Progress in Nonlinear Differential Equations and Their Applications
Volume 26

Algebraic Aspects of Integrable Systems

In Memory of Irene Dorfman

A. S. Fokas and I. M. Gelfand

Editors

Birkhäuser
Boston • Basel • Berlin

A. S. Fokas
Department of Mathematics
Imperial College
London SW7 2BZ
United Kingdom

I. M. Gelfand
Department of Mathematics
Rutgers University
New Brunswick, NJ 08903
United States

Library of Congress Cataloging-in-Publication Data

Algebraic aspects of integrable systems : in memory of Irene Dorfman /
  A. S. Fokas and I. M. Gelfand, editors.
      p.   cm. -- (Progress in nonlinear differential equations and
  their applications ; v. 26)
    Includes bibliographical references.
    ISBN-13:978-1-4612-7535-0   e-ISBN-13:978-1-4612-2434-1
    DOI: 10.1007/978-1-4612-2434-1
    1. Differentiable dynamical systems   2. Mathematical physics.
  I. Dorfman, Irene.   II. Fokas, A. S., 1952-   . III. Gelfand,
  I. M. (Izrail' Moiseevich)   IV. Series.
  QA614.8.A 425   1996   96-31350
  514'.74--dc20   CIP

Printed on acid-free paper
© 1997 Birkhäuser Boston   *Birkhäuser*
Softcover reprint of the hardcover 1st edition 1997

ISBN-13:978-1-4612-7535-0

Typeset by TeXniques, Boston, MA

9 8 7 6 5 4 3 2 1

Contents

Preface

Irene Dorfman died in Moscow on April 6, 1994, shortly after seeing her beautiful book on Dirac structures [1]. The present volume contains a collection of papers aiming at celebrating her outstanding contributions to mathematics. Her most important discoveries are associated with the algebraic structures arising in the study of integrable equations. Most of the articles contained in this volume are in the same spirit.

Irene, working as a student of Israel Gel'fand made the fundamental discovery that integrability is closely related to the existence of bi-Hamiltonian structures [2], [3]. These structures were discovered independently, and almost simultaneously, by Magri [4] (see also [5]). Several papers in this book are based on this remarkable discovery. In particular Fokas, Olver, Rosenau construct large classes on integrable equations using bi-Hamiltonian structures, Fordy, Harris derive such structures by considering the restriction of isospectral flows to stationary manifolds and Fuchssteiner discusses similar structures in a rather abstract setting.

Bi-Hamiltonian operators and their related Poisson brackets can be constructed via the so called Adler-Gelfand-Dickey [6], [7] approach. This construction is based on Lax operators [8] which are regarded as elements of the algebra of pseudo-differential symbols. Irene made important contributions in this direction (see for example [9]). Several papers in this book use extensively the algebra of pseudo-differential operators. In particular Oevel discusses such operators in connection with r-matrices and the modified Yang-Baxter equation, Dickey constructs the τ-function for large classes of systems of integrable equations, and Cohen, Manin, Zagier establish some beautiful relations between certain pseudo-differential operators and modular forms.

In recent years there has been much activity aiming at extending some of the above results from continuous to discrete systems. Several discrete versions of the equation $u_t + uu_x = 0$, of certain "Schwarzian" equations, and of the Newmann equation are discussed in the papers of Kupershmidt, of Nijhoff, and of Ragnisco, Suris respectively. Nonultralocal Poisson brackets for 1-dimensional lattice systems are discussed by Semenov-Tian-Shansky and Sevostyanov.

Irene dedicated a considerable part of her research studying the concept of mastersymmetries introduced in [10] (see chapter [7] of her book). This concept is used by Grunbaum and Haine to establish certain relations between the Toda lattice, Bocher's theorem, and orthogonal polynomials.

The study of the algebraic properties of integrable equations, and in particular the existence of infinitely many symmetries for such equations, has been used to study equations which are integrable only asymptotically (with respect to some small parameter ε). Two different approaches to asymptotic integrability are presented in the papers of Kodama, Mikhailov and of Santini.

Ferapontov and Mokhov use certain transformations to map the Witten-Dijkgraaf-Verlinde-Verlinde equations (i.e. the associativity equations) to a system of integrable equations of the hydrodynamic type. Schief and Rogers show that the $2+1$ Loewner-Konopelchenko-Rogers system is invariant under certain Laplace-Darboux transformations and then use this fact to obtain interesting integrable reductions.

Although most papers presented here are algebraic, some papers have a strong analytic flavor. In particular McKean uses some beautiful trace formulae associated with the Dirac and the Schrödinger periodic operators to study the canonical 1-forms of the symplectic geometry associated with the defocussing nonlinear Schrödinger and with the Korteweg-deVries equations. Alber, Luther and Marsden obtain interesting new classes of solutions of nonlinear PDE's by constructing certain finite dimensional integrable Hamiltonian systems on Riemannian manifolds.

A.S. Fokas and I. M. Gelfand, Editors

References

[1] I.Y. Dorfman, *Dirac Structures and Integrability of Nonlinear Evolution Equations*, Wiley and Sons, (1993).

[2] I.M. Gel'fand, I.Ya Dorfman, Funct. Anal. Appl. **13**:4, 13-30 (1979).

[3] I.M. Gel'fand, I.Ya Dorfman, Funct. Anal. Appl. **14**:3, 71-74 (1980).

[4] F. Magri, J. Math. Phys., **19**, 1156-1162 (1978).

[5] A.S. Fokas, B. Fuchssteiner, Lett. in Nuovo Cimento, **28** 299-303, (1980).

[6] I.M. Gel'fand, L.A. Dikii, in *I.M. Gel'fand, Collected Works*, Springer, NY (1990).

[7] M. Adler, Invent. Math **50**, 219 (1979).

[8] P. Lax, Comm. Pure Appl. Math. **21**, 467 (1968).

[9] I.Ya Dorfman, A.S. Fokas, J. Math. Phys. **33**, 2504-2514 (1992).

[10] A.S. Fokas, B. Fuchssteiner, Phys. Lett. A, **86**, 341 (1981).

*This volume is dedicated to the memory
of our dear friend and colleague
Irene Dorfman*

Complex Billiard Hamiltonian Systems and Nonlinear Waves

Mark S. Alber,[1] *Gregory G. Luther,*[2] *and Jerrold E. Marsden*[3]

In memory of Irene Dorfman

Abstract

The relationships between phase shifts, monodromy effects, and billiard solutions are studied in the context of Riemannian manifolds for both integrable ordinary and partial differential equations. The ideas are illustrated with the three wave interaction, the nonlinear Schrödinger equation, a coupled Dym system and the coupled non-linear Schrödinger equations.

1. Introduction

There is a deep connection between solutions of nonlinear equations and both geodesic flows and billiards on Riemannian manifolds. For example, in Alber, Camassa, Holm and Marsden [1995], a link between umbilic geodesics and billiards on n-dimensional quadrics and new soliton-like solutions of nonlinear equations in the Dym hierarchy was investigated. The geodesic flows provide the spatial x-flow, or the instantaneous profile of the solution of a partial differential equation. When combined with a prescription for advancing the solution in time, a t-flow, one is able to determine a class of solutions for the partial differential equation under study (see, for example, Alber and Alber [1985, 1987] and Alber and Marsden [1994]). New classes of solutions can be obtained using deformations of finite dimensional level sets in the phase space. In particular, to obtain soliton, billiard, and peakon solutions of nonlinear equations, one applies limiting procedures to the system of differential equations on the Riemann surfaces describing quasiperiodic solutions. To carry this out, one can use the method of *asymptotic reduction*—for details see Alber and Marsden [1992, 1994] and Alber, Camassa, Holm and Marsden [1994, 1995].

[1]Research partially supported by NSF grants DMS 9403861 and 9508711.
[2]GGL gratefully acknowledges support from BRIMS, Hewlett-Packard Labs and from NSF DMS under grant 9508711.
[3]Research partially supported by NSF grant DMS 9302992.

Elliptic and hyperbolic billiards can be obtained from the problem of geodesics on quadrics by collapsing along the shortest semiaxis—see Alber and Alber [1986] and Moser and Veselov [1991]. This process yields Hamiltonians and first integrals for the resulting billiard problem. A defining characteristic of a billiard solution is that some of the momentum variables experience a jump when a trajectory hits a boundary. In the context of generalized geodesic flows on Riemannian manifolds, these jumps are associated with a reflection map that is used to glue different pieces of the invariant variety in the phase space together to form one symplectic manifold (for generalized geodesic flows and completely integrable systems, see Lazutkin [1993]). These jumps on the invariant variety in the phase space correspond to discontinuities in the instantaneous profile of the solution of the partial differential equation.

In Hamiltonian systems, boundaries of Riemann surfaces are obtained by collapsing several branch points into one or by eliminating two or more branch points. Points obtained in this way are called *reflection points* or *boundaries* of Riemann surfaces. Collapsing branch points can change the type of the Riemann surface, for example from a hyperelliptic curve to a logarithmic surface. Boundary conditions must be imposed when reflection points are created. Introducing reflection conditions at boundaries as jumps from one sheet of a Riemann surface to another leads to billiard geodesic flows. The reflection condition then yields billiard-type weak solutions of nonlinear partial differential equations (See Alber, Camassa, Holm, Marsden [1994,1995]). The interpretation of the jump conditions depends on the particular physical problem and is different in each case. For example, in the case of a 3-wave interaction, the jump manifests itself as a shift of the relative phase of the three waves. We describe this example in some detail below to demonstrate a more general approach that can be used to analyze soliton phase functions.

The appearance of *monodromy* is prevalent in many Hamiltonian systems. A monodromy representation for a complex analytic Hamiltonian system X_H is a solution germ that has been analytically continued around a closed loop. The presence of holomorphic and meromorphic differentials in the action angle representations of the linearized Hamiltonian flow on the Jacobi variety of many integrable problems implies the existence of nontrivial fixed points for the action of homotopy classes represented by real periodic orbits (for details see Baider, Churchill and Rod [1990] and Magnus [1976].)

Below we find new classes of solutions of nonlinear partial differential equations with *monodromy* by constructing associated finite dimensional integrable Hamiltonian systems on Riemannian manifolds. Our approach to evolution equations with monodromy is illustrated for classes of solutions of the coupled Dym equations. Prior to this, such effects were recognized only in finite dimensional mechanical systems.

2. Phase Shifts on Riemann Surfaces

2.1. Phase Shifts and Monodromy of a Branch Point

Here we consider an example of a jump in the relative phase of a wave interaction caused by the presence of a branch point on a Riemann surface. The envelope equations for the resonant interaction of three waves can be written in the canonical form

$$\frac{d}{dz}A_1 = -iA_2 A_3 + i(\delta_1 + \lambda_{1\beta}|A_\beta|^2)A_1 , \tag{2.1}$$

$$\frac{d}{dz}A_2 = -iA_1 A_3^* + i(\delta_2 + \lambda_{2\beta}|A_\beta|^2)A_2 , \tag{2.2}$$

$$\frac{d}{dz}A_3 = -iA_1 A_2^* + i(\delta_3 + \lambda_{3\beta}|A_\beta|^2)A_3 , \tag{2.3}$$

where all quantities are measured in a frame moving with the wave and a repeated index implies summation (see for instance Guckenheimer and Mahalov [1992] or McKinstrie and Luther [1988] and references therein). The A_α's are the action flux amplitudes of each wave, normalized to the initial action flux amplitude of the pump wave. The δ_α's and the $\lambda_{\alpha\beta}$'s are the linear and nonlinear phase-mismatch coefficients of each wave. Note that the nonlinear phase-shift coefficients can be chosen to satisfy the symmetry relation $\lambda_{\alpha\beta} = \lambda_{\beta\alpha}$.

This system can be formulated in terms of the Hamiltonian

$$H = (A_1 A_2^* A_3^* + A_1^* A_2 A_3) - \left(\delta_\alpha + \frac{1}{2}\lambda_{\alpha\beta}|A_\beta|^2\right)|A_\alpha|^2 , \tag{2.4}$$

together with Hamilton's equations

$$i\frac{d}{dz}A_\alpha = \frac{\partial H}{\partial A_\alpha^*} . \tag{2.5}$$

In this formulation, the wave amplitudes A_α and A_α^* are canonically-conjugate variables. In addition to the Hamiltonian there are two other constants of the motion,

$$I = |A_1|^2 + |A_2|^2 , \quad J = |A_2|^2 - |A_3|^2 . \tag{2.6}$$

Writing

$$A_\alpha = F_\alpha^{1/2}\exp(i\phi_\alpha) , \tag{2.7}$$

the Hamiltonian takes the form

$$H = 2(F_1 F_2 F_3)^{1/2}\cos(\phi_1 - \phi_2 - \phi_3) - (\delta_\alpha + \lambda_{\alpha\beta}F_\beta)F_\alpha , \tag{2.8}$$

where the action flux densities F_α and the phases ϕ_α evolve according to

$$\frac{d}{dz}F_\alpha = -\frac{\partial H}{\partial \phi_\alpha} \ , \quad \frac{d}{dz}\phi_\alpha = \frac{\partial H}{\partial F_\alpha} \ . \tag{2.9}$$

Using the three constants H, I, and J, the evolution equation for each action flux density can be rewritten in the potential form

$$\left(\frac{d}{dz}F_\alpha\right)^2 + Q_\alpha(F_\alpha) = 0 \ , \tag{2.10}$$

where each Q_α is a fourth-order polynomial in F_α. These equations (2.10) can be transformed into the following problem of inversion:

$$\theta = \int_{z_0}^z \frac{dF_\alpha}{\sqrt{-Q_\alpha(F_\alpha)}} = z - z_0 \ , \tag{2.11}$$

which can be solved in terms of Riemann θ-functions on the Jacobi varieties of hyperelliptic curves

$$W_\alpha^2 + Q_\alpha(F_\alpha) = 0 \ . \tag{2.12}$$

Therefore, in this case the exchange of energy and momentum among the three waves is periodic. Once the spatial dependence of F_α has been obtained from the appropriate potential equation, ϕ_α is found by integrating the corresponding phase equation

$$\frac{d}{dz}\phi_\alpha = P_\alpha(F_\alpha) \ , \tag{2.13}$$

where each P_α is a simple algebraic function of F_α. For example, setting initial conditions in the form of

$$F_1(z) = 1, F_2(z) = \epsilon, F_3(0) = 0,$$

yields

$$F_3(z) =: F(z), F_2(0) = \epsilon + F(z), F_1(0) = 1 - F(z)$$

and

$$(d_z F)^2 = 4F((\epsilon + F)(1 - F) - F(\delta - \lambda F)^2) = C(F) \ . \tag{2.14}$$

This also results in the following expression for the relative phase $\Phi = \phi_1 - \phi_2 - \phi_3$:

$$\cos \Phi = -\frac{(\delta - \lambda F)\sqrt{F}}{\sqrt{(1 - F)(\epsilon + F)}} \ . \tag{2.15}$$

It follows from (2.14) that F is changing along a cycle on the Riemann surface

$$W^2 = C(F) \tag{2.16}$$

over the cut between two roots of the polynomial $C(F)$. In the process of investigating the dynamics of the phase function Φ one also has to deal with the multivaluedness of the expression on the righthand side of (2.15). The value of F is defined on the Riemann surface (2.16) and therefore $\cos \Phi$ should be considered on a covering of this Riemann surface. This yields relative phase shifts. For example, after going through the branch point at $F = 0$, $\cos \Phi$ goes from one sheet of the covering to another identified with $(\sqrt{F})$ and $(-\sqrt{F})$ respectively. This yields a shift of π in Φ.

When $\delta = \lambda$ the maximum energy transfer among the waves can be achieved. The point $F = 1$ becomes a branch point, and the expression for the relative phase becomes

$$\cos \Phi = -\frac{\delta\sqrt{F(1-F)}}{\sqrt{(\epsilon + F)}} \, . \tag{2.17}$$

Remark 1. We first notice that θ from (2.11) is an angle variable for our problem. The existence of the general phenomenon of a geometric phase (as in Berry and Hannay [1988] and Marsden, Montgomery and Ratiu [1990]) can be demonstrated as follows (see Alber and Marsden [1992]). For a periodic orbit we can write

$$\Delta_g \theta = \oint_\gamma dX \, \langle \partial_X \theta \rangle \tag{2.18}$$

where X denotes the parameters of the system, namely the coefficients of the polynomial $C(F)$, γ is a closed cycle in the space of parameters and where $\langle \, , \rangle$ denotes averaging along the periodic orbit, which is parametrized by the action. This action is the constant term in the polynomial $C(F)$. This will also result in a geometric phase for the ϕ.

Remark 2.   Variations in the medium that hosts the wave result in changes in the parameters δ and λ. Since these parameters along with ϵ determine the location of the fixed points and therefore the homoclinic point, their motion during the dynamics can significantly alter the dynamics of the waves. It is possible to vary δ and λ so that the phase point of a wave makes a circuit around one of the fixed points resulting in the shift associated with the monodromy.

2.2. Phase Shifts and Billiards

Now we demonstrate that in the limiting case $\epsilon \to 0$ the phase jump is generated by reflection on the boundary of a billiard system on a Riemann surface. When $\epsilon \to 0$ (infinite-period limit), the homoclinic orbit is recovered. This is associated with a singular leaf of the Jacobian foliation in the phase space. A homoclinic orbit is obtained when the daughter waves initially have exponentially small amplitudes for $\xi \to -\infty$. In this limit the three waves exchange their energy exactly once and the recurrence time is infinity. Note also that the position of the homoclinic is altered by changing the parameters δ and λ.

The potential for wave three reduces in the limit $\epsilon \to 0$ to

$$Q^2 = 4F^2((1-F) - (\delta - \lambda F)^2)$$

and equations for Φ and F become

$$z - z_0 \;=\; \int_{\xi^0}^{\xi} \frac{dF}{2F\sqrt{(1-F) - (\delta - \lambda F)^2}} \tag{2.19}$$

$$\cos \Phi \;=\; \frac{-(\delta - \lambda F)}{(1-F)^{1/2}} \;. \tag{2.20}$$

The Riemann surface for F has a logarithmic singularity at $F = 0$. At the same time the branch point at $F = 0$ in the expression for $\cos \Phi$ cancels out. Nevertheless, the sign of the expression changes every time F reaches $F = 0$ due to the *reflection* condition on the boundary $F = 0$. This yields a shift in the relative phase Φ.

3. Monodromy in Solutions of Nonlinear PDE's

In this section we proceed with our discussion by describing special solutions of nonlinear equations with monodromy. Examples include parametrized classes of solutions of completely integrable equations such as the coupled Dym and the NLS equations. The connection to Hamiltonian systems on Riemann surfaces in these examples is more complicated than in the case of the three-wave equations, but it can be established, for example, using the method of generating equations.

We begin by considering the spectral problem for an associated Schrödinger operator,

$$L = -\frac{\partial^2}{\partial x^2} + V(x, t, E), \tag{3.1}$$

where E is a parameter and

$$V(x, t, E) = \sum_{j=-l}^{m} V_j(x, t) E^j.$$

In some cases, such as the KdV equation, E appears as an eigenvalue and one ultimately equates the potential with a solution of the nonlinear equation itself. In the case of the nonlinear Schrödinger equation, the solution,

Q, and the potential, V, are related in a slightly more complicated way. To carry out the procedure, one looks for a solution A of the Lax system

$$\left. \begin{array}{c} L\psi = 0 \\ \left(\dfrac{\partial L}{\partial t} + [L, A] \right) \psi = 0 \end{array} \right\} \tag{3.2}$$

of the form

$$A = B_n \frac{\partial}{\partial x} - \frac{1}{2} \frac{\partial B_n}{\partial x}. \tag{3.3}$$

Substituting the given form of A into the Lax system, one gets

$$\frac{\partial V}{\partial t} = -\frac{B_n'''}{2} + 2B_n' V + B_n V' \tag{3.4}$$

where the prime denotes $\partial/\partial x$ and $B_n = \prod_{j=1}^n (E - \mu_j(x, t))$. Equation (3.4) is called the *generating equation*. Suppose $\partial V/\partial t = 0$, then the equation can be integrated resulting in the following stationary generating equation:

$$- B''B + \frac{B'^2}{2} + 2B^2 V = K(E) \tag{3.5}$$

where $K(E)$ is a rational function of E with constant coefficients. In the inverse scattering transform method it is called the *spectral polynomial* (see Moser [1981]).

To describe the dynamics of B_n we use the following procedure. We suppose that at each instant of time, B_n is a solution of the stationary generating equation. The evolution of B_n in time is then determined by

$$\dot{B}_n = B_n' B_l - B_l' B_n, \quad i.e., \quad \frac{\partial}{\partial t} \left(\frac{1}{B_n} \right) = \frac{\partial}{\partial x} \left(\frac{B_l}{B_n} \right), \tag{3.6}$$

where B_l is a solution of the dynamical generating equation (3.4). For particular choices of the form of the potential V, equations (3.4) and (3.6) generate a hierarchy of integrable equations.

Using (3.4) and (3.6), finite dimensional Hamiltonian systems in the μ-representation for solutions of the nonlinear equations in the hierarchy are obtained after substituting $E = \mu_j$, where $j = 1, ...n$. The x-flow is given by a system of equations for μ_j' and the t-flow is given by a system for $\dot{\mu}_j$. The μ-representations have complete sets of first integrals in the form of Riemann surfaces. The Abel map is frequently used for these Riemann surfaces, and it linearizes the Hamiltonian flows on the associated Jacobi varieties. Recall from Ercolani [1989] and Alber *et al.* [1994,1995] that the Abel map can also be viewed as a set of the complex angle variables called the *complex angle representation* of the problem. Continuous variation of the first integrals results in deformations of the level sets. Some of these deformations lead to singular cases for which one must introduce new types of angle representations on singular level sets. Examples of this situation are provided by homoclinic and soliton Hamiltonian flows; these systems usually have phase shifts produced by monodromies.

3.1. Phase Shifts for the NLS Equation

In this section we describe phase shifts produced by the interaction of fundamental solutions of the NLS equation. To do so we use the angle representation obtained in Alber and Marsden [1992, 1994].

Applying the general formalism of generating equations and asymptotic reduction (from Alber and Marsden [1992]) gives the angle representation that describes the collision of two standard solitons of the (f)NLS equation, namely

$$
\begin{aligned}
\theta_1 &= \frac{1}{4y_1} \int_{\mu_j^0}^{\mu_1} \left(\frac{1}{(\mu_1 - a_1)} - \frac{1}{(\mu_1 - \bar{a}_1)} \right) d\mu_1 + \phi_1(\mu_2, \mu_3) \\
&= x + v_1 t, \ v_1 = 2x_1
\end{aligned}
\tag{3.7}
$$

$$
\begin{aligned}
\theta_2 &= \frac{1}{4y_2} \int_{\mu_2^0}^{\mu_2} \left(\frac{1}{(\mu_2 - a_2)} - \frac{1}{(\mu_2 - \bar{a}_2)} \right) d\mu_2 + \phi_2(\mu_2, \mu_3) \\
&= x + v_2 t, \ v_2 = 2x_2,
\end{aligned}
\tag{3.8}
$$

and a third angle θ_3 which depends on a_1, a_2 and which is complex.

$a_1 = x_1 + iy_1$ and $a_2 = x_2 + iy_2$ are elements of the discrete spectrum of the differential operator L. We assume that $x_2 < x_1$ and $y_2 < y_1$, which implies that $v_2 < v_1$. Reality conditions are also imposed.

The complex phase functions associated with the angle variables are defined by

$$
\begin{aligned}
\phi_1(x, t) &= \frac{1}{4y_1} \left(\int_{\mu_2^0}^{\mu_2} \left(\frac{1}{(\mu_2 - a_1)} - \frac{1}{(\mu_2 - \bar{a}_1)} \right) d\mu_2 \right) \\
&\quad + \frac{1}{4y_1} \left(\int_{\mu_3^0}^{\mu_3} \left(\frac{1}{(\mu_3 - a_1)} - \frac{1}{(\mu_3 - \bar{a}_1)} \right) d\mu_3 \right)
\end{aligned}
\tag{3.9}
$$

and

$$
\begin{aligned}
\phi_2(x, t) &= \frac{1}{4y_2} \left(\int_{\mu_1^0}^{\mu_1} \left(\frac{1}{(\mu_1 - a_2)} - \frac{1}{(\mu_1 - \bar{a}_2)} \right) d\mu_1 \right) \\
&\quad + \frac{1}{4y_2} \left(\int_{\mu_3^0}^{\mu_3} \left(\frac{1}{(\mu_3 - a_2)} - \frac{1}{(\mu_3 - \bar{a}_2)} \right) d\mu_3 \right) .
\end{aligned}
\tag{3.10}
$$

The phase shifts of solitons are related to the monodromy of ϕ_1 and ϕ_2. The phase shifts are obtained by considering the asymptotic behavior of the relative phase function, $\phi(x,t) = \phi_1(x,t) - \phi_2(x,t)$. Notice that each of the μ-variables is defined on a logarithmic Riemann surface and μ_1, μ_2 and μ_3 describe a dynamical system with two homoclinic points. Phase shifts, or jumps in the values of phase functions, occur precisely when the system goes through one of these points. This situation is, therefore, very similar to the three-wave interaction.

3.2. Special Solutions for the Coupled Dym System

In what follows, we introduce a Hamiltonian system for the set of quasi-periodic solutions of the so-called coupled Dym system. When applied to this Hamiltonian system, different limiting procedures (involving the coalescence of roots of the basic polynomial of the spectral curve) yield special solutions with monodromy, billiard solutions, and umbilic geodesic flows (in the presence of potentials) on associated limiting Riemann surfaces.

The time evolution of functions of the generalized Schrödinger equations and the multi-Hamiltonian structures associated with the coupled Dym system were investigated in Antonowicz and Fordy [1988,1989] using an algebraic approach. Recall from these papers that the potential associated with the coupled Dym system

$$\left.\begin{aligned}
\frac{\partial u}{\partial t} &= \frac{1}{4}u''' - \frac{3}{2}uu' + v' \\[2mm]
\frac{\partial v}{\partial t} &= -u'v - \frac{1}{2}uv' ,
\end{aligned}\right\}
\tag{3.11}$$

has a pole in the spectral parameter. Using the method of Alber, Camassa, Holm and Marsden [1994, 1995] one obtains the stationary μ-flow,

$$\mu_j' = \frac{1}{\prod_{r \neq j}^{n}(\mu_j - \mu_r)}\sqrt{\frac{C(\mu_j)}{\mu_j}}, \quad j = 1, ..., n ,
\tag{3.12}$$

for this integrable problem after substituting $V = u(x) + \lambda + v(x)/\lambda$ into the generating equation (3.5). Each of the μ-variables is defined on a copy of the Riemann surface

$$\Re : P^2 = \frac{C(\mu)}{\mu} = -\frac{L_0^2}{\mu}\prod_{k=1}^{2n+2}(\mu - m_k) .
\tag{3.13}$$

Recall that the μ variables move along cycles on the corresponding Riemann surface (3.13) over the prohibited zones (that is, over the basic cuts between m_{2j} and m_{2j-1} on the Riemann surface). The system (3.12) is a Hamiltonian system with Hamiltonian

$$H = \sum_{j=1}^{n} \frac{(P_j^2 - \frac{C(\mu_j)}{\mu_j})}{\prod_{r \neq j}^{n}(\mu_j - \mu_r)}, \quad j = 1, ..., n , \tag{3.14}$$

and the set of first integrals

$$P_j^2 = \frac{C(\mu_j)}{\mu_j}, \quad j = 1, ..., n . \tag{3.15}$$

Here the system is degenerate because the genus of the Riemann surface (3.13) is $(n+1)$, yet we have only n μ variables. This difference produces a degeneracy in the problem of inversion. It can be resolved by introducing an additional μ variable, μ_{n+1}, and solving the problem of inversion in terms of Riemann θ-functions on the $(n+1)$-dimensional Jacobian of the hyperelliptic curve

$$\Re : P^2 = C(\mu)\mu , \tag{3.16}$$

which is a torus of genus $g = n + 1$. In this way the following problem of inversion is obtained in the complex angle representation,

$$\alpha_k = \sum_{j=1}^{n+1} \int_{\mu_j^0}^{\mu_j} \frac{\mu_j^{k+1} \, d\mu_j}{\sqrt{C(\mu_j)\mu_j}} = \delta_k^{n-1} x + \delta_k^{n-2} t + \alpha_k^0, \quad k = 0, ..., n , \tag{3.17}$$

where α_k^0 are constants, $\delta_k^{n-1}, \delta_k^{n-2}$ are Kronecker delta's and each μ_j is defined on a copy of the Riemann surface (3.16). The above integrals are taken along cycles a_j over basic cuts on the Riemann surface. The solution of the Hamiltonian system is obtained by setting $\mu_{n+1} = m_{2n+2}$.

After applying a limiting procedure similar to that described in Alber, Camassa, Holm and Marsden [1995] for the Dym equation and fixing one of the μ-variables, the system (3.17) leads to the so-called umbilic angle representation. Alber, Camassa, Holm and Marsden [1995] shows that these representations generate a class of umbilic solitons.

On the other hand, setting $m_{2n+2} = m_{2n+1} = d$ in the initial Hamiltonian system (without the additional μ variable) yields a well-defined system of inversion with monodromy. Namely, moving d along a certain closed loop in the space of parameters can lead to a nontrivial shift in action angle variables. This phenomenon is caused by a singularity on the associated Riemann surface and can be demonstrated as follows. In the 1-dimensional case, the limiting process $m_3, m_4 \to d$ yields the angle representation

$$\alpha_1 = \int_{\mu_1^0}^{\mu_1} \frac{\mu_1 d\mu_1}{(\mu_j - d)\sqrt{-\mu_1(\mu_1 - m_1)(\mu_1 - m_2)}} = L_0 x + \alpha_1^0. \tag{3.18}$$

Complex Billiard Hamiltonian Systems and Nonlinear Waves 11

In the case of a genus 3 initial Riemann surface, the limiting angle representation is as follows

$$
\left.
\begin{aligned}
\alpha_1 &= -\frac{\partial S}{\partial \beta_1} = \int_{\mu_1^0}^{\mu_1} \frac{\mu_1 d\mu_1}{(\mu_1 - d)\sqrt{C_5(\mu_1)}} + \int_{\mu_2^0}^{\mu_2} \frac{\mu_2 d\mu_2}{(\mu_2 - d)\sqrt{C_5(\mu_2)}} = \alpha_1^0. \\
\alpha_2 &= -\frac{\partial S}{\partial \beta_2} = \int_{\mu_1^0}^{\mu_1} \frac{\mu_1 d\mu_1}{\sqrt{C_5(\mu_1)}} + \int_{\mu_2^0}^{\mu_2} \frac{\mu_2 d\mu_2}{\sqrt{C_5(\mu_2)}} = L_0 x + \alpha_2^0
\end{aligned}
\right\}
$$

(3.19)

where

$$
S = \int_{\mu_1^0}^{\mu_1} \frac{\sqrt{C_5(\mu_1)}d\mu_1}{(\mu_1 - d)} + \int_{\mu_2^0}^{\mu_2} \frac{\sqrt{C_5(\mu_2)}d\mu_2}{(\mu_2 - d)}
$$

is an action function (the generating function of a canonical transformation).

The variables μ_1 and μ_2 move along cycles a_1 and a_2 over the cuts $[m_1, m_2]$ and $[m_3, m_4]$ on the Riemann surface $W^2 = C_5(\mu)$. There is also a singularity at $\mu = d$. Transport of a system of canonical action-angle variables, which linearize the Hamiltonian flow, along a certain loop in the space of parameters (d, m_j) in a way similar to the case of the spherical pendulum and some other integrable systems with monodromy (see, for example, Duistermaat [1980] and Bates and Zou [1993]) will result in a nontrivial shift, which is a manifestation of the monodromy phenomenon. This can be demonstrated as follows. Canonical actions are calculated in terms of periods of the holomorphic differential

$$
I_j = \oint_{a_j} dS = \oint_{a_j} \frac{\sqrt{C_5(\mu)}d\mu}{(\mu - d)}
$$

along cycles a_j on the Riemann surface (a torus); for details see Arnold [1978]. Now suppose initially that d does not belong to either of the cycles a_1 or a_2. By moving d along a closed loop on the Riemann surface such that it encircles one of the branch points, m_j, a shift in the action variable is produced that is given by the residue of the integrand at $\mu = d$.

Lastly, the method of Alber et al. [1994, 1995] of associating solutions of nonlinear partial differential equations with finite dimensional Hamiltonian systems on Riemann surfaces leads to the construction of a class of solutions of nonlinear partial differential equations with monodromy.

4. Coupled Nonlinear Schrödinger Equations

Our last example is provided by the coupled NLS equations. A pair of distinct wave packets propagating under the action of dispersion and cubic nonlinearity are modeled by the coupled nonlinear Schrödinger equation (CNLS),

$$\left[i(\partial_t + v_1\partial_z) + \lambda_{11}|A_1|^2 + \lambda_{12}|A_2|^2\right]A_1 + \Delta_1 A_2 = 0, \quad (4.1)$$

$$\left[i(\partial_t + v_2\partial_z) + \lambda_{21}|A_1|^2 + \lambda_{22}|A_2|^2\right]A_2 + \Delta_2 A_1 = 0. \quad (4.2)$$

The CNLS equations have several different forms and arise in fluid dynamics, plasma physics, and nonlinear optics. The CNLS models the coupling of wave envelopes propagating in two coupled nonlinear waveguides (dual-core fibers), in two separate polarization modes in a single waveguide, or at two distinct frequencies in the same waveguide and polarization mode. Understanding the integrable Manikov system, which is a special case of the system (4.1) and (4.2) as well as its near-integrable counterparts is of basic interest for the analysis of problems in optical communications and switching.

Stationary solutions of the CNLS are found by taking

$$(A, B) = (q_1(t), q_2(t)) \exp(i\Omega z).$$

That is, $(q_1(t), q_2(t))$ are the complex amplitude profiles of envelopes in two modes that couple nonlinearly as described above. Solutions $(q_1(t), q_2(t))$ can be found using algebraic geometric methods by choosing

$$\begin{aligned} B(E, x) &= 1 - \frac{q_1^2}{2(E - l_1)} - \frac{q_2^2}{2(E - l_2)} \\ &= \frac{(E - \mu_1)(E - \mu_2)}{2(E - l_1)(E - l_2)} \quad \text{and} \quad V = u(x) + E, \quad (4.3) \end{aligned}$$

in (3.5). This yields a system of two coupled ODE's for q_1 and q_2. Namely, substituting (4.3) into (3.5), multiplying by $(E - l_1)^2(E - l_2)^2$, and setting separately $E \to l_1$ and $E \to l_2$ yields

$$\left.\begin{aligned} q_1'' + (q_1^2 + q_2^2)q_1 &= l_1 q_1 \\ q_2'' + (q_1^2 + q_2^2)q_2 &= l_2 q_2. \end{aligned}\right\} \quad (4.4)$$

Solutions of (4.4) can be obtained using the μ representations on Riemann surfaces. Setting $E = \mu_1$ or $E = \mu_2$ in (3.5) yields

$$\left.\begin{aligned} \mu_1' &= \frac{\sqrt{-(\mu_1 - l_1)(\mu_1 - l_2)(\mu_1 - l_3)(\mu_1 - l_4)(\mu_1 - l_5)}}{\mu_1 - \mu_2} \\ \mu_2' &= \frac{\sqrt{-(\mu_2 - l_1)(\mu_2 - l_2)(\mu_2 - l_3)(\mu_2 - l_4)(\mu_2 - l_5)}}{\mu_2 - \mu_1} \end{aligned}\right\} \quad (4.5)$$

This coincides with the particular case of the quasiperiodic stationary KdV flow. In Christiansen *et al.* [1995] elliptic solutions of the Jacobi problem of inversion associated with (4.5) are linked with the Treibich-Verdier potentials for the Schrödinger equation.

The following limiting process $l_1 \to 0$, $l_2, l_3 \to a_1$ and $l_4, l_5 \to a_2$ when applied to (4.5), yields homoclinic orbits

$$\left.\begin{aligned}
\mu_1' &= \frac{\sqrt{-\mu_1}(\mu_1 - a_1)(\mu_1 - a_2)}{\mu_1 - \mu_2} \\[2mm]
\mu_2' &= \frac{\sqrt{-\mu_2}(\mu_2 - a_1)(\mu_2 - a_2)}{\mu_2 - \mu_1}
\end{aligned}\right\} \tag{4.6}$$

Lastly, solutions of the system (4.4) are related to μ-variables on Riemann surfaces as follows:

$$\left.\begin{aligned}
q_1 &= \sqrt{\frac{-(l_1 - \mu_1)(l_1 - \mu_2)}{l_1 - l_2}} \\[2mm]
q_2 &= \sqrt{\frac{-(l_2 - \mu_1)(l_2 - \mu_2)}{l_2 - l_1}}.
\end{aligned}\right\} \tag{4.7}$$

The finite dimensional invariant variety for q_1 and q_2 is a covering of the Riemann surface for the representation of the system in terms of the μ-variables. It's structure can be investigated using methods of algebraic geometry (see Christiansen et. al. [1995]); monodromy of the branch points plays central role in this approach.

References

[1] M.J. Ablowitz and H. Segur [1981], *Solitons and the Inverse Scattering Transform*, SIAM, Philadelphia.

[2] M.S. Alber and S.J. Alber [1986], On Hamiltonian formalism for finite-zone solutions of nonlinear integrable equations, Proc. VIII Int. Cong. on Math. Physics, Marseille, France (World Scientific) 447–462.

[3] M.S. Alber and S.J. Alber [1985], Hamiltonian formalism for finite-zone solutions of integrable equations, *C.R. Acad. Sc. Paris* **301**, 777–781.

[4] M.S. Alber and S. J. Alber [1987], Hamiltonian formalism for nonlinear Schrödinger equations and sine-Gordon equations, *J. London Math. Soc.* **36**, 176–192.

[5] M.S. Alber, R. Camassa, D.D. Holm and J.E. Marsden [1994], The geometry of peaked solitons and billiard solutions of a class of integrable partial differential equation's, *Lett. Math. Phys.* **32**, 137–151.

[6] M.S. Alber, R. Camassa, D.D. Holm and J.E. Marsden [1995], On the link between umbilic geodesics and soliton solutions of nonlinear partial differential equation's, *Proc. Roy. Soc. Lond. A* **450** 677-692.

[7] M.S. Alber and J.E. Marsden [1992], On geometric phases for soliton equations, *Commun. Math. Phys.*, **149**, 217–240.

[8] M.S. Alber and J.E. Marsden [1994a], Geometric phases and monodromy at singularities, N.M. Ercolani et al., eds., *NATO ASI Series B* (Plenum Press, New York) **320** 273-296.

[9] M.S. Alber and J.E. Marsden [1994b], Resonant Geometric Phases for Soliton Equations, *Fields Institute Commun.* **3** 1–26.

[10] M. Antonowicz and A.P. Fordy [1988], Coupled Harry Dym equations with multi-Hamiltonian structures, *J. Phys. A* **21** L269–L275.

[11] M. Antonowicz and A.P. Fordy [1989], Factorisation of energy dependent Schrödinger operators: Miura maps and modified systems, *Commun. Math. Phys.* **124** 465–486.

[12] V.I. Arnold [1989], *Mathematical Methods of Classical Mechanics*, (Springer-Verlag: New York, Heidelberg, Berlin).

[13] A. Baider, R. C. Churchill and D. L. Rod [1990], Monodromy and nonintegrability in complex Hamiltonian systems, *J. Dyn. Diff. Equations* **2**, 451–481.

[14] L. Bates and M. Zou [1993], Degeneration of Hamiltonian monodromy cycles, *Nonlinearity* **6**, 313–335.

[15] Berry, M. and J. Hannay [1988] Classical non-adiabatic angles, *J. Phys. A. Math. Gen.* **21**, 325–333.

[16] H. Braden [1982], A completely integrable mechanical system, *Lett. Math. Phys.* **6** 449–452.

[17] P.L. Christiansen, J.C. Eilbeck, V.Z. Enolskii, and N.A. Kostov [1995], Quasiperiodic solutions of the coupled nonlinear Schrödinger equations, preprint.

[18] J.J. Duistermaat [1980] On global action-angle coordinates, *Comm. Pure Appl. Math.* **23** 687-706.

[19] N. Ercolani [1989], Generalized theta functions and homoclinic varieties, *Proc. Symp. Pure Appl. Math.* **49** 87.

[20] J. Guckenheimer and A. Mahalov [1992], Resonant triad interactions in symmetric systems, *Physica D* **54**, 267-310.

[21] V. F. Lazutkin [1993], *KAM Theory and Semiclassical Approximations to Eigenfunctions*, A Series of Modern Surveys in Mathematics, Springer-Verlag Berlin Heidelberg.

[22] W. Magnus [1976], Monodromy groups and Hill's equations, *Commun. Pure Appl. Math.* **29**, 701-716.

[23] S.V. Manakov [1974], On the theory of two-dimensional stationary self-Focusing of electromagnetic waves, *Sov. Phys. JETP* **38**, 248-253.

[24] J. E. Marsden, R. Montgomery, and T.S. Ratiu [1990] *Reduction, symmetry, and phases in mechanics.* Memoirs AMS **436**.

[25] C.J. McKinstrie [1988], Relativistic solitary-wave solutions of the beat-Wave equations, *Phys. Fluids* **31**, 288-297.

[26] C.J. McKinstrie and G.G. Luther [1988], Solitary-wave solutions of the generalized three-wave and four-wave equations, *Phys. Lett. A* **127**, 14-18.

[27] C.J. McKinstrie and X.D. Cao [1993], The nonlinear detuning of three-wave interactions, *J. Opt. Soc. Am. B* **10**, 898-912.

[28] J. Moser [1981], Integrable Hamiltonian Systems and Spectral Theory, Lezioni Fermiane, Accademia Nazionale dei Lincei, Pisa.

[29] J. Moser and A.P. Veselov [1991], Discrete Versions of Some Classical Integrable Systems and Factorization of Matrix Polynomials, *Commun. Math. Phys.*, **139**, 217-243.

[30] A.C. Newell [1985], *Solitons in Mathematics and Physics*, Regional Conf. Series in Appl. Math. **48**, SIAM, Philadelphia.

Mark Alber
Department of Mathematics
University of Notre Dame
Notre Dame, IN 46556.

Gregory G. Luther
Department of Mathematics
University of Notre Dame
Notre Dame, IN 46556
and the Basic Research Institute in the Mathematical Sciences
Hewlett-Packard Laboratories
Filton Road
Stoke Gifford, Bristol BS12 6QZ, UK.

Jerrold E. Marsden
California Institute of Technology
Control and Dynamical Systems 104-44
Pasadena, CA 91125.

Submitted November, 1995
Revised January 1996

Automorphic Pseudodifferential Operators

Paula Beazley Cohen[1], Yuri Manin, and Don Zagier

To the memory of Irene Dorfman

Introduction

The theme of this paper is the correspondence between classical modular forms and pseudodifferential operators (ΨDO's) which have some kind of automorphic behaviour. In the simplest case, this correspondence is as follows. Let Γ be a discrete subgroup of $PSL_2(\mathbb{R})$, acting on the complex upper half-plane $\mathcal{H}$ in the usual way, and $f(z)$ a modular form of even weight k on Γ. Then there is a unique lifting from f to a Γ-invariant ΨDO with leading term $f(z)\,\partial^{-k/2}$, where ∂ is the differential operator $\frac{d}{dz}$. This lifting and the fact that the product of two invariant ΨDO's is again an invariant ΨDO imply a non-commutative multiplicative structure on the space of all modular forms whose components are scalar multiples of the so-called Rankin–Cohen brackets (canonical bilinear maps on the space of modular forms on Γ defined by certain bilinear combinations of derivatives; the definition will be recalled later). This was already discussed briefly in the earlier paper [Z], where it was given as one of several "raisons d'être" for the Rankin–Cohen brackets.

The basic lifting from modular forms to invariant ΨDO's can be interpreted and developed in many ways. We shall discuss some of them in this paper. The two main generalizations are as follows:

(I) Just as one generalizes the notion of a modular function to the notion of a modular form, one can consider ΨDO's which are not invariant with respect to Γ but instead transform with some automorphy factor. Because of the non-commutativity of ΨDO's, however, we have new possibilities which do not occur in the classical case: one can consider "conjugate-automorphic" ΨDO's which under the action of a fractional linear transformation $\left(\begin{smallmatrix} a & b \\ c & d \end{smallmatrix}\right) \in \Gamma$ are multiplied by a $(cz+d)^\kappa$ on the left and

---

[1]The first-named author was supported by the Institute for Advanced Study, Princeton, New Jersey, USA, NSF grant number DMS 9304580, by the Centre National de Recherche Scientifique, Lille, France and by the School MPCE, Macquarie University, NSW, Australia.

by $(cz + d)^{-\kappa}$ on the right for some κ, or "automorphic ΨDO's of mixed weight" which transform by different automorphy factors on the left and on the right. The first way leads to a whole family of multiplications on the space of modular forms on Γ, each of which can be expressed in terms of the Rankin–Cohen brackets, but with coefficients which turn out to be intricate combinatorial expressions having beautiful and surprising properties. The second way gives even more structure on the space of modular forms and provides the clearest conceptual framework for the Rankin–Cohen brackets.

(II) The whole theory has a supersymmetric analogue. This is a natural generalization for the following reason. One of the disadvantages of the usual theory is that the derivative of the fractional linear transformation $z \mapsto \frac{az+b}{cz+d}$ is $(cz + d)^{-2}$ and hence that there is no coupling between modular forms of even and odd weight: not only is the derivative of a modular form not quite modular (which is why the theory is so complicated), but its weight is larger than the weight of the original form by 2 rather than by 1. But in the supersymmetric context, one has available a superdifferentiation operator D with square equal to d/dz and super-fractional linear transformations whose automorphy factor reduces modulo nilpotents to $(cz + d)^{-1}$ and hence effectively raises the weight of (super)modular forms by 1. Specifically, in the supercomplex plane $\mathbb{C}^{1|1}$ one has one even coordinate z and one odd one ζ, with $z\zeta = \zeta z$, $\zeta^2 = 0$, so a superanalytic function has the form $F(z, \zeta) = f(z) + g(z)\zeta$ with f and g holomorphic functions of z. The differential operator $D = \frac{\partial}{\partial \zeta} + \zeta \frac{\partial}{\partial z}$ sends F to $g(z) + f'(z)\zeta$, so that $D^2 = \partial$ as claimed; and we get the desired theory by working with ΨDO's based on powers of D rather than of ∂.

The structure of the paper is as follows. In Section 1 we define ΨDO's and give the basic result about lifting modular forms to invariant ΨDO's. In Section 2 we describe other proofs and interpretations of that result and a generalization to ΨDO's with non-integral powers of ∂. The next few sections treat topic (I) above: in Sections 3–5 we define canonical liftings of modular forms to various kinds of automorphic ΨDO's and describe the induced multiplications on the space of modular forms explicitly in terms of Rankin–Cohen brackets, and in Section 6 we give a conceptual proof (in terms of the non-commutative residue map and the duality between modular forms of weights k and $2-k$) of the surprising symmetries exhibited by the numerical coefficients appearing in these formulas. Topic (II), the supersymmetric generalization of the theory, is treated in Section 7. We explain the superanalogues of modular forms and of ΨDO's and state and prove the superanalogue of the basic lifting property.

The last section contains some scattered remarks and questions. Whereas in the main body of the paper we described our constructions in the context of classical automorphic forms, here we try to put them in the framework of the theory of completely integrable Hamiltonian systems to which Irene Dorfman made a significant contribution (see e.g. [GD]).

1. Lifting Modular Forms to Pseudodifferential Operators

Let z be a local coordinate for $\mathbb{C}$. We have the associated differential operator $\partial = \frac{d}{dz}$ which transforms under a coordinate change $z \mapsto \tilde{z}$ as $\partial = \partial(\tilde{z}) \cdot \tilde{\partial}$. Let R be a ring of functions on $\mathbb{C}$ on which ∂ acts, so that the pair (R, ∂) is a ring with derivation. By a *pseudodifferential operator* (ΨDO) over R we will mean a formal Laurent series in the formal inverse ∂^{-1} of ∂ with coefficients in R, i.e. an element of the vector space

$$\Psi\mathrm{DO}(R) = \left\{ \sum_{n \in \mathbb{Z}} h_n \partial^n : h_n \in R, \quad h_n = 0 \text{ if } n \gg 0 \right\}. \qquad (1.1)$$

The subspace $\mathrm{DO}(R) = R[\partial]$ of *differential operators* over R, consisting of sums as in (1.1) but with $n \geq 0$, is a ring under composition, and the formula for the multiplication of differential operators implied by Leibniz's rule, viz.,

$$\left(\sum_n g_n(z)\, \partial^n \right) \left(\sum_m h_m(z)\, \partial^m \right) = \sum_{n,m} \sum_{r \geq 0} \binom{n}{r} g_n\, \partial^r (h_m)\, \partial^{n+m-r},$$

$$(1.2)$$

can be extended to the full space $\Psi\mathrm{DO}(R)$ if we remember that for $l \in \mathbb{Z}_{\geq 0}$ the binomial coefficient $\binom{w}{l} = w(w-1)\cdots(w-l+1)/l!$ is a polynomial in w and hence is defined for any integral (or even complex) value of w. We have an increasing filtration of $\Psi\mathrm{DO}(R)$ by the subspaces

$$\Psi\mathrm{DO}(R)_w = \left\{ \sum_{n=0}^{\infty} f_n \partial^{w-n}, \quad f_n \in R \right\} \qquad (1.3)$$

with $w \in \mathbb{Z}$. It follows from formula (1.2) that this filtration is compatible with the ring structure in the sense that

$$\Psi\mathrm{DO}(R)_{w_1} \cdot \Psi\mathrm{DO}(R)_{w_2} \subseteq \Psi\mathrm{DO}(R)_{w_1+w_2} \qquad \forall w_1, w_2. \qquad (1.4)$$

In particular, the subspace $\Psi\mathrm{DO}(R)_0$ of pure ΨDO's is a subring of $\Psi\mathrm{DO}(R)$, and $\Psi\mathrm{DO}(R)$ has an (additive) direct sum decomposition as $\Psi\mathrm{DO}(R)_{-1} \oplus \mathrm{DO}(R)$. We have a short exact sequence

$$0 \to \Psi\mathrm{DO}(R)_{w-1} \to \Psi\mathrm{DO}(R)_w \to R \to 0 \qquad (1.5)$$

for every w, where the final map sends $\sum_{m \geq 0} f_m \partial^{w-m}$ to f_0 (symbol map).

We shall be interested in the behavior of ΨDO's under (groups of) transformations of the coordinate z. Under a coordinate change $z \mapsto \tilde{z}$ the differentiation operator ∂ is transformed to $\tilde{\partial} = j^{-1}\partial$, where $j = d\tilde{z}/dz$ is the Jacobian of the transformation, and there is a corresponding action on ΨDO's (cf. [KZ1])

$$\tilde{\partial}^w = j^{-w}\partial^w - \binom{w}{2}j'j^{-w-1}\partial^{w-1} + [3\binom{w+1}{4}j'^2 + \binom{w}{3}jj'']j^{-w-2}\partial^{w-2} + \cdots \tag{1.6}$$

(prove this by induction on w for $w \in \mathbb{Z}_{\geq 0}$, and then extend to all w). In particular, the exact sequence (1.5) is equivariant with respect to coordinate transforms if we define the action on the last term by $f(z) \mapsto j^{-w}f(\tilde{z})$.

If the coordinate change is a fractional linear transformation $\tilde{z} = g(z) = \frac{az+b}{cz+d}$ with $g = \left(\begin{smallmatrix} a & b \\ c & d \end{smallmatrix}\right) \in \mathrm{SL}(\mathbb{C})$, then $j = (cz+d)^{-2}$, all the terms multiplying ∂^{w-n} in (1.6) become proportional, and the equation simplifies to

$$\tilde{\partial}^w = [(cz+d)^2\,\partial]^w = \sum_{n=0}^{\infty} n!\binom{w}{n}\binom{w-1}{n}c^n(cz+d)^{2w-n}\,\partial^{w-n}. \tag{1.7}$$

(Again one proves this by induction for $w \in \mathbb{Z}_{\geq 0}$ and then extends to other values of w.) The action on the symbol, and hence on the last term in the sequence (1.5), is the classical action $f \mapsto f|_{-2w}g$, where $f|_k$ is defined for $k \in \mathbb{Z}$ by $(f|_k g)(z) := (cz+d)^{-k}f\left(\frac{az+b}{cz+d}\right)$. If we have a group $\Gamma \subset \mathrm{SL}(2,\mathbb{C})$, acting on R via its fractional linear action on $\mathbb{C}$, then we will denote by $M_k(R,\Gamma)$ or simply by $M_k(\Gamma)$ the space of invariants of R under the action $f \mapsto f|_k$ of Γ. If we take for R the ring $\mathcal{F}$ of all holomorphic functions in the complex upper half-plane $\mathcal{H}$ which are bounded by a power of $(|z|^2+1)/\Im(z)$, and Γ is a discrete subgroup of $\mathrm{SL}(2,\mathbb{R})$ of finite covolume, then $M_k(\Gamma)$ is the usual space of holomorphic modular forms on Γ and is finite-dimensional for all $k \in \mathbb{Z}$ and zero for $k < 0$, but we can also take larger rings of functions (like the ring of all holomorphic functions in $\mathcal{H}$, or all those of at most exponential growth at the cusps) to allow modular forms of negative weight. By taking Γ-invariants in (1.5) we get (with $k = -w$) a sequence

$$0 \to \Psi\mathrm{DO}(R)^\Gamma_{-k-1} \to \Psi\mathrm{DO}(R)^\Gamma_{-k} \to M_{2k}(\Gamma) \to 0 \tag{1.8}$$

which is exact except perhaps for the last arrow. The basic fact studied in this paper is the following proposition, which says that (1.5) has a canonical equivariant splitting and hence that the sequence (1.8) is exact and splits canonically:

Proposition 1. *For $k \geq 1$ define an operator $\mathcal{L}_k : R \to \Psi\mathrm{DO}(R)_{-k}$ by*

$$\mathcal{L}_k(f) = \sum_{n=0}^{\infty}(-1)^n\frac{(n+k)!\,(n+k-1)!}{n!\,(n+2k-1)!}f^{(n)}\partial^{-k-n}$$

and an operator $\mathcal{L}_{-k} : R \to \mathrm{DO}(R)_k$ *by*

$$\mathcal{L}_{-k}(f) = \sum_{n=0}^{k-1} \frac{(2k-n)!}{n!\,(k-n)!\,(k-n-1)!}\, f^{(n)}\, \partial^{k-n}\,,$$

and set $\mathcal{L}_0(f) = f$. *Then* $\mathcal{L}_k(f|_{2k}g) = \mathcal{L}_k(f) \circ g$ *for all* $g \in \mathrm{PSL}(2,\mathbb{C})$ *and all* $k \in \mathbb{Z}$. *In particular, if* $f \in M_{2k}(\Gamma)$ *for some subgroup* $\Gamma \subset \mathrm{PSL}(2,\mathbb{C})$ *then* $\mathcal{L}_k(f) \in \Psi\mathrm{DO}(R)_{-k}^{\Gamma}$.

Proof. Write $g = \begin{pmatrix} a & b \\ c & d \end{pmatrix}$. By induction on n we obtain the formula

$$\frac{d^n}{dz^n}\left(f|_k g(z)\right) = \sum_{r=0}^{n} \frac{n!}{r!} \binom{k+n-1}{n-r} \frac{(-c)^{n-r}}{(cz+d)^{k+n+r}} f^{(r)}\left(\frac{az+b}{cz+d}\right) \quad (1.9)$$

for any $k \in \mathbb{Z}$ and any $n \geq 0$, where $f^{(r)}$ denotes $\partial^r f$ as usual. From this and (1.7) we find that for $k > 0$ both $\mathcal{L}_k(f|_{2k}g)(z)$ and $(\mathcal{L}_k(f) \circ g)(z)$ are equal to

$$\sum_{r,\,m \geq 0} \frac{(m+r+k)!\,(m+r+k-1)!}{m!\,r!\,(2k+r-1)!} \frac{(-1)^r\, c^m}{(cz+d)^{2k+m}} f^{(r)}\left(\frac{az+b}{cz+d}\right) \partial^{-k-m-r}.$$

The proof for $k < 0$, is similar, and the case $k = 0$ is of course trivial. $\quad\square$

2. Interpretations and Extensions of the Basic Lifting

In this section we discuss some further aspects of the proposition just proved. In particular we describe the relationship between modular forms, invariant $\Psi\mathrm{DO}$'s, and "Jacobi-like forms" (this was the point of view taken in [Z1]), give a different and more conceptual proof of Proposition 1 in terms of the Casimir operator for $\mathrm{sl}(2,\mathbb{C})$, and describe an extension to generalized $\Psi\mathrm{DO}$'s where one allows non-integral powers of ∂.

Jacobi-like forms. One interpretation of the lifting from modular forms with respect to Γ to Γ-invariant $\Psi\mathrm{DO}$'s is to identify both spaces with the space $\mathcal{J}(\Gamma)$ of *Jacobi-like forms*, namely power series $\Phi(z,X) \in R[[X]]$ satisfying the transformation law

$$\Phi\left(\frac{az+b}{cz+d}, \frac{X}{(cz+d)^2}\right) = e^{cX/(cz+d)}\, \Phi(z,X) \qquad \forall \begin{pmatrix} a & b \\ c & d \end{pmatrix} \in \Gamma. \quad (2.1)$$

(Here Γ is a subgroup of $\mathrm{PSL}(2,\mathbb{R})$ and R a Γ-invariant ring of functions in $\mathcal{H}$, e.g., the ring $\mathcal{F}$ defined in Section 1.) This space is filtered by the subspaces $\mathcal{J}(\Gamma)_k = \mathcal{J}(\Gamma) \cap X^k R[[X]]$. Clearly, if $\Phi(z,X)$ belongs to $\mathcal{J}(\Gamma)_k$

and has leading term $f(z)X^k$, then $f|_{2k}\gamma = f$ for all $\gamma \in \Gamma$, so we have a sequence

$$0 \to \mathcal{J}(\Gamma)_{k+1} \to \mathcal{J}(\Gamma)_k \to M_{2k}(\Gamma) \to 0 \qquad (2.2)$$

which is exact except possibly at the last place. The following proposition, which is a sharpening of Prop. 1 for the case of positive weights, says that this sequence splits and is canonically isomorphic to the split short exact sequence (1.8).

Proposition 2. *Let $\phi_k = \phi_k(z)$ $(k = 1, 2, \ldots)$ be elements of R. Then the following are equivalent:*

(1) $\displaystyle \Phi(z, X) := \sum_{k=1}^{\infty} \phi_k(z)\,X^k \in \mathcal{J}(\Gamma)$;

(2) $\displaystyle \psi(z) := \sum_{k=1}^{\infty} (-1)^k\, k!\,(k-1)!\,\phi_k(z)\,\partial^{-k} \in \Psi\mathrm{DO}(R)^{\Gamma}$;

(3) $\displaystyle (\phi_k|_{2k}\gamma)(z) = \sum_{n=0}^{k-1} \frac{1}{n!}\left(\frac{c}{cz+d}\right)^n \phi_{k-n}(z)$ *for all* $k \geq 1$; $\gamma = \left(\begin{smallmatrix} a & b \\ c & d \end{smallmatrix}\right) \in \Gamma$;

(4) $\displaystyle \sum_{r=0}^{k-1} (-1)^r\, \frac{(2k-2-r)!}{r!}\, \phi_{k-r}^{(r)}(z) \in M_{2k}(\Gamma)$ *for all* $k \geq 1$;

(5) $\displaystyle \phi_n(z) = \sum_{r=0}^{n-1} \frac{1}{r!\,(2n-r-1)!}\, f_{n-r}^{(r)}(z)$ *where* $f_k \in M_{2k}(\Gamma)$ $(\forall k \geq 1)$.

Proof. One checks that each of the properties in question is equivalent to the transformation law (3). For (1) this is obvious from the definition (2.1), for (2) it follows directly from (1.7), and for (5) it follows from (1.9). Property (4) can be checked the same way or else we can note that by a simple binomial coefficient identity it is equivalent to (5) if we define f_k to be $2k - 1$ times the sum in (4). $\qquad\square$

We can restate the result of Proposition 2 in the following way. Denote by $\mathcal{M}(\Gamma)_+ = \prod_{n>0} M_{2n}(\Gamma)$ the space of sequences of modular forms of positive weights, with the trivial filtration by the subspaces $\mathcal{M}(\Gamma)_k = \prod_{n \geq k} M_{2n}(\Gamma)$, with successive quotients $\mathcal{M}(\Gamma)_k/\mathcal{M}(\Gamma)_{k-1} = M_{2k}(\Gamma)$. Proposition 2 says that $\mathcal{M}(\Gamma)_1$ is canonically isomorphic as a filtered vector space to both the space $\mathcal{J}(\Gamma)_1$ of Jacobi-like forms with no constant term and the space $\Psi\mathrm{DO}(R)_1^{\Gamma}$, the correspondence sending the sequence $(f_1, f_2, \ldots)$ $(f_k \in M_{2k}(\Gamma))$ to the elements $\Phi \in \mathcal{J}(\Gamma)_1$ and $\psi \in \Psi\mathrm{DO}(R)_1^{\Gamma}$ defined by (1) and (2), respectively. Note also that, by linearity, the Jacobi-like property of Φ and the Γ-invariance of ψ need only be checked in the case when there is only a single non-zero f_k. In this case, writing f for f_k, we find that the Φ is simply the *Cohen–Kuznetsov lifting*

$$\tilde{f}(z, X) = \sum_{n=0}^{\infty} \frac{f^{(n)}(z)}{n!(n + 2k - 1)!} X^{n+k} \qquad (2.3)$$

of f whose Jacobi-like property was discovered in [Ku] and [Co], while ψ is precisely the lifting $\mathcal{L}_k(f)$ of Proposition 1.

The reason for the name "Jacobi-like," by the way, is that the space $J(\Gamma)_k$ can be identified via $\phi(z, 2\pi i m u^2) = u^{2k}\phi(z, u)$ with the set of all $\phi(z, u) \in R[[u^2]]$ satisfying

$$\phi(\frac{az + b}{cz + d}, \frac{u}{cz + d}) = (cz + d)^{2k} e^{2\pi i c m u^2/(cz+d)} \phi(z, X)$$

for all $\left(\begin{smallmatrix} a & b \\ c & d \end{smallmatrix}\right) \in \Gamma$, and this is one of the two transformation laws characterizing Jacobi forms of weight k and index m in the sense of [EZ].

The Casimir operator. The proof of Proposition 1 by direct computation as given in Section 1 is very short, but not particularly enlightening. We now describe another way to see the existence (and uniqueness) of the equivariant splitting map $\mathcal{L}_k$ which was pointed out to us by Beilinson. Let $SL(2, \mathbb{C})$ act by fractional linear transformations as usual. The action of its Lie algebra $sl(2, \mathbb{C})$ is then given by the three vector fields $L_j = z^{j+1}\partial$ ($j = -1, 0, 1$), with Lie bracket given by commutation. There is an induced operation of $sl(2, \mathbb{C})$ on ΨDO's by commutation (adjoint representation). Explicitly, we have $L_{-1}(f\partial^w) = [\partial, f\partial^w] = f'\partial^w$ and similarly $L_0(f\partial^w) = (zf' - wf)\partial^w$, $L_1(f\partial^w) = (z^2 f' - 2wzf)\partial^w - w(w+1)zf\partial^{w-1}$, so a short computation shows that the Casimir operator

$$C = L_0^2 - \frac{1}{2}(L_1 L_{-1} + L_{-1} L_1),$$

which acts trivially on functions, acts on ΨDO's by

$$C(f\partial^w) = w(w + 1)f\partial^w + w(w - 1)f'\partial^{w-1}.$$

In particular, the induced action of C on the quotient

$$\Psi DO(R)_w / \Psi DO(R)_{w-1} \cong R$$

in (1.5) is multiplication by $w(w + 1)$, so if there is any equivariant splitting of this sequence then the lift ψ of $f \in R$ to $\Psi DO(R)_w$ must be an eigenvector of C with eigenvalue $w(w + 1)$. Writing $w = -k$ and $\psi(z) = \sum_{n=0}^{\infty} f_n(z) \partial^{-k-n}$, we find

$$[C - k(k - 1)]\psi = \sum_{n=1}^{\infty} [n(n + 2k - 1) f_n$$

$$+ (n + k)(n + k - 1) f_{n-1}'] \partial^{-k-n},$$

and equating all coefficients of this to 0 we find by induction that each f_n is a multiple of the nth derivative $f^{(n)}$ with coefficients as given in Proposition 1. (To get exactly the lift $\mathcal{L}_k$ rather than a multiple of it we must normalize by taking $f_0 = \lambda_k f$ with $\lambda_k = \binom{2k-1}{k}^{-1}$ if $k \geq 0$ and $\lambda_k = \frac{(2|k|)!}{|k|!(|k|-1)!}$ if $k < 0$.)

Generalized pseudodifferential operators. Since a ΨDO is defined as a formal expression anyway, one can allow symbols ∂^w with arbitrary complex powers w. Both the transformation property (1.6) of ΨDO's under changes of variables and the rule (1.2) for multiplying ΨDO's involve together with each power ∂^w all lower powers ∂^{w-n} with n a positive integer, so we again define ΨDO$(R)_w$ for any $w \in \mathbb{C}$ by equation (1.3) and define a generalized ΨDO as an element of any such space or a finite sum of such elements [KZ1]. Because formula (1.2) involves only binomial coefficients whose lower index is a nonnegative integer, and hence makes (formal) sense even for non-integral m and n, the space ΨDO$(R)_{\mathbb{C}}$ of generalized ΨDO's is a ring just as before, formula (1.4) still holds, and there is a direct sum decomposition

$$\Psi\text{DO}(R)_{\mathbb{C}} = \bigoplus_{w\in\mathbb{C}/\mathbb{Z}} \Psi\text{DO}(R)_{w+\mathbb{Z}}, \qquad \Psi\text{DO}(R)_{w+\mathbb{Z}} := \bigcup_{k\in\mathbb{Z}} \Psi\text{DO}(R)_{w+k}.$$

The summand ΨDO$(R)_{\mathbb{Z}}$ is the ring ΨDO(R) previously considered and each other summand ΨDO$(R)_{w+\mathbb{Z}}$ is a module over this ring and is filtered by the subspaces ΨDO$(R)_{w+n}$ ($n \in \mathbb{Z}$), and we again have the exact sequence (1.5).

    Formula (1.6) defines the behavior of the generalized ΨDO's under coordinate changes (again the binomial coefficients make sense even for w non-integral), and formula (1.7) their behavior under the action of $SL(2,\mathbb{C})$. Of course there is now a problem because the quantity j^{-w} or $(cz+d)^{2w}$ is not uniquely defined for w non-integral. This can be overcome in several ways. In the case when R is a space of functions on the upper half-plane $\mathcal{H}$, we replace the group $SL(2,\mathbb{R})$ by its universal covering, consisting of matrices $\begin{pmatrix} a & b \\ c & d \end{pmatrix} \in SL(2,\mathbb{R})$ together with a choice of logarithm of $cz+d$ in $\mathcal{H}$, and take for Γ a subgroup of this covering which maps isomorphically onto a discrete co-finite volume subgroup of $SL(2,\mathbb{R})$. In this case the elements of $M_{-2w}(\Gamma)$ are essentially what are classically known as modular forms with multiplier systems. This does not work for $SL(2,\mathbb{C})$ acting on $\mathbb{P}^1$ because then there is no global logarithm of $cz+d$. But actually, as we could have pointed out even when looking at the case of integral weight, there is no reason that we have to work with a ring R of functions defined globally on all of $\mathcal{H}$ or all of $\mathbb{P}^1$: all of the considerations in Section 1 were local, so in the formulas of that section we could always have considered ΨDO's $\psi(z) = \sum f_n(z)\partial^{w-n}$ defined for z in some open subset of $\mathbb{C}$, and coordinate changes $z \mapsto \tilde{z}$ mapping this set to some possibly different open subset. On a simply connected open set on which j or $cz + d$ has no zeros or poles, we can choose a branch of j^{-w} or $(cz + d)^{2w}$ and make sense of all the formulas we have been writing. The correct language to describe all of this is actually that of sheaves of D-modules over Riemann surfaces with a projective structure (i.e., having an atlas such that the coordinate transformation maps between charts are fractional linear), as will be discussed in Section 8. For now we will ignore this issue and use the same terminology as before, with the understanding that the results have to be interpreted in one of the ways just indicated. Proposition 1 then generalizes to the following result.

Proposition 3. *Let $w \in \mathbb{C}$, $2w$ not a nonnegative integer. Then the map*

$$\mathcal{D}_w : R \to \Psi\mathrm{DO}(R)_w\,, \qquad \mathcal{D}_w(f) = \sum_{n=0}^{\infty} \frac{\binom{w}{n}\binom{w-1}{n}}{\binom{2w}{n}} f^{(n)}\, \partial^{w-n} \qquad (2.4)$$

satisfies $\mathcal{D}_w(f|_{-2w}g) = \mathcal{D}_w(f) \circ g$ for all $g \in \mathrm{SL}(2, \mathbb{C})$, so $\mathcal{D}_w$ gives an equivariant splitting of the exact sequence (1.5). If w is a nonnegative integer, then the same assertion remains true if the sum in the definition of $\mathcal{D}_w$ is replaced by a sum from $n = 0$ to $n = w$. If w is a positive half-integer, then there is no equivariant splitting of the sequence (1.5).

Proof. For $w \in \mathbb{Z}$ this is the same as the statement of Proposition 1, since one easily checks that $\mathcal{D}_{-k} = \lambda_k \mathcal{L}_k$ for $k \in \mathbb{Z}$, with λ_k defined as at the end of the last subsection. Both the proof by direct computation given in Section 1 and the proof given above using the Casimir operator given above apply unchanged for general w (with the change of notation that we again use w instead of $k = -w$, which was more convenient before because classical modular forms have positive weight). The proof using the Casimir operator showed the uniqueness of the lift and hence also its non-existence in the case when w is a positive half-integer (corresponding to modular forms of odd negative weight), since the recursive relation $n(n - 2w - 1) f_n = -(n - w)(n - w - 1) f'_{n-1}$ cannot be solved in general for $n = 2w + 1$. In this case there is a lifting if and only if f is a polynomial of weight $\leq 2w$. $\qquad\square$

3. A Non-Commutative Multiplication of Modular Forms

Let Γ be a discrete subgroup of $\mathrm{PSL}(2, \mathbb{R})$. As explained in the last section, one interpretation of Proposition 1 is that $\Psi\mathrm{DO}(R)^{\Gamma}$ is canonically isomorphic to the space $\mathcal{M}(\Gamma) = \prod_{k \gg -\infty} M_{2k}(\Gamma)$ of semi-infinite sequences of modular forms on Γ (i.e., sequences $f_k \in M_{2k}(\Gamma)$ with $f_k = 0$ for all but finitely many negative k; in the first subsection of Section 2 we looked only at the subspace $\mathcal{M}(\Gamma)_+$ of sequences of forms of positive weight). On the other hand, the product of Γ-invariant ΨDO's is again Γ-invariant, so there is an induced non-commutative ring structure on $\mathcal{M}(\Gamma)$. In this section we describe it explicitly in terms of the "Rankin–Cohen brackets." These are the bilinear maps

$$[\, , \,]_n = [\, , \,]_n^{(k,l)} : M_{2k} \otimes M_{2l} \to M_{2k+2l+2n} \qquad (k, l \in \mathbb{Z}, \quad n \in \mathbb{Z}_{\geq 0})$$

defined by the formula

$$[f,g]_n^{(k,l)}(z) = \sum_{m=0}^{n} (-1)^m \binom{2k+n-1}{n-m} \binom{2l+n-1}{m} f^{(m)}(z)\, g^{(n-m)}(z).$$

$$(3.1)$$

(Here $\phi^{(m)} = \partial^m \phi$ as usual, and we have dropped the Γ in the notation for spaces of modular forms; we will also usually omit the superscripts "(k,l)" on the brackets except when necessary for clarity, since we will always apply them with superscripts equal to half the weights of the arguments.) They were introduced and shown to be modular in 1974 by H. Cohen [Co], this result being a special case of a general theorem of Rankin [Ra] describing all multilinear differential operators which send modular forms to other modular forms. The easiest proof of the modularity of $[f,g]_n$ is to use the Cohen–Kuznetsov lifting (2.3) from modular forms to Jacobi-like forms: the transformation law (2.1) shows that the product $\tilde{f}(z, -X)$ and $\tilde{g}(z, X)$ is invariant under $(z, X) \mapsto \left(\frac{az+b}{cz+d}, \frac{X}{(cz+d)^2} \right)$, which means that the coefficient of X^{k+l+n} in this product is modular of weight $2k + 2l + 2n$, and this coefficient is just a scalar multiple of $[f,g]_n$. It is also easy to see that the combination (3.1) is the only universal bilinear combination of derivatives of f and g which goes from $M_{2k} \otimes M_{2l}$ to $M_{2k+2l+2n}$.

Proposition 4. *For integers* n, k, $l \geq 0$ *define coefficients* $t_n(k,l)$ *by*

$$t_n(k,l) = \frac{1}{\binom{-2l}{n}} \sum_{r+s=n} \frac{\binom{-k}{r}\binom{-k-1}{r}}{\binom{-2k}{r}} \frac{\binom{n+k+l}{s}\binom{n+k+l-1}{s}}{\binom{2n+2k+2l-2}{s}}. \qquad (3.2)$$

Then the multiplication μ *on* $\mathcal{M}(\Gamma)$ *defined by*

$$\mu(f,g) = \sum_{n=0}^{\infty} t_n(k,l)\, [f,g]_n^{(k,l)} \qquad (f \in M_{2k}(\Gamma),\ g \in M_{2l}(\Gamma))$$

is associative and the lifting map $\mathcal{D} = \prod_w \mathcal{D}_w : \mathcal{M}(\Gamma) \to \Psi\mathrm{DO}(R)^{\Gamma}$ *is a ring homomorphism with respect to this multiplication.*

Proof. As already mentioned, the isomorphism between $\mathcal{M}(\Gamma)$ and $\Psi\mathrm{DO}(R)^{\Gamma}$ permits us to transfer the non-commutative structure on the latter space to the former one, i.e., to associate to $f \in M_{2k}$ and $g \in M_{2l}$ a unique sequence of elements $h_n \in M_{2k+2l+2n}$ $(n = 0, 1, \ldots)$ such that $\mathcal{D}_{-k}(f)\mathcal{D}_{-l}(g) = \sum_{n=0}^{\infty} \mathcal{D}_{-k-l-n}(h_n)$. The map $(f, g) \mapsto h_n$ from $M_{2k} \otimes M_{2l}$ to $M_{2k+2l+2n}$ is expressed by a universal formula as a linear combination of products of the first n derivatives of f and g, so by the uniqueness mentioned above it must be a multiple of the Rankin–Cohen bracket, i.e., we have $h_n = t_n [f, g]_n$ for all $n \geq 0$, where the coefficient t_n depends only on n and on the weights k and l. Substituting the definitions of the Rankin–Cohen brackets and of $\mathcal{D}$ and multiplying everything out, we obtain a rather complicated identity which overdetermines the coefficients t_n: for each $m \geq 0$ the comparison of the coefficients of $f^{(n)}g^{(m)}\partial^{-k-l-n-m}$ on the two sides of the equation for $n = 0, 1, 2, \ldots$ gives an infinite sequence of equations which inductively determine the coefficients t_n. For $m = 0$ these equations are

$$\frac{\binom{n+k}{n}\binom{n+k-1}{n}}{\binom{n+2k-1}{n}} = \sum_{r+s=n} \frac{\binom{2l+r-1}{r}\binom{n+k+l}{s}\binom{n+k+l-1}{s}}{\binom{n+r+2k+2l-1}{s}} t_r,$$

and this can easily be inverted to yield the formula for $t_n = t_n(k, l)$ given in the proposition.   □

Computing the first three coefficients t_n from (3.2), we find

$$t_0 = 1, \qquad t_1 = -\frac{1}{4}, \qquad t_2 = \frac{1}{16}\left(1 + \frac{3}{(2k+1)(2l+1)(2k+2l+1)}\right),$$
$$(3.3)$$

and computing a few more coefficients we are led to conjecture the formula

$$t_n(k, l) = \left(-\frac{1}{4}\right)^n \sum_{j \geq 0} \binom{n}{2j} \frac{\binom{-\frac{3}{2}}{j}\binom{-\frac{1}{2}}{j}\binom{\frac{1}{2}}{j}}{\binom{-k-\frac{1}{2}}{j}\binom{-l-\frac{1}{2}}{j}\binom{n+k+l-\frac{3}{2}}{j}}. \qquad (3.4)$$

(Note that the sum on the right is finite since $\binom{n}{2j}$ vanishes for $j > n/2$.) The equivalence of (3.2) and (3.4) is a special case of the following result, whose fairly complicated proof will be given in a separate paper [Z2].

Identity. *For an integer $n \geq 0$ and variables X, Y, Z satisfying $X + Y + Z = n - 1$, we have*

$$\frac{(-4)^n}{\binom{2X}{n}} \sum_{r+s=n} \frac{\binom{Y}{r}\binom{Y-1}{r}}{\binom{2Y}{r}} \frac{\binom{Z}{s}\binom{Z+1}{s}}{\binom{2Z}{s}} = \sum_{j \geq 0} \binom{n}{2j} \frac{\binom{-\frac{3}{2}}{j}\binom{-\frac{1}{2}}{j}\binom{\frac{1}{2}}{j}}{\binom{X-\frac{1}{2}}{j}\binom{Y-\frac{1}{2}}{j}\binom{Z-\frac{1}{2}}{j}}. \qquad (3.5)$$

A first corollary of the identity (3.4) is that the coefficient $t_n(k, l)$ is symmetric in k and l, a property which is not at all obvious from the definition (the product of ΨDO's is neither commutative nor anti-commutative) and is also not at all obvious from the closed formula (3.2). But in fact there is an even less obvious three-fold symmetry which is seen best in the formulation (3.5): even though the expression on the left apparently has a slightly different dependence on Y and Z, and a totally different dependence on X, the identity shows that it is in fact symmetric in all three variables. Going back to the special case (3.4), we can rewrite these properties as

$$t_n(k,l) = t_n(l,k) = t_n(k, 1 - n - k - l) \qquad \forall k, l \in \mathbb{Z}, \quad n \in \mathbb{Z}_{\geq 0}. \quad (3.6)$$

(This makes sense because the denominators in (3.4) do not vanish for any integral values of k and l and the sum is finite, so that $t_n(k,l)$ is defined for all $k, l \in \mathbb{Z}$.) An explanation of these symmetries in terms of residues will be given in Section 6.

4. Conjugate-Automorphic ΨDO's and New Multiplications on $\mathcal{M}(\Gamma)$

In this section we will show how to generalise the above discussion to produce a whole family of new associative multiplications. The starting point for this was an observation by W. Eholzer, who discovered (and verified for the first few terms of the expansion) that the anti-commutative bracket

$$[f,g]_E := \sum_{n \text{ odd}} [f,g]_n \qquad (4.1)$$

satisfies the Jacobi identity and hence equips $\mathcal{M}(\Gamma)$ with the structure of a Lie algebra. Since the nth Rankin–Cohen bracket is $(-1)^n$-symmetric, the bracket $[f,g]_E$ is just the odd part $\frac{1}{2}(f*g - g*f)$ of the *Eholzer product*

$$f * g := \sum_{n=0}^{\infty} [f,g]_n. \qquad (4.2)$$

So Eholzer's observation suggested the following result:

Proposition 5. *The multiplication $*$ on $\mathcal{M}(\Gamma)$ defined by (4.2) is associative.*

Comparing this statement with Proposition 4, we see that both have the same form, except that the complicated coefficients $t_n = t_n(k,l)$ defined by (3.2) or (3.4) are replaced simply by 1. On the other hand, from the special cases in (3.3) we see that the coefficients $(-4)^n t_n$ (where the factor $(-4)^n$ of course does not affect the associativity of the product $\sum t_n[f,g]_n$) are a kind of "small deformation" of 1. This suggested that there might be a whole family of multiplications of $\mathcal{M}(\Gamma)$ of which both Propositions 4 and 5 are specializations, and after a fair amount of experimentation a formula which worked empirically was discovered:

Theorem 1. *For $\kappa \in \mathbb{C}$ define coefficients $t_n^\kappa(k,l)$ $(n = 0, 1, 2, \ldots)$ by*

$$t_n^\kappa(k,l) = \left(-\frac{1}{4}\right)^n \sum_{j \geq 0} \binom{n}{2j} \frac{\binom{-\frac{1}{2}}{j}\binom{\kappa-\frac{3}{2}}{j}\binom{\frac{1}{2}-\kappa}{j}}{\binom{-k-\frac{1}{2}}{j}\binom{-l-\frac{1}{2}}{j}\binom{n+k+l-\frac{3}{2}}{j}} \cdot \tag{4.3}$$

Then the multiplication μ^κ on $\mathcal{M}(\Gamma)$ defined by

$$\mu^\kappa(f,g) = \sum_{n=0}^{\infty} t_n^\kappa(k,l)\, [f,g]_n^{(k,l)} \qquad (f \in M_{2k}(\Gamma),\ g \in M_{2l}(\Gamma)) \tag{4.4}$$

is associative.

The first few coefficients $t_n = t_n^{(s)}(k,l)$ are

$$t_0 = 1, \qquad t_1 = -\frac{1}{4}, \qquad t_2 = \frac{1}{16}\left(1 + \frac{(1-2\kappa)(3-2\kappa)}{(2k+1)(2l+1)(2k+2l+1)}\right).$$

From these special cases or from the formula (4.3) we again see non-trivial symmetries, namely

$$t_n^\kappa(k,l) = t_n^\kappa(l,k) = t_n^\kappa(k, 1-n-k-l) \tag{4.5}$$

(generalizing (3.6)) and

$$t_n^\kappa(k,l) = t_n^{2-\kappa}(k,l) \tag{4.6}$$

(which says that the multiplications μ^κ and $\mu^{2-\kappa}$ coincide). We will discuss both of these equations in Section 6 in terms of the residue map and the duality between automorphic forms of weights κ and $2 - \kappa$. We note that Proposition 4 is the special case $\kappa = 0$ (or $\kappa = 2$) of Theorem 1 and Proposition 5 (up to a harmless rescaling of t_n by $(-4)^n$) is the special case $\kappa = 1/2$ or $\kappa = 3/2$. Another interesting special case is given by taking $\kappa = 1/\varepsilon$, multiplying t_n by a factor $(-4\varepsilon)^n$ (again, this does not affect the statement about associativity), and letting ε tend to 0. The resulting coefficient $t_n^{(\infty)}(k,l)$ is simpler than the general coefficient t_n^κ, since in the limit all terms in (4.3) except the one (if any) with $2j = n$ vanish and we have

$$t_{2j}^{(\infty)}(k,l) = \binom{-\frac{1}{2}}{j} \Big/ j!^2 \binom{k+j-\frac{1}{2}}{j}\binom{l+j-\frac{1}{2}}{j}\binom{k+l+2j-\frac{3}{2}}{j},$$

$$t_{2j+1}^{(\infty)}(k,l) = 0.$$

The vanishing of $t_n^{(\infty)}(k,l)$ for n odd means that the corresponding multiplication $\mu^{(\infty)}$, unlike the multiplications μ^κ for κ finite, is commutative.

Problem.   Find a natural interpretation for the ring structure $\mu^{(\infty)}$ on $\mathcal{M}(\Gamma)$.

We now turn to the proof of Theorem 1. One can prove it by direct combinatorial manipulation of the sums of binomial coefficients involved. However, this proof is not only very laborious, but also does not explain where the new multiplications come from. Instead we give a proof using pseudodifferential operators with a new invariance property. Namely, we can use the non-commutativity of the ring of ΨDO's to define a "twisted" action of $SL(2, \mathbb{C})$ by

$$(\psi|_\kappa g)(z) = (cz+d)^{-\kappa}\, \psi\left(\frac{az+b}{cz+d}\right)(cz+d)^{\kappa} \qquad \left(\kappa \in \mathbb{C}, \quad g = \begin{pmatrix} a & b \\ c & d \end{pmatrix}\right).$$
$$(4.7)$$

Note that this makes sense even for non-integral κ since any two determinations of the factor $(cz+d)^{\kappa}$ differ by a scalar factor and scalars commute with ΨDO's. If $\Gamma \subset PSL(2, \mathbb{C})$ is a group acting on the ring R as usual, then we call an element of $\Psi DO(R)$ which is Γ-invariant with respect to the action (4.7), i.e., which satisfies

$$\psi\left(\frac{az+b}{cz+d}\right) = (cz+d)^{\kappa}\psi(z)(cz+d)^{-\kappa} \qquad \text{for all } \left(\begin{smallmatrix} a & b \\ c & d \end{smallmatrix}\right) \in \Gamma, \qquad (4.8)$$

a *conjugate-automorphic pseudodifferential operator of weight κ with respect to* Γ. We denote the space of such elements by $\Psi DO(\Gamma)^{\kappa}$ and write $\Psi DO(\Gamma)_w^{\kappa}$ for its intersection with $\Psi DO(R)_w$. (We omit R from the notation; usually we think of the case when Γ is a discrete subgrop of $PSL(2, \mathbb{R})$ and $R = \mathcal{F}$.) Since conjugation of a ΨDO by a function does not change the leading term (symbol), we see that the exact sequence (1.5) is equivariant with respect to the action of Γ on the first two terms by (4.7) and on the last term by $|_{-2w}$, so taking invariants we get a sequence

$$0 \to \Psi DO(\Gamma)_{w-1}^{\kappa} \to \Psi DO(\Gamma)_w^{\kappa} \to M_{-2w}(\Gamma) \to 0 \qquad (4.9)$$

which is exact except possibly for the final term. We then have the following generalization of Proposition 3:

Proposition 6. *The map $\mathcal{D}_w^{\kappa} : R \to \Psi DO(R)_w$ defined by*

$$\mathcal{D}_w^{\kappa}(f) = \sum_{n=0}^{\infty} \frac{\binom{w}{n}\binom{w+\kappa-1}{n}}{\binom{2w}{n}} f^{(n)}\, \partial^{w-n} \qquad (4.10)$$

(where the sum must be replaced by $\sum_{n=0}^{w}$ if w is a nonnegative integer and is not defined if w is a positive half-integer) satisfies $\mathcal{D}_w^{\kappa}(f|_{-2w}g) = \mathcal{D}_w^{\kappa}(f)|_{\kappa}g$ for all $g \in SL(2, \mathbb{C})$. In particular, the sequence (4.9) is exact and splits canonically.

Proof. The proof, either by direct calculation, via Jacobi-like forms, or using the Casimir operator, is exactly the same as before. $\qquad\square$

Now we proceed just as in Section 3. The lifting $\mathcal{D}^\kappa = \prod_k \mathcal{D}_w^\kappa$ gives an isomorphism from $\mathcal{M}(\Gamma)$ to $\Psi\mathrm{DO}(R,\Gamma)^\kappa$, the inverse map being given explicitly by

$$\sum_{n\ll\infty} g_n\,\partial^n \mapsto \{f_k\}_{k\gg-\infty}, \qquad f_k = \sum_{r\geq 0} \frac{\binom{k-1}{r}\binom{k-\kappa}{r}}{\binom{2k-2}{r}}\,g_{r-k}^{(r)} \in M_{2k}, \quad (4.11)$$

generalizing (4) of Proposition 2, Section 2. On the other hand, it is clear that the product of two conjugate-automorphic $\Psi\mathrm{DO}$'s of weight κ is again conjugate-automorphic of the same weight, so by transporting the multiplication of $\Psi\mathrm{DO}$'s to $\mathcal{M}(\Gamma)$ by $\mathcal{D}^\kappa$ we get a new ring structure μ^κ on $\mathcal{M}(\Gamma)$. Again the uniqueness of the Rankin–Cohen brackets says that we must have $\mathcal{D}^\kappa(f)\mathcal{D}^\kappa(g) = \sum_n t_n[f,g]_n$ for all $f \in M_{2k}$, $g \in M_{2l}$ for some universal coefficients $t_n = t_n^\kappa(k,l)$, and by substituting all definitions and multiplying out what this says we get an infinite sequence of equations for the t_n of which the simplest is

$$\frac{\binom{n+k-\kappa}{n}\binom{n+k-1}{n}}{\binom{n+2k-1}{n}} = \sum_{r+s=n} \frac{\binom{2l+r-1}{r}\binom{n+k+l-\kappa}{s}\binom{n+k+l-1}{s}}{\binom{n+r+2k+2l-1}{s}}\,t_r\,.$$

Inverting this as in the previous case we find the closed formula

$$t_n^\kappa(k,l) = \frac{1}{\binom{-2l}{n}} \sum_{r+s=n} \frac{\binom{-k}{r}\binom{-k-1+\kappa}{r}}{\binom{-2k}{r}}\frac{\binom{n+k+l-\kappa}{s}\binom{n+k+l-1}{s}}{\binom{2n+2k+2l-2}{s}}\,. \quad (4.12)$$

That this is equivalent to (4.3) follows from the following generalization of the identity given in Section 3, and whose proof again will be postponed to the paper [Z2]:

Identity. *For an integer $n \geq 0$ and variables a, x, y, z satisfying $x + y + z = n - 1$, we have*

$$\frac{(-4)^n}{\binom{2x}{n}} \sum_{r+s=n} \frac{\binom{y}{r}\binom{y-a}{r}}{\binom{2y}{r}}\frac{\binom{z}{s}\binom{z+a}{s}}{\binom{2z}{s}} = \sum_{j\geq 0} \binom{n}{2j}\frac{\binom{-\frac{1}{2}}{j}\binom{a-\frac{1}{2}}{j}\binom{-a-\frac{1}{2}}{j}}{\binom{x-\frac{1}{2}}{j}\binom{y-\frac{1}{2}}{j}\binom{z-\frac{1}{2}}{j}}\,. \quad (4.13)$$

(In our case $x = -l$, $y = -k$, $z = n + k + l - 1$, and $a = 1 - \kappa$.) Again this identity reveals surprising "hidden symmetries": the left-hand side is symmetric under interchanging y and z and simultaneously replacing a by $-a$ and has no other evident symmetries, but the identity shows that it is in fact symmetric in all three variables x, y, z and at the same time an even function of a. In terms of the coefficients $t_n^\kappa(k, l)$, these symmetries become the equations (4.5) and (4.6) mentioned above.

As a final remark, we observe that in the special case $\kappa = 1/2$ corresponding to the Eholzer multiplication (4.2), not only the multiplication but also the formula for the lifting map $\mathcal{D}^\kappa$ simplifies, since (4.10) becomes simply

$$\mathcal{D}_w^{1/2}(f) = \sum_{n=0}^{\infty} 4^n \binom{2w - n}{n} f^{(n)} \partial^{w-n} . \qquad (4.14)$$

5. Automorphic ΨDO's of Mixed Weight

We can generalize still further by considering the action of Γ defined by

$$(\psi|_{\kappa_1,\kappa_2}\gamma)(z) = (cz + d)^{-\kappa_1} \psi\left(\frac{az + b}{cz + d}\right)(cz + d)^{\kappa_2} \qquad \left(\gamma = \begin{pmatrix} a & b \\ c & d \end{pmatrix} \in \Gamma\right)$$

where κ_1 and κ_2 are complex constants. If κ_1 and κ_2 differ by an integer, then this makes sense independently of the branch of $\log(cz + d)$ chosen; if not, then we either have to pick a lifting from Γ to the universal cover of $\mathrm{SL}(2, \mathbb{R})$ or else work with locally defined functions, as discussed at the end of Section 2. We call the elements of $\Psi\mathrm{DO}(R)$ which are Γ-invariant with respect to this action, i.e., which satisfy the transformation law

$$\psi\left(\frac{az + b}{cz + d}\right) = (cz + d)^{\kappa_1} \psi(z) (cz + d)^{-\kappa_2} \qquad \forall \begin{pmatrix} a & b \\ c & d \end{pmatrix} \in \Gamma, \qquad (5.1)$$

automorphic pseudodifferential operators of mixed weight (κ_1, κ_2) with respect to Γ. We denote the space of such operators by $\Psi\mathrm{DO}(\Gamma)^{\kappa_1,\kappa_2}$ and its intersection with $\Psi\mathrm{DO}(R)_w$ by $\Psi\mathrm{DO}(\Gamma)_w^{\kappa_1,\kappa_2}$. If $\psi(z) = \sum_{n\geq0} f_n(z)\partial^{w-n}$ belongs to this latter space, then its leading coefficient f_0 is Γ-invariant with respect to the action $|_{\kappa_1-\kappa_2-2w}$, so the sequence (4.9) generalizes to

$$0 \to \Psi\mathrm{DO}(\Gamma)_{w-1}^{\kappa_1,\kappa_2} \to \Psi\mathrm{DO}(\Gamma)_w^{\kappa_1,\kappa_2} \to M_{\kappa_1-\kappa_2-2w}(\Gamma) \to 0 \qquad (5.2)$$

and the liftings described in the previous sections to the following proposition:

Proposition 7. *The map $\mathcal{D}_w^{\kappa_1,\kappa_2} : R \to \Psi\mathrm{DO}(R)_w$ defined by*

$$\mathcal{D}_w^{\kappa_1,\kappa_2}(f) = \sum_{n=0}^{\infty} \frac{\binom{w}{n}\binom{w+\kappa_2-1}{n}}{\binom{\kappa_2-\kappa_1+2w}{n}} f^{(n)} \partial^{w-n} , \qquad (5.3)$$

where the upper index in the sum must be replaced by $\sum_{n=0}^{w}$ if w is a non-negative integer and values of w for which the denominator of any of the coefficients vanishes must be excluded, satisfies

$$(\mathcal{D}_w^{\kappa_1,\kappa_2} f)|_{\kappa_1,\kappa_2}g = \mathcal{D}_w^{\kappa_1,\kappa_2}(f|_{\kappa_1-\kappa_2-2w}g) \qquad \forall g \in \mathrm{SL}(2,\mathbb{C}). \qquad (5.4)$$

In particular, the sequence (5.2) *is exact and splits canonically.*

Just as before, we could prove the proposition by direct computations as in Section 1 or else by an argument using Jacobi-like forms or the Casimir operator as in Section 2. Now, however, there is a new argument which is perhaps the simplest of all. In the special case when $w = n$ is a nonnegative integer, the lifting $\mathcal{D}_w^{\kappa_1,\kappa_2}(f)$ of an element $f \in M_{\kappa_1-\kappa_2-2n}$ is a differential rather than a pseudodifferential operator, and hence acts on functions. Moreover, it is clear from the transformation law (5.1) that if $g \in M_{\kappa_2}(\Gamma)$ and $\psi \in DO(\Gamma)^{\kappa_1,\kappa_2}$, then the image $\psi(z)g(z)$ belongs to $M_{\kappa_1}(\Gamma)$. Hence, changing notation from κ_1, κ_2 to $2k = \kappa_1 - \kappa_2 - 2n$, $2l = \kappa_2$, we see that the map $f \otimes g \mapsto (\mathcal{D}_n^{2k+2l+2n,2l} f)(g)$ goes from $M_{2k}(\Gamma) \otimes M_{2l}(\Gamma)$ to $M_{2k+2l+2n}(\Gamma)$, and comparing the definition (5.3) with the definition (3.1), we see that this map is, up to a scalar, nothing else than the Rankin–Cohen bracket (as indeed it must be by the uniqueness of the latter). Turning this around, the fact that the Rankin–Cohen bracket is given in terms of derivatives means that, for a fixed $f \in M_{2k}(\Gamma)$, the operator $[f, \cdot]_n^{(k,l)}$ is a differential operator which sends $M_{2l}(\Gamma)$ to $M_{2k+2l+2n}(\Gamma)$ and hence satisfies the transformation law (5.2) (with $\kappa_1 = 2k + 2l + 2n$, $\kappa_2 = 2l$), so that the modularity property of the bracket implies the equivariance property of the lifting (5.3) in this case. Since this equivariance property is at each level equivalent to a finite number of binomial coefficient identities, and since this argument shows that these identities (which are polynomial in κ_1 and κ_2) are true for infinitely many values κ_1, κ_2, this special case is enough to prove the proposition.

Now just as in the previous cases $\kappa_1 = \kappa_2 = 0$ and $\kappa_1 = \kappa_2 = \kappa$, this proposition induces an isomorphism between $\mathcal{M}(\Gamma)$ and $\Psi DO(\Gamma)^{\kappa_1,\kappa_2}$. However, the latter space is no longer a ring, so this does not directly induce a single multiplication on the space of modular forms. Instead, we clearly have

$$\Psi DO(\Gamma)_{w_1}^{\kappa_1,\kappa_2} \Psi DO(\Gamma)_{w_2}^{\kappa_2,\kappa_3} \subseteq \Psi DO(\Gamma)_{w_1+w_2}^{\kappa_1,\kappa_3} \tag{5.5}$$

(if we restrict this to differential operators rather than ΨDO's then $DO(R)^{\kappa_1,\kappa_2}$ can be thought of as giving homomorphisms from M_{κ_2} to M_{κ_1} as just explained, and this is just the composition of homomorphisms) and combining this with the lifting of Proposition 7 we get a corresponding collection of multiplication maps $\mu^{\kappa_1,\kappa_2,\kappa_3}$ on $\mathcal{M}(\Gamma)$ which satisfy the evident associativity property (groupoid structure). These multiplications must again be expressible in terms of Rankin–Cohen brackets, i.e., we must have

$$\mathcal{D}_{w_1}^{\kappa_1,\kappa_2}(f)\,\mathcal{D}_{w_2}^{\kappa_2,\kappa_3}(g) = \sum_{n=0}^{\infty} t_n^{\kappa_1,\kappa_2,\kappa_3}(k,l)\,\mathcal{D}_{w_1+w_2-n}^{\kappa_1,\kappa_3}([f,g]_n) \tag{5.6}$$

for some numerical coefficients $t_n^{\kappa_1,\kappa_2,\kappa_3}(k,l)$, where $2k$ and $2l$ are the weights of f and g and $w_1 = \frac{1}{2}(\kappa_1 - \kappa_2) - k$, $w_2 = \frac{1}{2}(\kappa_2 - \kappa_3) - l$. These coefficients can be evaluated as before to give the formula

$$t_n^{\kappa_1,\kappa_2,\kappa_3}(k,l) = \left(-\frac{1}{4}\right)^n T_n\left(1-\kappa_1, 1-\kappa_3, 1-\kappa_2; -l, -k, k+l+n-1\right),$$
(5.7)

where $T_n(a,b,c;x,y,z)$ is defined for a non-negative integer n and variables a, b, c, x, y, z with $x+y+z=n-1$ by the formula

$$T_n(a,b,c;x,y,z) = \frac{(-4)^n}{\binom{2x}{n}} \sum_{r+s=n} \frac{\left(y-\frac{a+c}{2}\right)\left(y+\frac{c-a}{2}\right)}{\binom{2y}{r}} \frac{\left(z+\frac{a-b}{2}\right)\left(z+\frac{a+b}{2}\right)}{\binom{2z}{s}}, \quad (5.8)$$

which reduces to the left-hand side of (4.13) in case $a=b=c$.

In Section 6 we will use the interpretation (5.7) of the numbers T_n to prove the following purely combinatorial result.

Theorem 2. *The coefficient $T_n(a,b,c;x,y,z)$ is symmetric in the three pairs of variables (a,x), (b,y), and (c,z) and is an even function of a, b and c.*

As examples of the theorem, we found (with effort!) the symmetric expressions

$$T_0 = 1,$$

$$T_1 = 1 + \frac{1}{4}\left(\frac{a^2}{yz} + \frac{b^2}{xz} + \frac{c^2}{xy}\right) \qquad (x+y+z=0),$$

$$\begin{aligned}
T_2 = {} & 1 + \frac{1}{2}\left(\frac{a^2}{yz} + \frac{b^2}{xz} + \frac{c^2}{xy}\right) - \frac{1}{(2x-1)(2y-1)(2z-1)} \\
& + \frac{1}{4}\left(\frac{a^2(a^2-2)}{yz(2y-1)(2z-1)} + \frac{b^2(b^2-2)}{xz(2x-1)(2z-1)}\right. \\
& \left. + \frac{c^2(c^2-2)}{xy(2x-1)(2y-1)}\right) \\
& + \frac{1}{4xyz}\left(\frac{b^2c^2}{2x-1} + \frac{a^2c^2}{2y-1} + \frac{a^2b^2}{2z-1}\right) \qquad (x+y+z=1).
\end{aligned}$$

The expression for T_1 clearly simplifies to 1 if $a=b=c$, but already for $n=2$ the verification that T_2 reduces to $1+(4a^2-1)/(2x-1)(2y-1)(2z-1)$ when $a=b=c$, as it must by (4.13), requires the non-obvious identities

$$\begin{aligned}
& \frac{1}{zx} + \frac{1}{xy} + \frac{1}{yz} - \frac{8}{(2x-1)(2y-1)(2z-1)} \\
& = \frac{-1}{xyz}\left(\frac{1}{2x-1} + \frac{1}{2y-1} + \frac{1}{2z-1}\right) \\
& = \frac{1}{zx(2z-1)(2x-1)} + \frac{1}{xy(2x-1)(2y-1)} + \frac{1}{yz(2y-1)(2z-1)}
\end{aligned}$$

for variables x, y, z with $x + y + z = 1$. It would be nice to find a direct combinatorial proof of Theorem 2 or, even better, a closed formula for $T_n(a, b, c; x, y, z)$ which

(a) makes the symmetries stated in Theorem 2 evident, and

(b) reduces term-by-term to the right-hand side of (4.13) when $a = b = c$,

but so far we could not find a formula having either one of these properties.

6. Residues, Duality, and Symmetry

In the previous section we found a striking symmetry among the *three* weights k, l, and $m := 1 - k - l - n$ in the formulas giving the coefficients of the nth bracket $[f, g]_n$ ($f \in M_{2k}$, $g \in M_{2l}$) in the various multiplications on $\mathcal{M}(\Gamma)$. To explain it, we use the *non-commutative residue map*

$$\mathrm{Res}_\partial : \sum_m h_m(z)\partial^m \mapsto h_{-1}(z)dz \in H(R) := \Omega^1(R)/d\Omega^0(R),$$

where $\Omega^1(R) = R\,dz$ denotes the space of formal differentials $f(z)\,dz$ ($f \in R$) and $d\Omega^0(R) = dR$ the subspace of exact differentials $f'(z)dz$, $f \in R$. This residue map was introduced in [Ma2] and shown to have the properties

$$\mathrm{Res}_\partial(\psi \circ g) = \mathrm{Res}_\partial(\psi) \circ g \tag{6.1}$$

for any ψ and any holomorphic map $z \mapsto g(z)$ (invariance under holomorphic change of coordinates) and

$$\mathrm{Res}_\partial(\psi_1(z)\psi_2(z)) = \mathrm{Res}_\partial(\psi_2(z)\psi_1(z)) \tag{6.2}$$

for any two ΨDO's ψ_1 and ψ_2 (trace property). The invariance under changes of variables implies in particular that Res_∂ maps $\Psi\mathrm{DO}(R)^\Gamma$ to $H(R)^\Gamma$ if Γ is a group of fractional linear transformations acting on R, and the conjugacy-invariance property (6.2) implies that the same is true for the space $\Psi\mathrm{DO}(R, \Gamma)^\kappa$ of conjugate-automorphic ΨDO's of arbitrary (complex) weight κ. The space $H(R)^\Gamma$ is isomorphic via $f(z)\,dz \mapsto f(z)$ to the space $H(R, \Gamma) = M_2(\Gamma)/\partial(M_0(\Gamma))$, and by abuse of notation we will simply identify these spaces and write Res_∂ for the corresponding map $\Psi\mathrm{DO}(R, \Gamma)^\kappa \to H(R, \Gamma)$. We must choose R large enough that there are plenty of modular forms of positive and negative weight, so that we can test an identity in $M_{2k}(\Gamma)$ by checking whether its product with an arbitrary element of $M_{2-2k}(R, \Gamma)$ is 0 in $H(R, \Gamma)$. For instance, we could take R to be the set of all functions which are meromorphic in the upper half-plane including the cusps, or the subspace of those which are holomorphic outside some specified non-empty Γ-invariant set.

We also define a projection map P from $M_*(\Gamma) = \oplus_k M_{2k}(\Gamma)$ to $H(R,\Gamma)$ by sending $f \in M_{2k}(\Gamma)$ to 0 if $k \neq 1$ and to its natural image in $H(R,\Gamma)$ if $k = 1$.

Proposition 8. (i) *The map* P *annihilates all higher Rankin–Cohen brackets, i.e.* $P([f,g]_n) = 0$ *for all* $f, g \in M_*(\Gamma)$ *and all* $n > 0$.

(ii) *The "triple bracket"* $\{f,g,h\}_n := P([f,g]_n h)$ $(f, g, h \in M_*(\Gamma),$ $n \geq 0)$ *is invariant under cyclical permutation of its three arguments.*

Proof. Suppose $f \in M_{2k}(\Gamma)$ and $g \in M_{2l}(\Gamma)$. If $k+l+n \neq 1$, then $P([f,g]_n)$ vanishes by definition. If $k + l + n = 1$ then a one-line computation shows that

$$n[f,g]_n = (k-l)\,\partial\left([f,g]_{n-1}\right)$$

and hence that $[f,g]_n$ vanishes in $H(R,\Gamma)$ if $n \neq 0$. This proves (i). To prove (ii), let $h \in M_{2m}(\Gamma)$ be a third modular form, and suppose that $k+l+m+n = 1$ (otherwise $\{f,g,h\}_n$ and $\{g,h,f\}_n$ are zero by definition). Let $\equiv$ denote congruence modulo dR. From $f'g \equiv -fg'$ we get $(-1)^p f^{(p)}g \equiv g^{(p)}f$ by induction and hence

$$(-1)^p\, f^{(p)}\, g^{(q)}\, h \equiv (g^{(q)}h)^{(p)}\, f = \sum_{r+s=p} \binom{p}{s} g^{(q+r)}\, h^{(s)}\, f$$

by Leibniz's rule, so

$$
\begin{aligned}
[f,g]_n h &= \sum_{p+q=n} (-1)^p \binom{2k+n-1}{q}\binom{2l+n-1}{p} f^{(p)}\, g^{(q)}\, h \\
&\equiv \sum_{q+r+s=n} \binom{2k+n-1}{q}\binom{2l+n-1}{r+s}\binom{r+s}{s} g^{(q+r)}\, h^{(s)}\, f \\
&= \sum_{s=0}^{n} \binom{2l+n-1}{s} \\
&\quad \left\{ \sum_{q+r=n-s} \binom{2k+n-1}{q}\binom{2l+n-s-1}{r} \right\} g^{(n-s)}\, h^{(s)}\, f.
\end{aligned}
$$

But the term in braces is given by

$$
\begin{aligned}
\{\cdots\} &= \binom{2k+2l+2n-s-2}{n-s} = \binom{-2m-s}{n-s} \\
&= (-1)^{n-s}\binom{2m+n-1}{n-s},
\end{aligned}
$$

so this last expression equals $[g,h]_n f$, proving the claim. We also note that (i) is a special case of (ii), since $P([f,g]_n) = \{f,g,1\}_n = \{g,1,f\}_n = 0$ if $n > 0$. $\qquad\square$

We can now give the promised explanation of the cyclic symmetry property (3.6) of the coefficients $t_n(k, l)$. Let k, l, m, n be integers with $n \geq 0$ and $k + l + m + n = 1$, and let f, g, and h be modular forms of weight $2k$, $2l$, and $2m$, respectively. (For the application to (3.6) we imagine that k and l are positive and hence that m is negative, but the signs play no role.) Write $\mathcal{D}$ for the lifting map from $\mathcal{M}(\Gamma)$ to $\Psi\mathrm{DO}(R)^\Gamma$ (so $\mathcal{D}(F) = \mathcal{D}_{-K}(F)$ for F modular of weight K) and μ for the multiplication on $\mathcal{M}(\Gamma)$ defined in Section 3, Proposition 4, so that $\mathcal{D}(F)\mathcal{D}(G) = \mathcal{D}(\mu(F, G))$ for any F and G in $\mathcal{M}(\Gamma)$. Also $\mathrm{Res}_\partial(\mathcal{D}(F)) = P(F)$ for any modular form F, because the coefficient of ∂^{-1} in $\mathcal{D}(F)$ is F if F has weight 2 and is either 0 or else a higher derivative of F if F has any other weight. Hence for any two modular forms F and G we have

$$\mathrm{Res}_\partial\left(\mathcal{D}(F)\mathcal{D}(G)\right) = \mathrm{Res}_\partial\left(\mathcal{D}\left(\mu(F, G)\right)\right) = P\left(\mu(F, G)\right) = P(FG),$$

where the last line follows from part (i) of Proposition 8 and the fact that $\mu(F, G)$ is the sum of FG plus a linear combination of higher Rankin–Cohen brackets. Applying this with $F = \mu(f, g)$ and $G = h$ we find

$$\begin{aligned} \mathrm{Res}_\partial\left(\mathcal{D}(f)\mathcal{D}(g)\mathcal{D}(h)\right) &= \mathrm{Res}_\partial\left(\mathcal{D}(\mu(f, g))\mathcal{D}(h)\right) \\ &= P\left(\mu(f, g)h\right) = t_n(k, l)\left\{f, g, h\right\}_n. \end{aligned}$$

The expression on the left is invariant under cyclic permutation of f, g and h by the trace property (6.1), and the triple bracket $\{f, g, h\}_n$ is invariant under cyclic permutations by part (ii) of Proposition 8, so the coefficient $t_n(k, l)$ must have the same symmetry, i.e., $t_n(k, l) = t_n(l, m) = t_n(l, 1 - n - k - l)$.

The same argument works unchanged if we replace $\Psi\mathrm{DO}(R)^\Gamma$ by the group $\Psi\mathrm{DO}(R, \Gamma)^\kappa$ of conjugate-invariant $\Psi\mathrm{DO}$'s of weight κ and $\mathcal{D}$ by the lifting map $\mathcal{M}(\Gamma) \to \Psi\mathrm{DO}(R, \Gamma)^\kappa$ constructed in Section 4, so we also get an explanation of the analogous cyclic symmetry property of the more general coefficients $t_n^\kappa(k, l)$.

Everything also goes through in the case of mixed weights introduced in the last section. Choose three complex numbers κ_1, κ_2, and κ_3, and consider equation (5.6). Multiplying this equation on the right by $\mathcal{D}_{w_3}^{\kappa_3, \kappa_1}(h)$, where h is a modular form of weight $2m$ with $k + l + m = 1 - n$ for some integer $n \geq 0$ and w_3 is defined as $\frac{1}{2}(\kappa_3 - \kappa_1) - m$, we find by a second application of the same equation that

$$\mathcal{D}_{w_1}^{\kappa_1, \kappa_2}(f)\, \mathcal{D}_{w_2}^{\kappa_2, \kappa_3}(g)\, \mathcal{D}_{w_3}^{\kappa_3,\ \kappa_1}(h)$$
$$= \sum_{r, s \geq 0} t_r^{\kappa_1, \kappa_2, \kappa_3}(k, l)\, t_s^{\kappa_1, \kappa_3, \kappa_1}(k + l + r, m)\, \mathcal{D}_{w_1 + w_2 + w_3 - r - s}^{\kappa_1}([[f, g]_r, h]_s).$$

Now applying Res_∂ to both sides and arguing as before we find

$$\mathrm{Res}_\partial\left(\mathcal{D}_{w_1}^{\kappa_1, \kappa_2}(f)\, \mathcal{D}_{w_2}^{\kappa_2, \kappa_3}(g)\, \mathcal{D}_{w_3}^{\kappa_3, \kappa_1}(h)\right) = t_n^{\kappa_1, \kappa_2, \kappa_3}(k, l) \cdot \{f, g, h\}_n, \quad (6.3)$$

and this implies just as before the invariance of $t_n^{\kappa_1,\kappa_2,\kappa_3}(k,l)$ with respect to simultaneous cyclic permutation of $(\kappa_1,\kappa_2,\kappa_3)$ and of $(k,l,-k-l-n+1)$.

Proof of Theorem 2. Formula (5.7) together with the cyclic symmetry just proved implies that $T_n(a,b,c;x,y,z)$ is invariant with respect to cyclic permutations of the three pairs of variables (a,x), (b,y), and (c,z). On the other hand, it is clear from the defining formula (5.8) that $T_n(a,b,c;x,y,z)$ is an even function of b and of c. From the cyclic invariance it follows that the three variables a, b, and c play equal roles, so it is also an even function of a. On the other hand, by interchanging the roles of r and s in (5.8) we see that T_n is unchanged if we interchange (b,y) and (c,z) and simultaneously replace a by $-a$, so we obtain also the invariance of T_n under odd permutations of (a,x), (b,y), and (c,z).   □

In terms of the coefficients t_n of the multiplications of ΨDO's of mixed weights, Theorem 2 says that these coefficients are invariant not only under cyclic permutations of the indices, but also under interchange of k and l (and simultaneously of κ_1 and κ_3), as well as under each of the three involutions $\kappa_i \mapsto 2 - \kappa_i$. We have given an intrinsic explanation of the first symmetry in terms of the residue map, but this is only a subgroup of order 3 out of a total symmetry group $\mathfrak{S}_3 \ltimes (\mathbb{Z}/2\mathbb{Z})^3$ of order 48. We now explain where the other symmetries come from. For this we will use both a duality and an isomorphism between the (abstract) spaces of modular forms of weight κ and weight $2 - \kappa$.

We first give an argument which shows that

$$t_n^{\kappa_1,\kappa_2,\kappa_3}(k,l) = t_n^{2-\kappa_3,2-\kappa_2,2-\kappa_1}(l,k) \qquad (6.4)$$

and hence that the coefficients t_n are invariant if we subject $(k,l,m = 1-n-k-l)$ to any odd permutation, apply the corresponding permutation to the κ_i's, and simultaneously replace each κ_i by $2 - \kappa_i$.

There is a canonical involution $A \mapsto A^*$ on the ring $\mathrm{DO}(R)$ of differential operators over R defined by the property that $A(f) \cdot g \equiv f \cdot A^*(g) \pmod{dR}$ for all $f, g \in R$. This involution is the identity on functions, sends ∂ to $-\partial$ (formula for integration by parts!), and satisfies $(AB)^* = B^*A^*$, so it must be given by

$$\left(\sum_n f_n \, \partial^n \right)^* = \sum_n (-1)^n \, \partial^n \, f_n \,. \qquad (6.5)$$

We can now use this formula to extend $*$ to all of ΨDO(R), and all its formal properties (like being a ring anti-automorphism) must remain true, since all such properties are equivalent to binomial coefficient identities which hold identically if they hold for positive integers. We also find the further property

$$\mathrm{Res}_\partial \left(\psi^* \right) = -\mathrm{Res}_\partial \left(\psi \right) \qquad \forall \psi \in \Psi\mathrm{DO}(R) \,. \qquad (6.6)$$

Indeed, any ψ can be decomposed as $\psi_1 + \psi_2$ with $\psi_1 \in \mathrm{DO}(R)$ and $\psi_2 = \sum_{n=1}^{\infty} h_n \partial^{-n}$; then ψ_1 and ψ_1^* are differential operators and hence map to 0 under Res_∂ while $\psi_2 = h_1 \partial^{-1} + O(\partial^{-2})$ and $\psi_2^* = -\partial^{-1} h_1 + O(\partial^{-2}) = -h_1 \partial^{-1} + O(\partial^{-2})$ have opposite images under Res_∂. Finally, one can check either from the defining property of $*$ or else by direct computation that

$$(\psi \circ \gamma)^* (z) = (cz+d)^{-2} \, \psi^*(\gamma z)(cz+d)^2 \qquad \forall \gamma = \begin{pmatrix} a & b \\ c & d \end{pmatrix}.$$

In particular, if ψ belongs to $\Psi\mathrm{DO}(\Gamma)^{\kappa_1,\kappa_2}$ then ψ^* lies in $\Psi\mathrm{DO}(\Gamma)^{2-\kappa_2,2-\kappa_1}$. We also have:

$$\mathcal{D}_w^{\kappa_1,\kappa_2}(f) = (-1)^w \, \mathcal{D}_w^{2-\kappa_2,2-\kappa_1}(f)^* \qquad \text{if } f \in M_{\kappa_1-\kappa_2-2w}(\Gamma). \qquad (6.7)$$

Indeed, the map $f \mapsto \mathcal{D}_w^{2-\kappa_2,2-\kappa_1}(f)^*$ is an equivariant splitting of (5.2) and hence by uniqueness is a multiple of $\mathcal{D}_w^{\kappa_1,\kappa_2}(f)$, and the multiple is $(-1)^w$ because of (6.5) and the fact that the leading term of $\mathcal{D}_w^*(f)$ is $f \, \partial^w$.

Combining (6.3), (6.6) and (6.7) and noting that $w_1 + w_2 + w_3 = n-1$, we find

$$\begin{aligned}
t_n^{\kappa_1,\kappa_2,\kappa_3}(k,l)\,\{f,g,h\}_n &= -\mathrm{Res}_\partial \left(\left(\mathcal{D}_{w_1}^{\kappa_1,\kappa_2}(f) \, \mathcal{D}_{w_2}^{\kappa_2,\kappa_3}(g) \, \mathcal{D}_{w_3}^{\kappa_3,\kappa_1}(h) \right)^* \right) \\
&= -\mathrm{Res}_\partial \left(\mathcal{D}_{w_3}^{\kappa_3,\kappa_1}(h)^* \, \mathcal{D}_{w_2}^{\kappa_2,\kappa_3}(g)^* \, \mathcal{D}_{w_1}^{\kappa_1,\kappa_2}(f)^* \right) \\
&= (-1)^n \, \mathrm{Res}_\partial \left(\mathcal{D}_{w_3}^{2-\kappa_1,2-\kappa_3}(h) \, \mathcal{D}_{w_2}^{2-\kappa_3,2-\kappa_2}(g) \right. \\
&\qquad\qquad\qquad \left. \mathcal{D}_{w_1}^{2-\kappa_2,2-\kappa_1}(f) \right) \\
&= (-1)^n \, t_n^{2-\kappa_3,2-\kappa_2,2-\kappa_1}(m,l)\,\{h,g,f\}_n\,.
\end{aligned}$$

But $(-1)^n \{h,g,f\}_n = \{f,g,h\}_n$ by Proposition 8 and the $(-1)^n$-symmetry of the nth Rankin–Cohen bracket, so this equation (after one more cyclic permutation of its arguments) implies (6.4).

Finally, we have to see why each κ_i can be replaced by $2 - \kappa$; this will give the rest of our symmetry group (so far we have explained only 6 out of a total of 48 symmetries) and in particular show why the original coefficients $t_n(k,l)$ of Section 3 are symmetric in k and l. Consider the case when the w of $\mathcal{D}_w^{\kappa_1,\kappa_2}$ is a positive integer, so that $\mathcal{D}_w^{\kappa_1,\kappa_2}(f)$ is a differential rather than a pseudo-differential operator. Then, as discussed in Section 5, it maps the space of modular forms of weight κ_2 on Γ to the space of modular forms of weight κ_1. Suppose that $\kappa_1 = 2 - h$ for some positive even integer h. (As usual, these restrictions on w and κ_1 are not important since in proving formal identities it is enough to prove them for infinitely many special cases.) A classical identity from the theory of modular forms says that

$$\frac{d^{h-1}}{dz^{h-1}}\,(f|_{2-h}g) = \frac{d^{h-1}}{dz^{h-1}}\,(f)\,|_h g \qquad \forall\, g \in \mathrm{SL}(2,\mathbb{C}), \quad h \in \mathbb{N}.$$

(This formula, due to G. Bol, is the basis of Eichler cohomology and the theory of periods of modular forms.) Hence ∂^{h-1} maps $M_{2-h}(\Gamma)$ to $M_h(\Gamma)$, so if $\psi \in DO(R,\Gamma)^{2-h,\kappa_2}$ and $f \in M_{\kappa_2}(\Gamma)$ then $\partial^{h-1}(\psi(f)) \in M_h$. This says that the product in $\Psi DO(R)$ of ∂^{h-1} and ψ belongs to $\Psi DO(R,\Gamma)^{h,\kappa_2}$. Replacing $\kappa_1 = 2 - h$ by an arbitrary value of κ_1, we see that we have proved that $\Psi DO(R,\Gamma)^{\kappa_1,\kappa_2}$ is canonically isomorphic to $\Psi DO(R,\Gamma)^{2-\kappa_1,\kappa_2}$ by left multiplication with $\partial^{1-\kappa_1}$. The same argument shows that it is also canonically isomorphic to $\Psi DO(R,\Gamma)^{\kappa_1,2-\kappa_2}$ by right multiplication with ∂^{κ_2-1}. It follows that the equations

$$\partial^{1-\kappa_1} \circ \mathcal{D}_w^{\kappa_1,\kappa_2}(f) = \mathcal{D}_{w+1-\kappa_1}^{2-\kappa_1,\kappa_2}(f), \qquad \mathcal{D}_w^{\kappa_1,\kappa_2}(f) \circ \partial^{\kappa_2-1} = \mathcal{D}_{w+\kappa_2-1}^{\kappa_1,2-\kappa_2}(f)$$

$$(6.8)$$

must be true up to scalar factors, and by looking at the leading term one sees that these factors equal 1. (As a check, note that the second of these equations implies that the coefficient of $f^{(n)}\partial^{-n}$ in (5.3) must be invariant under $(\kappa_1,\kappa_2,w) \mapsto (\kappa_1, 2-\kappa_2, w+\kappa_2-1)$, and this is indeed true.) These identities let one replace κ by $2-\kappa$ wherever they occur as superscripts, which was the observed symmetry.

7. Supermodular Forms and Superpseudodifferential Operators

We work on the supercomplex plane $\mathbb{C}^{1|1}$ with local coordinate (z,ζ), where $\zeta^2 = 0$, and with the canonical supersymmetric (SUSY) structure given by the maximal non-integrable structure distribution of rank $0|1$ generated by the vector field $D = \frac{\partial}{\partial\zeta} + \zeta\frac{\partial}{\partial z}$ satisfying $D^2 = \frac{\partial}{\partial z}$. This is, up to isomorphism, the unique such SUSY structure extendible to $\mathbb{P}^{1|1}$. (For an exposition of those aspects of the theory of supersymmetry needed for the present paper see [Ma2].) If $(\tilde{z},\tilde{\zeta})$ is another local coordinate defining the same SUSY structure, then $D = J \cdot \tilde{D}$ where $J = D(\tilde{\zeta})$ is the superanalogue of the usual Jacobian. We let R be a $\mathbb{Z}/2$-graded ring of functions on $\mathbb{C}^{1|1}$ on which D acts; these will have the form $F(z,\zeta) = f(z) + \zeta g(z)$ where the functions $f(z)$ and $g(z)$ can themselves have coefficients in a superring (or $\mathbb{Z}/2$-graded ring) of constants Λ. By convention, even coordinates or constants will always be denoted by Latin letters $a, b, c, d \ldots$ and odd coordinates or constants by Greek letters $\alpha, \beta, \gamma, \delta \ldots$. Even constants and variables commute with even and odd constants and variables, while odd constants and variables anti-commute with odd constants and variables and in particular have square zero.

    The superanalogue of the group $\mathrm{PSL}(2,\mathbb{C})$ is the group $\mathrm{PC}(2,\mathbb{C}^{1|1})$ whose elements are matrices

$$\mathbf{A} = \begin{pmatrix} a & b & \gamma \\ c & d & \delta \\ \alpha & \beta & e \end{pmatrix} \qquad (7.1)$$

satisfying

$$ad - bc - \alpha\beta = 1, \qquad e^2 + 2\gamma\delta = 1, \quad \alpha e = a\delta - c\gamma, \qquad \beta e = b\delta - d\gamma$$

together with the condition that e reduces to 1 modulo nilpotent elements. (The last condition prevents both $\mathbf{A}$ and $-\mathbf{A}$ from belonging to the group.) The matrix (7.1) acts on $\mathbb{C}^{1|1}$ by the "fractional linear SUSY-compatible transformation"

$$(z, \zeta) \mapsto (\tilde{z}, \tilde{\zeta}) = \left(\frac{az + b + \gamma\zeta}{cz + d + \delta\zeta}, \frac{\alpha z + \beta + e\zeta}{cz + d + \delta\zeta} \right) \tag{7.2}$$

and on R by sending $F(z, \zeta) = f(z) + \zeta g(z)$ to $F^{\mathbf{A}}(z, \zeta) = f(\tilde{z}) + \tilde{\zeta} g(\tilde{z})$. A calculation shows that the superjacobian $J(\mathbf{A}) = D(\tilde{\zeta})$ of the transformation (7.2) is equal to $(cz + d + \delta\zeta)^{-1}$. We will use this as the automorphy factor to define supermodular forms (notice that it becomes the square root of the classical automorphy factor $d\tilde{z}/dz$ when $\delta = 0$). For an integer k and a (discrete) subgroup $\Gamma \subset PC(2, \mathbb{C}^{1|1})$, we denote by $SM_k(\Gamma, R)$ the space of supermodular forms of weight k, i.e., elements of R satisfying, for $\mathbf{A} \in \Gamma$ as in (7.1),

$$F\left(\frac{az + b + \gamma\zeta}{cz + d + \delta\zeta}, \frac{\alpha z + \beta + e\zeta}{cz + d + \delta\zeta} \right) = (cz + d + \delta\zeta)^k F(z, \zeta).$$

By direct calculation we find that this is equivalent to the two equations

$$(1 - k\alpha\beta) f(z) - (cz + d)^{-k} f\left(\frac{az + b}{cz + d} \right) = e (\alpha z + \beta) g(z),$$

$$e\, g(z) - (cz + d)^{-k-1} g\left(\frac{az + b}{cz + d} \right) = (\alpha z + \beta) f'(z) + k \frac{c(\alpha z + \beta) + \delta e}{cz + d} f(z).$$

Notice that when $\alpha = \beta = \gamma = \delta = 0$ and $e = 1$ the element $\mathbf{A}$ corresponds to an element of $PSL(2, \mathbb{C})$ and these two equations give the separate transformation laws

$$f\left(\frac{az + b}{cz + d} \right) = (cz + d)^k f(z)$$

$$g\left(\frac{az + b}{cz + d} \right) = (cz + d)^{k+1} g(z)$$

corresponding to the transformation law for M_k and M_{k+1} respectively, so the new theory automatically combines the cases of modular forms of even and odd weight.

We next turn to the definition of *superpseudodifferential operators*, see also [MR]. We first need the analogue of the Leibniz formula. The usual Leibniz formula $\partial(fg) = \partial(f)g + f\partial g$ is replaced in the supercase by

$$D(FG) = D(F)G + \sigma(F)D(G), \qquad F, G \in R.$$

where the involution σ is the *grading automorphism* of R, equal to 1 on the even part and to -1 on the odd part of R (in other words, D is a superderivation). This formula generalises by induction on m to the graded Leibniz formula

$$D^m(FG) = \sum_{k=0}^{\infty} \binom{m}{r}_S D^r(\sigma^{m-r}(F)) D^{m-r}(G) \qquad (7.3)$$

for all integers $m \geq 0$, where the *supersymmetric binomial coefficients* $\binom{m}{r}_S$ are defined by

$$\binom{m}{r}_S = \begin{cases} \binom{[m/2]}{[r/2]} & \text{if } r \text{ is even or } m \text{ is odd,} \\ 0 & \text{if } r \text{ is odd and } m \text{ is even,} \end{cases}$$

with $[x]$ as usual denoting the integral part of a real number x, so we can define a multiplication on the space $S\Psi DO(R)$ of super-ΨDO's (Laurent series in D^{-1}) by

$$FD^m \cdot GD^n = \sum_{r \geq 0} \binom{m}{r}_S FD^r(\sigma^{m-r}G) D^{m+n-r} \qquad (m, n \in \mathbb{Z}),$$

and with respect to this multiplication the subspace $SDO(R) = R[D] \subset S\Psi DO(R)$ of superdifferential operators is a subring. As before, we have a filtration of $S\Psi DO(R)$ by the subspaces

$$S\Psi DO(R)_w = \left\{ \sum_{m=0}^{\infty} F_m D^{w-m}, \quad F_m \in R \right\}$$

and this filtration is compatible with the ring structure.

In the supercase, the group Γ acts on $S\Psi DO(R)$ via its action on R and on D. The element $\mathbf{A}$ of Γ as in (7.1) transforms D into $(cz+d+\delta\zeta)D$. The ring $S\Psi DO(R)^\Gamma$ denotes the Γ-invariant elements of $S\Psi DO(R)$. We have the filtration $S\Psi DO(R)_k^\Gamma, k \in \mathbb{Z}$, inherited from the filtration of $S\Psi DO(R)$. The analogue of (1.8) for the supercase is the sequence, which is split short exact by Theorem 3 below, involving supermodular forms of weight k (for all parities of k)

$$0 \to S\Psi DO(R)^\Gamma_{-k-1} \to S\Psi DO(R)^\Gamma_{-k} \to SM_k(\Gamma) \to 0. \qquad (7.4)$$

The analogue of Proposition 1 for supermodular forms is as follows.

Theorem 3. *For $k > 0$ define an operator $S\mathcal{L}_k : R \to S\Psi DO(R)_{-k}$ by*

$$S\mathcal{L}_k(F) = \sum_{n=0}^{\infty} (-1)^{[n/2]} \frac{[\frac{n+k}{2}]! \, [\frac{n+k-1}{2}]!}{[\frac{n}{2}]! \, [\frac{n+2k-1}{2}]!} D^n(F) D^{-k-n}$$

and an operator $S\mathcal{L}_{-k} : R \to \mathrm{SDO}(R)_k$ by

$$S\mathcal{L}_{-k}(F) = \sum_{n=0}^{k-1} \frac{[\frac{2k-n}{2}]!}{[\frac{n}{2}]!\,[\frac{k-n}{2}]!\,[\frac{k-n-1}{2}]!}\, D^n(F)\, D^{k-n} \,,$$

and set $S\mathcal{L}_0(F) = F$. Then $S\mathcal{L}_k(F^{\mathbf{A}}J(\mathbf{A})^k) = S\mathcal{L}_k(F) \circ \mathbf{A}$ for any $\mathbf{A} \in \mathrm{PC}(2, \mathbb{C}^{1|1})$ and any $k \in \mathbb{Z}$. In particular, if $F \in SM_k(\Gamma)$ for any $k \in \mathbb{Z}$ then $S\mathcal{L}_k(F)$ is a Γ-invariant superpseudodifferential operator.

Remark.   Just as in Section 2, if we denote by $S\mathcal{D}_w : R \to S\Psi\mathrm{DO}(R)_w$ the lifting map renormalized to have leading coefficient FD^w then we can write the formulas for positive and negative w uniformly using binomial coefficients as

$$S\mathcal{D}_w(F) = \sum_{n\geq 0} \frac{\binom{[\frac{w}{2}]}{[\frac{n+1}{2}]}\binom{[\frac{w-1}{2}]}{[\frac{n}{2}]}}{\binom{w}{[\frac{n+1}{2}]}}\, D^n(F)\, D^{w-n} \qquad (w \in \mathbb{Z}), \qquad (7.5)$$

where the sum goes only up to $n = w$ if $w \geq 0$.

Proof of Theorem 3.   We imitate the proof of Proposition 1 in Section 1. The analogues of (1.7) and (1.9), both proved by induction, are

$$[(cz+d+\delta\zeta) \circ D]^w = (cz+d+\delta\zeta)^w \sum_{r=0}^{\infty} \alpha_r(w)\, \Phi_r(\mathbf{A})\, D^{w-r} \qquad (7.6)$$

and

$$D^n\left(F^{\mathbf{A}}\,(cz+d+\delta\zeta)^w\right)$$
$$= \sum_{r=0}^{n} \beta_r(w,n)\,(D^{n-r}F)^{\mathbf{A}}\,\Phi_r(\mathbf{A})\,(cz+d+\delta\zeta)^{w-n+r}$$

where $\Phi_r(\mathbf{A})$ is defined for $\mathbf{A}$ as in (7.1) by

$$\Phi_r(\mathbf{A}) = \left(\frac{c}{cz+d+\delta\zeta}\right)^{[\frac{r}{2}]}\left(\frac{c\zeta-\delta}{cz+d+\delta\zeta}\right)^{[\frac{r+1}{2}]-[\frac{r}{2}]}$$

and the numerical coefficients $\alpha_r(w)$ and $\beta_r(w,n)$ are given by

$$\alpha_r(w) = \frac{[\frac{w}{2}]!\,[\frac{w-1}{2}]!}{[\frac{r}{2}]!\,[\frac{w-r}{2}]!\,[\frac{w-r-1}{2}]!}\,, \qquad \beta_r(n,w) = \frac{[\frac{n}{2}]!\,[w-\frac{n-r}{2}]!}{[\frac{r}{2}]!\,[\frac{n-r}{2}]!\,[w-\frac{n}{2}]!}\,.$$

(The last two formulas are written for $w > 0$; there are similar formula for $w = -k < 0$ and again a uniform formula using binomial coefficients). Now letting $\gamma_n(w)$ denote the coefficient of $D^n(F)\,D^{w-n}$ in the definition of $\mathcal{L}_{-w}(F)$ (or of $S\mathcal{D}_w(F)$), we find that the desired equality is equivalent to the trivially verified identity $\gamma_s(w)\,\alpha_r(w-s) = \gamma_n(w)\,\beta_r(w,n)$. $\square$

One can also give proof of Theorem 3 along the lines of the one in Section 2 using the superanalogue of the Casimir operator. The other results of this paper can also all be generalized to the SUSY case, but we will not do this here. We say a few words about the super-version of the generalized ΨDO's mentioned in Section 2. The obvious idea of taking complex powers of D does not work. Instead, we must take linear combinations over R of formal symbols ∂^u and $\partial^u D$ with $u \in \mathbb{C}$, the multiplication being defined by $D^2 = \partial$ and by Leibniz's rule and its superextension (7.3). The transformation behavior under changes of coordinates (7.2) is given by the same formula (7.6) except that when one replaces D^w by $\partial^u D^p$ with $u \in \mathbb{C}$ and $p \in \{0, 1\}$ one must reinterpret the formula

$$\alpha_r(w) = [\tfrac{r+1}{2}]! \begin{pmatrix} [\tfrac{w}{2}] \\ [\tfrac{r+1}{2}] \end{pmatrix} \begin{pmatrix} [\tfrac{w-1}{2}] \\ [\tfrac{r}{2}] \end{pmatrix}$$

which was valid for $w \in \mathbb{Z}$ by replacing $[\tfrac{w}{2}]$ and $[\tfrac{w-1}{2}]$ by u and $u + p - 1$, respectively, i.e., by the unique expression which is correct when $u \in \mathbb{Z}$ and $w = 2u + p$, and similarly for the lifting formula (7.5). The considerations of Section 3–6 about the multiplications of modular forms induced by the multiplications of various kinds of automorphic ΨDO's can be generalized in the more or less obvious way (thus an automorphic super-ΨDO of mixed weight is just a super-ΨDO which is multiplied on the left and on the right by some powers of $J(\mathbf{A})$ under the action of $\mathbf{A} \in \Gamma$), and the arguments given in the last section can also be generalized using the superversion of the non-commutative residue map given in [MR]. Some of these things may be carried out in more detail in a later paper.

8. Concluding Remarks

The study of formal ΨDO's in the last two decades was primarily motivated by the needs of the theory of completely integrable systems of nonlinear differential equations like the Korteweg–de Vries equation and the Kadomtsev–Petviashvili hierarchy: see e.g. [KZ2] for some recent developments and extensive references. A few remarks added here may help the interested reader to put our constructions in this framework.

A sheaf-theoretic version. Let X be a complex Riemannian surface, not necessarily compact. For any open subset $U \subset X$ let $\mathcal{O}_X(U)$ be the ring of holomorphic functions in U. If U admits a local coordinate z, put $\partial_z = \partial/\partial z$ and form the ring $\mathcal{E}_{X,z}(U) = \{\sum_m h_m \partial_z^{-m} \mid h_m \in \mathcal{O}_X(U)\}$. A change of local coordinate induces a canonical isomorphism of the respective rings compatible with restriction to smaller sets, so that we get a sheaf of rings $\mathcal{E}_X$. It is naturally filtered by the subsheaves $\mathcal{E}_X^{(-m)}$, and the associated sheaf of graded algebras is $\oplus_m \omega_X^{\otimes m}$ where ω_X is the sheaf of holomorphic differentials. Assume now that X is additionally endowed with a projective structure p i.e., with a maximal atlas (U_α, z_α) whose transition functions $z_\alpha = f_{\alpha,\beta}(z_\beta)$ are fractional linear. Define the local lifting maps $\Lambda_{m,z_\alpha} : w_{X,z_\alpha}^{\otimes m} \to \mathcal{E}_X^{(-m)}$ by the same formulas as in Section 1. They will be automatically compatible on the intersections and therefore define a sheafified lifting map $\Lambda_m(p) : w_X^{\otimes m} \to \mathcal{E}_X^{(-m)}$ depending only on the flat structure p. This must be evident from the Beilinson construction of the lifting using Casimir operators discussed in Section 2. In fact, p determines a sheaf of $sl(2)$-algebras on X consisting of projectively flat tangent fields, and the local Casimirs in the relevant sheaf of universal enveloping algebras glue to form a global section $C(p)$. Then $\Lambda_m(p)$ is a differential operator (of infinite order for $m \geq 1$) identifying $w_X^{\otimes m}$ with a subsheaf of $\mathcal{E}_X^{(-m)}$ of operators with the same top symbol consisting of the eigenvectors of $C(p)$ with eigenvalue $m(m-1)$.

In the context of automorphic forms we considered essentially a modular curve $X_\Gamma = \mathcal{H}/\Gamma$ with a fixed projective structure coming from $\mathcal{H}$. Now we can vary p and ask how $\Lambda(p) = \oplus \Lambda_m(p)$ varies with p. Formally, $C(p)$ varies isospectrally so that for any pair of flat structures p, p' we have

$$C(p') = T(p',p)C(p)T(p',p)^{-1}, \quad \Lambda(p') = T(p',p)\Lambda(p)$$

for some $T(p',p)$ (acting e.g. upon $\Gamma(X, \mathcal{E}_X^{(-1)})$) for compact X of genus ≥ 2).

Now, all p's on X form an affine space associated with the vector space of quadratic holomorphic differentials on X: locally we have $p' - p = S_z^{z'}(dz)^2$ where p (resp. p') corresponds to a local flat coordinate z (resp. z'), and $S_z^{z'}$ is the Schwarz derivative (see e.g. A. Tyurin's report [Ty]).

Question. Is it true that $T(p',p)$ depends only on $p' - p$?

For example, a direct calculation shows that $\Lambda_m(p) = \Lambda_m(p')$ for $m = 1, 0, -1, -2$, whereas

$$(\Lambda_3(p) - \Lambda_3(p'))(f(dz)^{-3}) = \frac{1}{5}f\left(3\frac{(\partial_z j)^2}{j^2} - 2\frac{\partial_z^2 j}{j}\right)\partial_z = \frac{1}{5}f\,S_z^{z'}\,\partial_z$$

where $j = \partial z'/\partial z$. This means that $\Lambda_3(p) - \Lambda_3(p')$ is essentially multiplication by $p' - p$, if one writes $(dz)^{-1}$ instead of ∂_z at the last place of the right hand side.

Question. Can $T(p',p)$ be described in terms of derivations and multiplication in $\Gamma(X, \mathcal{E}_X)$?

All of this has a straightforward supersymmetric version.

Complex powers and $\mathcal{D}$-modules. The complex powers of ∂ were treated in [KZ1] in the Hamiltonian context. If one attempts to sheafify them, then one has to make some sense of complex powers of holomorphic functions because they appear already on the level of coordinate change for principal symbols. A well known way to interpret f^w for complex w is to treat it as a section of a $\mathcal{D}_X$-module. This of course incorporates the formal rule of derivation which we used to define ΨDO_w. This problem deserves further investigation. Let us mention in addition that the complex eigenvalues of the Casimir operator were recently used to define so-called "matrices of complex size" which are infinite-dimensional algebras $U(sl_2)/(C - w(w - 1))$ where C is the Casimir (cf. [KM].)

ΨDO as a Lie algebra and its central extension. In the context of the automorphic forms, we related via liftings the multiplication in ΨDO with Cohen-Kuznetsov brackets. We could have looked at the Lie bracket in ΨDO instead. The point is that this Lie algebra admits a nontrivial central extension which can be suggestively described by introducing the formal expression $\log \partial$ and the commutator

$$\left[\log \partial, \sum h_m \partial^{-m}\right] := \sum_m \sum_{k \geq 1} \frac{(-1)^{k+1}}{k} \partial^k h_m \partial^{-m-k}$$

which is then used to define a cocycle $c(A, B) = tr\left([\log \partial, A] \circ B\right)$ for an appropriate trace functional tr. This construction is important for clarifying the Poisson–Lie structure of ΨDO. Does it admit a sensible descent to modular forms?

References

[Co] H. Cohen, Sums involving the values at negative integers of L functions of quadratic characters, *Math. Ann.* **217** (1977), 81–94.

[EZ] M. Eichler and D. Zagier, *The Theory of Jacobi Forms*, Prog. Math. **55**, Birkhäuser, Boston-Basel-Stuttgart (1985).

[GD] I.M. Gelfand and I. Ya. Dorfman, The Schouten bracket and Hamiltonian operators, *Funct. Anal. Appl.* **14** (1980), 223–226.

[KM] B. Khesin and F. Malikov, Universal Drinfeld–Sokolov reduction and matrices of complex size, preprint hep–th/9405116.

[KZ1] B. Khesin and I. Zakharevich, Poisson-Lie group of pseudo-differential symbols and fractional KP-KdV hierarchies, *C. R. Acad. Sci.* **316** (1993), 621–626.

[KZ2] B. Khesin and I. Zakharevich, Lie-Poisson group of pseudo-differential symbols, *Comm. Math. Phys.* **171:3** (1995), 475–530.

[Ku] N.V. Kuznetsov, A new class of identities for the Fourier coefficients of modular forms, (in Russian), *Acta Arith.* **27** (1975), 505–519.

[Ma1] Yu.I. Manin, Algebraic aspects of differential equations, *J. Sov. Math.* **11** (1979), 1–128.

[Ma2] Yu.I. Manin, *Topics in Non-commutative Geometry*, Princeton Univ. Press, Princeton (1991).

[MR] Yu.I. Manin and A. O. Radul, A supersymmetric extension of the Kadomtsev-Petviashvili hierarchy, *Commun. Math. Phys.* **98** (1985), 65–77.

[Ra] R.A. Rankin, The construction of automorphic forms from the derivatives of a given form, *J. Indian Math. Soc.* **20** (1956), 103–116.

[Ty] A. Tyurin, On periods of quadratic differentials, *Russian Math. Surveys* **33:6** (1978), 169–221.

[Z1] D. Zagier, Modular forms and differential operators, *Proc. Indian Acad. Sci. (Math. Sci.)* **104** (1994), 57–75.

[Z2] D. Zagier, Some combinatorial identities occurring in the theory of modular forms, in preparation.

Paula Beazley Cohen
URA Géométrie-Analyse-Topologie,
UFR de Mathématiques,
Université des Sciences et Technologies de Lille
59655 Villeneuve d'Ascq cedex, France

Yuri Manin and Don Zagier
Max-Planck Institut für Mathematik
Gottfried-Claren str. 26,
53225 Bonn, Germany

Received October 1995; additions December 1995

On $\mathcal{T}$-Functions of Zakharov–Shabat and Other Matrix Hierarchies of Integrable Equations

L. A. Dickey

Abstract

Matrix hierarchies are: multi-component KP, general Zakharov–Shabat (ZS) and its special cases, e.g., AKNS. The ZS comprises all integrable systems having a form of zero-curvature equations with rational dependence of matrices on a spectral parameter. The notion of a τ-function is introduced here in the most general case along with formulas linking τ-functions with wave Baker functions. The method originally invented by Sato et al. for the KP hierarchy is used. This method goes immediately from definitions and does not require any assumption about the character of a solution, being the most general. Applied to the matrix hierarchies, this involves considerable sophistication. The paper is self-contained and does not expect any special prerequisite from a reader.

1. Introduction

Integrable systems of differential equations exist not isolated but united in large communities called hierarchies. All equations inside a hierarchy are commuting with each other. The first known hierarchies were generalized Korteweg-de Vries (KdV) hierarchies, one for every natural number n. (For detail, see, e.g., [9]). Then an immense Kadomtsev-Petviashvili (KP) hierarchy was found which united all the KdV's. Those hierarchies consisted of scalar equations. Almost immediately they were generalized to matrix equations and formed "multi-component" KdV's and KP.

All the above mentioned hierarchies are generated by linear differential (KdV) or pseudo-differential (KP) operators of arbitrary orders. Equations of another type are generated by matrix first order differential operators linearly depending on a spectral parameter. These are AKNS (for Ablowitz, Kaupp, Newell and Segur) with 2×2 matrix first order operators; they were generalized by Dubrovin to $n \times n$ matrices; we call the latter AKNS-D hierarchies. The next generalization is when linear operators depend on a parameter as polynomials of any degree. Finally, the most general case involves arbitrary rational dependence on a parameter. These equations are called general Zakharov–Shabat (ZS) equations. They also form a hierarchy (see [7]). The hierarchy with polynomial dependence on a parameter is a special case of the general ZS when there is a single pole at infinity; we call this hierarchy s-p ZS. All KdV's and AKNS's are nothing but reductions of the general ZS hierarchy. The exact definitions will be given below.

The importance of the theory of integrable system was essentially enhanced with the invention of the "tau"-function by mathematicians of the Kyoto school, see [1], [2]. This one single function of infinitely many "time" variables replaces infinitely many dynamical variables, coefficients of linear differential or pseudo-differential operators. It happened that this function linked integrable systems to Lie algebra representations and to many problems of modern physics, such as conformal field theory, matrix models in the statistical physics, 2-dimensional gravity and string theory. Up to now, all these achievements applied solely to scalar hierarchies, nth KdV and KP. There are also some published results about the multi-component KP. Concerning the general ZS equations and their τ-functions, we know the only work [10] done in very abstract terms; it is difficult to extract concrete formulas from it. The aim of the present article is to fill this gap. Physicists have not turned yet their attention to the general ZS hierarchy (except some particular equations of these hierarchy). We believe that its time will come sooner or later.

The paper is self-contained and, formally speaking, does not require a special prerequisite (see also [9]). It was easier not to start with the most complicated case of the general ZS but to pass gradually from the simplest to the most difficult model referring when needed to what was proven before.

In the first part of the paper we deal with the multi-component KP (mcKP), as defined in [1]. In [3] and [4] there are formulas written for its τ-function, without proofs in both the articles. Therefore it is difficult to guess what the method and thinking were. Most probably, they used the techniques of free fermion representations. Meanwhile, the authors had suggested their own excellent method which was invented in [2] for KP, based on nothing but the bilinear identity, i.e., being close to very first definitions. The advantage of this approach is its full generality, independence of the origin and the nature of a solution. Our first goal was to adjust this method to the mcKP hierarchy. Basically, the method remains the same as in [2], however, it becomes a little tricky. (In [5] we derived the τ-function in terms of the Grassmannian, in [6] found it for special, algebraic geometrical solutions; in contrast to that, we now discuss the general case).

The next part is devoted to the simplest special case of the ZS hierarchy, namely, the single-pole hierarchy (s-p ZS). It is closely connected with the mcKP since it is proven below that a Baker function of the s-p ZS is at the same time that of mcKP, and the s-p ZS is a subhierarchy of the mcKP. A similar statement was made before in [5].

Then we introduce a "not normalized" s-p ZS hierarchy which differs from the previous one by the fact that the expansion of its Baker function in powers of a spectral parameter starts with a matrix of a general form, not from the unity. It can be reduced to the normalized ZS. Nevertheless, it is convenient to study this case separately because it provides a good preparation for the general ZS where one cannot normalize Baker functions simultaneously at all poles.

Finally, and this is the main point, we treat the general ZS hierarchy. It is discussed in [7] in a way that one can understand the totality of all ZS equations as a hierarchy, i.e., as a set of commuting vector fields. There are definitions of a Baker function, of a corresponding Grassmannian, etc. in that paper. However, it is lacking a concept of the τ-function. We are doing this now. The main results of the present article are contained in the theorems of Sections 4 and 7 and Proposition 3 and its Corollary of section 5.

Despite the absence of a general definition and of a proof of the existence of the τ-function, there were a few examples of that function found earlier. In [7] this is done for soliton-type solutions and quite recently, in [8], for algebraic geometrical solutions that can be expressed in terms of θ-functions. Those examples were stimulating for the present study.

2. Multi-Component KP

Let
$$L = A\partial + u_0 + u_1\partial^{-1} + \cdots, \quad \partial = d/dx$$

be a pseudo-differential operator where u_i are $n \times n$ matrices, $A = \mathrm{diag}(a_1, ..., a_n)$, a_i are distinct nonzero constants. Diagonal elements of u_0 are assumed to be zero.

Let $R_\alpha = \sum_{j=0}^{\infty} R_{j\alpha}\partial^{-j}$, $\alpha = 1, ..., n$, where $R_{0\alpha} = E_\alpha$, E_α is a matrix having only one nonzero element on the (α, α) place which is equal to 1; R_α is supposed to satisfy
$$[L, R_\alpha] = 0.$$

It is shown below that such matrices exist being
$$R_\alpha R_\beta = \delta_{\alpha\beta}R_\alpha, \quad \sum_{\alpha=1}^{n} R_\alpha = I$$

(i.e. this is a spectral decomposition of the unity). The mcKP hierarchy (multi-component KP) is

$$
\begin{aligned}
\partial_{k\alpha} L &= [(L^k R_\alpha)_+, L], \\
\partial_{k\alpha} R_\beta &= [(L^k R_\alpha)_+, R_\beta] \\
\partial_{k\alpha} &= \partial/\partial t_{k\alpha}, \quad k = 0, 1, \ldots; \ \alpha = 1, \ldots, n
\end{aligned}
$$

and $t_{k\alpha}$ are the "time variables" of the hierarchy. The subscript $+$ refers, as usual, to a purely differential part of a pseudo-differential operator, $(\sum a_k \partial^k)_+ = \sum_{k \geq 0} a_k \partial^k$, $A_- = A - A_+$.

It can be shown that the equations for different k, α commute. The variables x and $t_{k\alpha}$ are not independent:

$$
\partial = \sum_\alpha a_\alpha^{-1} \partial_{1\alpha}
$$

(Greek indices always run from 1 to n).

Let

$$
L = \hat{w} A \partial \hat{w}^{-1}, \quad \text{where} \ \hat{w} = \hat{w}(A\partial) = \sum_0^\infty w_i (A\partial)^{-i}, \ w_0 = I;
$$

Then $R_\alpha = \hat{w} E_\alpha \hat{w}^{-1}$ has all needed properties. Put

$$
w = \hat{w}(A\partial) \exp \xi(t, z) = \hat{w}(z) \exp \xi(t, z);
$$

where

$$
\xi(t, z) = \sum_{k=0}^\infty \sum_{\alpha=1}^n z^k E_\alpha t_{k\alpha}.
$$

This is the *Baker function*, satisfying the equations

$$
Lw = zw, \quad \text{and} \ \partial_{k\alpha} w = (L^k R_\alpha)_+ w.
$$

The latter equation is equivalent to

$$
\partial_{k\alpha} \hat{w} = -(L^k R_\alpha)_- \hat{w}.
$$

Remark 1. It is very important to note that the series $\hat{w}$ are defined up to a multiplication on the right by series $\sum_0^\infty c_i \partial^{-i}$ with constant diagonal matrices c_i where $c_0 = I$. Correspondingly, the Baker function is defined up to a multiplication by $\sum_0^\infty a_i z^{-i}$. Two functions which differ by such a factor are said to be equivalent. For two equivalent Baker functions the Lax operator L is the same. All the formulas below will be obtained up to equivalence.

We have $\partial_{0\alpha}\hat{w} = -(R_\alpha - E_\alpha)\hat{w} = -\hat{w}E_a + E_\alpha\hat{w} = [E_\alpha, \hat{w}]$. Symmetries related to "zero" time variables $t_{0\alpha}$ are similarity transformations with constant matrices.

The *adjoint Baker function* is

$$w^a = (\hat{w}^*(A\partial))^{-1}\exp(-\xi(t, z))$$

where the star means the conjugation: for every matrix X the equality $(X\partial)^* = -\partial X^*$ holds where X^* is the transpose of X. The equations

$$L^* w^a = z w^a, \quad \text{and} \quad \partial_{k\alpha} w^a = -(L^k R_\alpha)^*_+ w^a$$

hold.

Remark 2. Our definition of the mcKP differs from that in [1],[2] and [3] where $u_0 = 0$ and $A = I$. It is easy to show that, in our definition, the coefficients of the equations are local in terms of u_i's, i.e., differential polynomials. Indeed, the dressing formula $L = \hat{w}A\partial\hat{w}^{-1}$ permits us to express every differential polynomial in elements of w_i's as a differential polynomial in elements of u_i's which is also an ordinary polynomial in w_i's (i.e., it does not depend on derivatives of w_i's). Then the elements of R_α's are such polynomials as well. Let us show that, in fact, they do not depend on w_i's at all. Let us give to $\hat{w}$ an infinitesimal deformation $\delta\hat{w}$ such that L is not changed. This means that $\delta L = [\delta\hat{w}\cdot\hat{w}^{-1}, A\partial] = 0$, which easily implies that the matrix $K = \delta\hat{w}\cdot\hat{w}^{-1}$ is constant and diagonal. Now, $\delta R_\alpha = [K, E_\alpha] = 0$. The rest is clear. The fact that all diagonal elements of A are distinct is crucial. It is easy to compute that otherwise the R_α are not local. If one is only interested in the hierarchy in terms of Baker functions, and not the operator L, then this distinction is not important.

The significance of the mcKP as well as KP is in their universality.

Proposition. Universality of the mcKP hierarchy. *If an expression of the form*

$$w = \hat{w}(A\partial)\exp\xi(t, z) = \hat{w}(z)\exp\xi(t, z);$$

where

$$\hat{w}(A\partial) = \sum_0^\infty w_i(A\partial)^{-i}, \ w_0 = I$$

satisfies arbitrary equations $\partial_{k\alpha}w = \overline{B}_{k\alpha}w$ *with some differential operators* $\overline{B}_{k\alpha}$, *then this is nothing but mcKP.*

Indeed, the given equations yield

$$0 = \partial_{k\alpha}\hat{w}\cdot e^\xi + \hat{w}E_\alpha z^k e^\xi - \overline{B}_{k\alpha}w = \partial_{k\alpha}\hat{w}\cdot e^\xi + \hat{w}E_\alpha(A\partial)^k e^\xi - \overline{B}_{k\alpha}w.$$

Letting $L = \hat{w}A\partial\hat{w}^{-1}$ and $R_\alpha = \hat{w}E_\alpha\hat{w}^{-1}$ we have $\partial_{k\alpha}\hat{w}\cdot\hat{w}^{-1} + R_\alpha L^k - \overline{B}_{k\alpha} = 0$. Taking the positive part of this equation, we get $\overline{B}_{k\alpha} = (R_\alpha L^k)_+$

and the negative part is $\partial_{k\alpha}\hat{w} = -(R_\alpha L^k)_-\hat{w}$. This is the equation of the hierarchy. □

3. Bilinear Identity

The so-called bilinear identity is basic for Sato's theory.

Lemma. *Let $\Phi = \sum \Phi_i(A\partial)^i$ and $\Psi = \sum \Psi_i(A\partial)^i$ be two pseudo-differential operators (ΨDO). Then the equality*

$$\mathrm{res}_\partial \Phi\Psi^* = \mathrm{res}_z(\Phi e^\xi) A^{-1}(\Psi e^{-\xi})^*$$

holds.

The notations res_∂ and res_z mean, as usual, coefficients of ∂^{-1} and z^{-1}.

Proof. It is easy to check that both the left- and the right-hand side are equal to $\sum \Phi_i A^{-1}\Psi^*_{-i-1}(-1)^{i+1}$.   □

Proposition. *If $\Phi = \sum \Phi_i(A\partial)^i$ is a ΨDO, and $w = \Phi \exp \xi(t, z)$, $w^a = (\Phi^*)^{-1}\exp(-\xi(t, z))$ then*

$$\mathrm{res}_z(\partial^i w) A^{-1}(w^a)^* = 0. \tag{1a}$$

Moreover, if w depends on infinitely many variables $t_{i\alpha}$, $i = 0, 1, ..., \alpha = 1, ..., n$ and satisfies a system of differential equations of the form $\partial_{i\alpha} w = B_{i\alpha} w$ where $B_{i\alpha}$ are any differential (matrix) operators in $\partial = \sum a_\alpha^{-1}\partial_{1\alpha}$ then

$$\mathrm{res}_z(\partial_{i_1\alpha_1}\partial_{i_2\alpha_2}...\partial_{i_s\alpha_s} w) A^{-1}(w^a)^* = 0 \tag{1b}$$

for an arbitrary set of indices $i_1, \alpha_1, i_2, \alpha_2, ..., i_s, \alpha_s$. This happens, e.g., when w is a Baker function of the mcKP hierarchy.

Conversely, if there are two expressions of the form $w = \sum_0^\infty w_i(t, z)z^{-i} \exp \xi$ and $w^a = \sum v_i(t, z)z^{-i}\exp(-\xi)$ with $w_0 = v_0 = I$, and Eq.(1a) holds for them, then letting $\Phi = \sum w_i(A\partial)^{-i}$ we will have $w = \Phi \exp \xi$ and $w^a = (\Phi^*)^{-1}\exp(-\xi)$.

Moreover, if the stronger equality (1b) holds, then w and w^a are the Baker and the adjoint Baker functions of the mcKP.

Proof. We have

$$\mathrm{res}_z(\partial^i w) A^{-1}(w^a)^* = \mathrm{res}_z(\partial^i \Phi e^\xi) A^{-1}((\Phi^*)^{-1}e^{-\xi})^*$$

$$= \mathrm{res}_\partial \partial^i \Phi((\Phi^*)^{-1})^* = \mathrm{res}_\partial \partial^i \Phi\Phi^{-1} = \mathrm{res}_\partial \partial^i = 0.$$

This proves the first statement. Now the equations $\partial_{i\alpha} w = B_{i\alpha} w$ allow us to express all the derivatives $\partial_{i\alpha}$ in terms of ∂ and then to apply (1a) which proves (1b).

The first statement of the converse proposition can be obtained in the following way. Let $\Phi = \sum w_i (A\partial)^{-i}$ and $\Psi = \sum v_i (-A\partial)^{-i}$. Then $w = \Phi \exp \xi$ and $w^a = \Psi \exp(-\xi)$. We have

$$0 = \mathrm{res}_z (\partial^i \Phi e^{\xi}) A^{-1} (\Psi e^{-\xi})^* = \mathrm{res}_{\partial} \partial^i \Phi \Psi^*$$

for all $i \geq 0$. The operator $\Phi \Psi^*$ is $I + O(\partial^{-1})$, and the last equality implies that the negative part is zero. Hence $\Phi \Psi^* = I$ and $\Psi = (\Phi^*)^{-1}$.

Now let (1b) hold. Put $L = \Phi A \partial \Phi^{-1}$. We have

$$
\begin{aligned}
((\partial_{k\alpha} \Phi) + (L^k R^{\alpha})_- \Phi) e^{\xi} &= (\partial_{k\alpha} \cdot \Phi - \Phi (A\partial)^k E_{\alpha} \\
&\quad + (L^k R^{\alpha})_- \Phi) e^{\xi} \\
&= (\partial_{k\alpha} - (L^k R^{\alpha})_+) \Phi e^{\xi}.
\end{aligned}
$$

Then, applying the assumption and the lemma,

$$
\begin{aligned}
0 &= \mathrm{res}_z \partial^i (\partial_{k\alpha} - (L^k R^{\alpha})_+) w A^{-1} (w^a)^* \\
&= \mathrm{res}_z \partial^i (\partial_{k\alpha} - (L^k R^{\alpha})_+) \Phi e^{\xi} A^{-1} ((\Phi^*)^{-1} e^{-\xi})^* \\
&= \mathrm{res}_{\partial} \partial^i ((\partial_{k\alpha} \Phi) + (L^k R^{\alpha})_- \Phi) \Phi^{-1}.
\end{aligned}
$$

This yields $(\partial_{k\alpha} \Phi) + (L^k R^{\alpha})_- \Phi = 0$, i.e., the equation of the hierarchy. $\square$

The bilinear identity can also be written in a dual form

$$\mathrm{res}_z w A^{-1} (\partial_{i_1 \alpha_1} \partial_{i_2 \alpha_2} \ldots \partial_{i_s \alpha_s} w^a)^* = 0.$$

The proof is similar.

Very often the identity in the form

$$\mathrm{res}_z w(t, z) A^{-1} (w^a(t', z))^* = 0$$

where t' is another set of values $t_{k\alpha}$ is used. This identity makes sense as a formal expansion in powers of $t'_{k\alpha} - t_{k\alpha}$.

4. τ-Function

Let $G_{\alpha}(\varsigma)$ be an operator of translation acting as

$$G_{\alpha}(\varsigma) f(t, z) = f(\ldots, t_{k\gamma} - \delta_{\alpha\gamma} \frac{1}{k\varsigma^k}, \ldots, z).$$

Let

$$N_{\alpha}(\varsigma) = -\sum_{j=0}^{\infty} \varsigma^{-j-1} \partial_{j\alpha} + \partial_{\varsigma}, \quad \partial_{\varsigma} = \partial/\partial \varsigma.$$

It is easy to see that $N_{\alpha}(\varsigma) G(\varsigma) f(t, z) = 0$.

According to the bilinear identity,

$$\operatorname{res}_z w(t,z) A^{-1} G_\beta(\zeta)(w^a(t,z))^* = 0.$$

We have

$$G_\beta(\zeta) \exp(-\sum_{k\gamma} t_{k\gamma} E_\gamma z^k)$$
$$= (I - E_\beta + (1 - \frac{z}{\zeta})^{-1} E_\beta) \exp(-\sum_{k\gamma} t_{k\gamma} E_\gamma z^k)$$

which is easy to check. If $w(z) = \hat{w}(z) \exp \xi$ and $w^a(z) = \hat{w}^a(z) \exp(-\xi)$ then

$$\operatorname{res}_z \hat{w}(z) A^{-1}(I - E_\beta + (1 - \frac{z}{\zeta})^{-1} E_\beta) G_\beta(\zeta)(\hat{w}^a(z))^* = 0.$$

It is easy to see that if $f(z) = \sum f_i z^i$ then $\operatorname{res}_z f(z)(1 - z/\zeta)^{-1} = \zeta f_-(\zeta)$ where the subscript "$-$" symbolizes the negative part of the series. We have $\hat{w} = I + w_1 z^{-1} + \dots$ and, as a simple calculation shows, $(\hat{w}^a)^* = I - A w_1 A^{-1} z^{-1} + \dots$. The identity becomes

$$w_1(I - E_\beta) A^{-1} - (I - E_\beta) G_\beta w_1 A^{-1}$$
$$+ \zeta[\hat{w}(\zeta) A^{-1} E_\beta G_\beta(\zeta)(\hat{w}^a(\zeta))^*]_- = 0.$$

The (β, β)th element of this matrix identity is

$$\hat{w}_{\beta\beta}(\zeta) a_\beta^{-1} G_\beta(\zeta)(\hat{w}^a(\zeta))^*_{\beta\beta} - a_\beta^{-1} I = 0.$$

Thus, we have

$$\hat{w}_{\beta\beta}(\zeta) G_\beta(\zeta)(\hat{w}^a(\zeta))_{\beta\beta} = I. \tag{2}$$

The shifted $(\hat{w}^a(\zeta))_{\beta\beta}$ happens to be just the inverse of $\hat{w}_{\beta\beta}(\zeta)$.

Let us take now the (α, β)th element of the matrix identity:

$$-a_\beta^{-1} G_\beta(\zeta) w_{1,\alpha\beta} + \zeta \hat{w}_{\alpha\beta}(\zeta) a_\beta^{-1} G_\beta(\zeta)(\hat{w}^a(\zeta))_{\beta\beta} = 0.$$

Using (2), transform this to

$$G_\beta(\zeta) w_{1,\alpha\beta} = \zeta \hat{w}_{\alpha\beta}(\zeta)(\hat{w}_{\beta\beta}(\zeta))^{-1}. \tag{3}$$

Now consider a more complicated relation which also follows from the bilinear identity

$$\operatorname{res}_z w(z) A^{-1} G_\alpha(\zeta_1) G_\beta(\zeta_2)(w^a(z))^* = 0.$$

In the case when $\alpha = \beta$, this reduces to

$$\text{res}_z \hat{w}(z) A^{-1} [I - E_\beta + (1 - \frac{z}{\zeta_1})^{-1}(1 - \frac{z}{\zeta_2})^{-1} E_\beta]$$
$$G_\beta(\zeta_1) G_\beta(\zeta_2)(\hat{w}^a(z))^* = 0.$$

Taking the (β, β)th element we have

$$\text{res}_z \hat{w}_{\beta\beta}(z)(1 - \frac{z}{\zeta_1})^{-1}(1 - \frac{z}{\zeta_2})^{-1} G_\beta(\zeta_1) G_\beta(\zeta_2)(\hat{w}^a(z))_{\beta\beta} = 0$$

or

$$\text{res}_z \hat{w}_{\beta\beta}(z)[\zeta_1^{-1}(1 - \frac{z}{\zeta_1})^{-1} - \zeta_2^{-1}(1 - \frac{z}{\zeta_2})^{-1}]$$
$$G_\beta(\zeta_1) G_\beta(\zeta_2)(\hat{w}^a(z))_{\beta\beta} = 0$$

which yields

$$\hat{w}_{\beta\beta}(\zeta_1) G_\beta(\zeta_1) G_\beta(\zeta_2)(\hat{w}^a(\zeta_1))^*$$
$$= \hat{w}_{\beta\beta}(\zeta_2) G_\beta(\zeta_1) G_\beta(\zeta_2)(\hat{w}^a(\zeta_2))_{\beta\beta}.$$

Using (2), we obtain

$$\frac{G_\beta(\zeta_2) \hat{w}_{\beta\beta}(\zeta_1)}{\hat{w}_{\beta\beta}(\zeta_1)} = \frac{G_\beta(\zeta_1) \hat{w}_{\beta\beta}(\zeta_2)}{\hat{w}_{\beta\beta}(\zeta_2)}.$$

Taking a logarithm and denoting $\ln \hat{w} = f$ we get

$$(G_\beta(\zeta_2) - 1) f_{\beta\beta}(\zeta_1) = (G_\beta(z_1) - 1) f_{\beta\beta}(\zeta_2). \tag{4}$$

In the case when $\alpha \neq \beta$ the identity is

$$\text{res}_z \hat{w}(z) A^{-1} [I - E_\alpha - E_\beta + (1 - \frac{z}{\zeta_1})^{-1} E_\alpha + (1 - \frac{z}{\zeta_2})^{-1} E_\beta]$$
$$G_\alpha(\zeta_1) G_\beta(\zeta_2)(\hat{w}^a(z))^* = 0.$$

The (α, α)th element of this matrix identity is

$$\zeta_1 \hat{w}_{\alpha\alpha}(\zeta_1) a_\alpha^{-1} G_\alpha(\zeta_1) G_\beta(\zeta_2)(\hat{w}^a(\zeta_1))_{\alpha\alpha}$$
$$+ \zeta_2 \hat{w}_{\alpha\beta}(\zeta_2) a_\beta^{-1} G_\alpha(\zeta_1) G_\beta(\zeta_2)(\hat{w}^a(\zeta_2))^*_{\beta\alpha} - \zeta_1 a_\alpha^{-1} I = 0.$$

The (β, α)th element is

$$\zeta_2 \hat{w}_{\beta\beta}(\zeta_2) a_\beta^{-1} G_\alpha(\zeta_1) G_\beta(\zeta_2)(\hat{w}^a(\zeta_2))^*_{\beta\alpha}$$
$$+ \zeta_1 \hat{w}_{\beta\alpha}(\zeta_1) a_\alpha^{-1} G_\alpha(\zeta_1) G_\beta(\zeta_2)(\hat{w}^a(\zeta_1))_{\alpha\alpha} = 0.$$

Eliminating $(\hat{w}^a)^*_{\beta\alpha}$ from two equations and applying (2), we obtain

$$-\hat{w}_{\beta\beta}(\zeta_2) + (\hat{w}_{\alpha\alpha}(\zeta_1)\hat{w}_{\beta\beta}(\zeta_2)$$
$$-\hat{w}_{\beta\alpha}(\zeta_1)\hat{w}_{\alpha\beta}(\zeta_2))G_\beta(\zeta_2)(\hat{w}_{\alpha\alpha}(\zeta_1))^{-1} = 0.$$

Take a logarithm

$$\ln\hat{w}_{\beta\beta}(\zeta_2) = \ln(\hat{w}_{\alpha\alpha}(\zeta_1)\hat{w}_{\beta\beta}(\zeta_2) - \hat{w}_{\beta\alpha}(\zeta_1)\hat{w}_{\alpha\beta}(\zeta_2))$$
$$-G_\beta(\zeta_2)\ln\hat{w}_{\alpha\alpha}(\zeta_1)$$

and subtract this equation from the one obtained by the permutation of α and β, ζ_1 and ζ_2. The result is

$$(G_\beta(\zeta_2) - 1)f_{\alpha\alpha}(\zeta_1) = (G_\alpha(\zeta_1) - 1)f_{\beta\beta}(\zeta_2), \quad f = \ln\hat{w}. \qquad (5)$$

Equation (4) is a special case of this one when $\alpha = \beta$.

Now we have to prove the existence of a function $\tau(t)$ such that $f_{\alpha\alpha}(\zeta) = (G_\alpha(\zeta) - 1)\ln\tau$. If the operator $(G_\alpha(\zeta) - 1)$ had an inverse, this would immediately follow from (5). This operator has a kernel consisting of constants (with respect to $\{t_{k\alpha}\}$). Let us apply the operator $N_\alpha(\zeta_1)$ to equation (5):

$$G_\beta(\zeta_2)N_\alpha(\zeta_1)f_{\alpha\alpha}(\zeta_1) - N_\alpha(\zeta_1)f_{\alpha\alpha}(\zeta_1) = \sum_{j=0}^{\infty} \zeta_1^{-j-1}\partial_{j\alpha}f_{\beta\beta}(\zeta_2).$$

Then multiply this by ζ_1^i and take res_{ζ_1}:

$$b_{i\alpha} \equiv \mathrm{res}_{\zeta_1}\zeta_1^i N_\alpha(\zeta_1)f_{\alpha\alpha}(\zeta_1)$$
$$= G_\beta(\zeta_2)\mathrm{res}_{\zeta_1}\zeta_1^i N_\alpha(\zeta_1)f_{\alpha\alpha}(\zeta_1) + \partial_{i\alpha}f_{\beta\beta}(\zeta_2),$$

i.e.,

$$b_{i\alpha} = G_\beta(\zeta_2)b_{i\alpha} + \partial_{i\alpha}f_{\beta\beta}(\zeta_2). \qquad (6)$$

Here (i, α) is an arbitrary pair of indices; one can replace them by (j, γ):

$$b_{j\gamma} = G_\beta(\zeta_2)b_{j\gamma} + \partial_{j\gamma}f_{\beta\beta}(\zeta_2).$$

Differentiating the first equality with respect to $t_{j\gamma}$, the second with respect to $t_{i\alpha}$ and subtracting, we have $(G_\beta(\zeta_2) - 1)(\partial_{j\gamma}b_{i\alpha} - \partial_{i\alpha}b_{j\gamma}) = 0$, whence $\partial_{j\gamma}b_{i\alpha} - \partial_{i\alpha}b_{j\gamma}$ is a constant. It is not difficult to see from the definition of $b_{i\alpha}$ that this constant can be only zero. Thus, $\partial_{j\gamma}b_{i\alpha} = \partial_{i\alpha}b_{j\gamma}$. This implies the existence of a function of the variables $\{t_{i\alpha}\}$, called $\ln\tau(t)$, such that $b_{i\alpha} = \partial_{i\alpha}\ln\tau$:

$$\mathrm{res}_\zeta\zeta^i(-\sum_0^{\infty} z^{-j-1}\partial_{j\alpha} + \partial_\zeta)\ln\hat{w}_{\alpha\alpha}(\zeta) = \partial_{i\alpha}\ln\tau. \qquad (7)$$

The equation (6) yields that $\partial_{i\alpha} f_{\beta\beta}(\zeta) = (G_\beta(\zeta) - 1)b_{i\alpha} = (G_\beta(\zeta) - 1)\partial_{i\alpha}\ln\tau$ and $f_{\beta\beta}(\zeta) = (G_\beta(\zeta) - 1)\ln\tau+\text{const}$. In more detail, this formula looks like

$$\hat{w}_{\beta\beta}(\zeta) = c_\beta(\zeta)\frac{\tau(..., t_{k\gamma} - \delta_{\beta\gamma}\cdot 1/(k\zeta^k), ...)}{\tau(t)}. \tag{8}$$

In the numerator only the variables $t_{k\gamma}$ with $\gamma = \alpha$ are shifted. The constant $c_\beta(\zeta)$ is a series $c_\beta(\zeta) = \sum_{i=0}^{\infty} c_{i\beta}z^{-i}$ with $c_{0\beta} = 1$.

We have obtained this formula only for diagonal elements of $\hat{w}$ yet. Equation (7) is a conversion of equation (8). Let $C =\text{diag } c_\beta(\zeta)$, a constant diagonal matrix. Then the Baker function wC^{-1} is equivalent to w. For this function (8) holds with $c_\beta = 1$.

Let us return to equation (3). We find $\hat{w}_{\alpha\beta}(\zeta) = \zeta^{-1}G_\beta(\zeta)\, w_{1,\alpha\beta}\cdot\hat{w}_{\beta\beta}$, substituting $\hat{w}_{\beta\beta}$ from (8) and denoting

$$\tau_{\alpha\beta}(t) = \tau(t)w_{1,\alpha\beta}, \quad \alpha \neq \beta, \tag{9}$$

this becomes $\hat{w}_{\alpha\beta}(\zeta) = \zeta^{-1}G_\beta(\zeta)\tau_{\alpha\beta}\cdot(\tau(t))^{-1}$, or

$$\hat{w}_{\alpha\beta}(\zeta) = \zeta^{-1}c_\beta(\zeta)\frac{\tau_{\alpha\beta}(..., t_{k\gamma} - \delta_{\beta\gamma}\cdot 1/(k\zeta^k), ...)}{\tau(t)}, \quad \alpha \neq \beta. \tag{10}$$

Thus, only those variables $t_{k,\gamma}$ are shifted whose index γ coincides with the number of the column, β. Thus, we have

Theorem. *For any Baker function there are functions $\tau(t)$ and $\tau_{\alpha\beta}(t)$ and constant series $c_\beta(\zeta)$ such that equations (8) and (10) hold. Coefficients $c_\beta(\zeta)$ are insignificant if a Baker functions is considered to within the equivalence.*

The formulas (8) and (10) are the main formulas of the theory of the τ-function.

Remark. The definition of the τ-function and the derivation of the formulas (8) and (10) based on the bilinear identity do not depend on the property of the matrix A to have distinct elements on the diagonal. The definition remains valid even if $A = I$. This will be used in the next section.

5. Single-Pole Zakharov–Shabat Hierarchy

The general Zakharov–Shabat equation is $[I\partial+U(z), I\partial_t+V(z)] = 0$ where matrices U and V are rational functions of a parameter z. In [7] it was explained in what sense the totality of all possible equations of this form can be considered as one hierarchy. Now we are interested in the case when both the functions $U(z)$ and $V(z)$ have a single pole which is at infinity, i.e., they are polynomials in z.

Let $\hat{w} = \sum_{i=0}^{\infty} w_i z^{-i}$ be a formal series, $w_0 = I$.

Definition.   The single-pole ZS hierarchy is the totality of all the equations

$$\partial_{l\alpha}\hat{w} = -(z^l R_\alpha)_-\hat{w} \quad \text{where} \quad R_\alpha = \hat{w}E_\alpha\hat{w}^{-1}. \tag{11}$$

The subscript "$-$" refers to the negative part of an expansion in powers of z.

Letting $w = \hat{w}\exp\xi(t,z)$, where ξ is as before, we get an equivalent form of the equations of the hierarchy

$$\partial_{l\alpha}w = B_{l\alpha}w, \quad B_{l\alpha} = (z^l R_\alpha)_+. \tag{12}$$

The same equation can also be expressed as

$$w\partial_{l\alpha}\cdot w^{-1} = \hat{w}(\partial_{l\alpha} - z^l E_\alpha)\hat{w}^{-1} = I\partial_{l\alpha} - B_{l\alpha}. \tag{13}$$

Thus, dressing $I\partial_{l\alpha}$ yields a first-order differential operator (13) that is an l-th degree polynomial in z. The expression w is called a *formal Baker function*. It can be proven that the operators $\partial_{l\alpha}$ commute. This fact and equation (13) imply that operators $I\partial_{l\alpha} - B_{l\alpha}$ commute, i.e.,

$$\partial_{l\alpha}B_{m\beta} - \partial_{m\beta}B_{l\alpha} - [B_{l\alpha}, B_{m\beta}] = 0. \tag{14}$$

Let λ_l, $l = 0, ..., m+1$ be a sequence of constant diagonal matrices, $\lambda_l = \text{diag}\,(\lambda_{l\alpha})$, $\lambda_{m+1,\alpha} = a_\alpha$ being distinct, and $\partial = -\sum_{l=0}^{m+1}\sum_{\alpha=1}^{n}\lambda_{l\alpha}\partial_{l\alpha}$. Set

$$L = -\sum_{l=0}^{m+1}\sum_{\alpha=1}^{n}\lambda_{l\alpha}(I\partial_{l\alpha} - B_{l\alpha}) = I\partial + U \tag{15}$$

where $U = \sum_{i=0}^{m+1}\sum_{\alpha=1}^{n}\lambda_{i\alpha}B_{l\alpha}$. Then

$$L = w\partial w^{-1} = I\partial + U = I\partial + u_0 + u_1 z + ... + u_m z^m - A z^{m+1},$$

$$A = \text{diag}\,a_\alpha. \tag{16}$$

The hierarchy equations imply

$$\partial_{m\beta}L = [B_{m\beta}, L].$$

If M is another operator defined in the same way as L with other matrix diagonal coefficients, $\mu_{l\alpha}$ instead of $\lambda_{l\alpha}$, then $[L, M] = 0$. This is exactly the ZS equation with a single pole.

The notion of the *equivalence* is the same as for the mcKP: two Baker functions are equivalent if they differ by a factor on the right which is a constant diagonal matrix series. Then $B_{l\alpha}$'s remain the same along with all differential operators L.

Proposition 1 (Universal property). *Let $\hat{w}$ be a series $\hat{w} = \sum_{i_0}^{\infty} w_i z^{-i}$, $w_0 = I$ and $w = \hat{w} \exp \xi$. All the functions depend on variables $t_{k\alpha}$. If w satisfies an equation of the form $\partial_{k\alpha} w = \overline{B}_{k\alpha} w$ where $\overline{B}_{k\alpha} w$ is a polynomial in z, then this is an equation of the hierarchy, i.e. $\overline{B}_{k\alpha} = (z^k R_\alpha)_+$.*

Proof. We have

$$0 = \partial_{k\alpha} \hat{w} \cdot e^\xi + \hat{w} E_\alpha z^k e^\xi - \overline{B}_{k\alpha} w$$

and

$$0 = \partial_{k\alpha} \hat{w} \cdot \hat{w}^{-1} + \hat{w} E_\alpha z^k \hat{w}^{-1} - \overline{B}_{k\alpha}.$$

Taking the positive part, we obtain $\overline{B}_{k\alpha} = (\hat{w} E_\alpha z^k \hat{w}^{-1})_+ = (z^k R_\alpha)_+$. $\quad\square$

Proposition 2. *Let $\hat{w}$ be a series $\hat{w} = \sum_{i_0}^{\infty} w_i z^{-i}$, $w_0 = I$ and $w = \hat{w} \exp \xi$. All the functions depend on variables $t_{k\alpha}$. Thus if w satisfies the hierarchy equations (12), then the following bilinear identity*

$$\text{res}_z z^i \partial_{k_1 \alpha_1} ... \partial_{k_s \alpha_s} w \cdot w^{-1} = 0 \tag{17}$$

holds for arbitrary sets of indices, $i \geq 0$.
    Conversely, if there is another series $\hat{v} = \sum_{i_0}^{\infty} v_i z^{-i}$, $v_0 = I$, $v = \exp(-\xi)\hat{v}$ and

$$\text{res}_z z^i \partial_{k_1 \alpha_1} ... \partial_{k_s \alpha_s} w \cdot v = 0$$

for all sets of indices, then $v = w^{-1}$ and w is a Baker function of the hierarchy.

Proof. Let w be a Baker function of the hierarchy. Then, by virtue of the equation (12), the left-hand side of (12) is a residue of a polynomial which is zero.
    Conversely, $\text{res}_z z^i wv = 0$ for all i implies that $(wv)_- = 0$, $\hat{w}\hat{v} = I$, and $v = w^{-1}$. We have further

$$(\partial_{k\alpha} \hat{w} + (z^k R_\alpha)_- \hat{w})e^\xi = (\partial_{k\alpha} - (z^k R_\alpha)_+)w$$

where R_α is defined as in (11). Using the assumption, one gets

$$0 = \text{res}_z z^i (\partial_{k\alpha} - (z^k R_\alpha)_+)w \cdot w^{-1} = \text{res}_z z^i (\partial_{k\alpha} \hat{w} + (z^k R_\alpha)_- \hat{w})\hat{w}^{-1}.$$

This implies $\partial_{k\alpha} \hat{w} + (z^k R_\alpha)_- \hat{w} = 0$ which is the hierarchy equation (11). $\quad\square$

Proposition 3. *Baker functions of the s-p ZS are those of mcKP, more-over, the action of the operators $\partial_{k\alpha}$ on them is the same in both the hier-archies. In other words, the s-p ZS hierarchy is a restriction of the mcKP.*

Proof. The first statement follows from the fact that a Baker function of the s-p ZS hierarchy satisfies a bilinear identity (17) stronger than (1b). The converse part of the proposition of Section 3 can be applied (letting $A^{-1}(w^a)^* = w^{-1}$). It also follows from the second statement. Let us prove the latter. Let A be an arbitrary constant diagonal matrix with distinct diagonal elements. Put $\partial = \sum_\alpha a_\alpha^{-1}\partial_{1\alpha}$. Let w be a Baker function of the single-pole ZS hierarchy. Then w satisfies (17) for every multi-index. It suffices to show that w satisfies the set of equations of the form $\partial_{k\alpha}w = \overline{B}_{k\alpha}w$ for all k and α where $\overline{B}_{k\alpha}$ are differential operators in ∂ (see Proposition, Section 2). We have obvious relations:

$$\begin{aligned}
\partial_{k\alpha}w &= (E_\alpha z^k + O(z^{k-1}))e^\xi, \\
A^q\partial^q w &= (z^q + O(z^{q-1}))e^\xi.
\end{aligned}$$

whence

$$\begin{aligned}
\partial_{k\alpha}w - E_\alpha A^k \partial^k w &= O(z^{k-1})\exp\xi \\
&= (V_{k-1}z^{k-1} + O(z^{k-2}))\exp\xi.
\end{aligned}$$

The process can be prolonged:

$$\partial_{k\alpha}w - E_\alpha A^k \partial^k w - V_{k-1}A^{k-1}\partial^{k-1}w = O(z^{k-2})\exp\xi$$

etc. In the end we have

$$\begin{aligned}
\partial_{k\alpha}w - \overline{B}_{k\alpha}w &\equiv \partial_{k\alpha}w - E_\alpha A^k\partial^k w - V_{k-1}A^{k-1}\partial^{k-1}w \\
&\quad - \cdots - V_0 w = O(z^{-1})e^\xi
\end{aligned}$$

where $\overline{B}_{k\alpha}$ is a differential operator. Now, the bilinear identity

$$\mathrm{res}_z z^i (\partial_{k\alpha} - \overline{B}_{k\alpha})w \cdot w^{-1} = 0$$

where $(\partial_{k\alpha} - \overline{B}_{k\alpha})w \cdot w^{-1} = O(z^{-1})$ implies that $(\partial_{k\alpha} - \overline{B}_{k\alpha})w \cdot w^{-1} = 0$, and $(\partial_{k\alpha} - \overline{B}_{k\alpha})w = 0$ as required. $\qquad\square$

Remark. It can seem strange that $\overline{B}_{k\alpha}w$ whose elements are differential polynomials with respect to ∂ coincides with $B_{k\alpha}w$ where only ordinary polynomials are involved. The explanation is that one of the equations of the s-p ZS hierarchy is $\partial w \equiv \sum_\alpha a_\alpha^{-1}\partial_{1\alpha}w = B_{1\alpha}w$ or, in detail, $(\partial + [A^{-1}, w_1] - A^{-1}z)w = 0$. It enables us to eliminate all the derivatives. It is spectacular and instructive (though cumbersome) to verify the statement of the last proposition directly even in the simplest case of $\partial_{2\alpha}$.

Corollary. *The τ-functions for the s-p ZS hierarchy exist and they are a special case of those for the mcKP.*

6. Not Normalized s-p ZS Hierarchy

The hierarchy in the last section was normalized in the sense that $w_0 = I$. Now w_0 also will depend on time variables, $w_0(t)$. The definition of the hierarchy (11) must be adjusted to this requirement since (11) implies that $w_0 =$const.

Let $A_{(+)}$ symbolize the purely positive part of an expansion in powers of z, i.e., without the constant term, and $A_{(-)}$ negative part with the constant term, i.e. the constant term passes from the positive part to the negative one. Equation (11) will be replaced by

$$\partial_{l\alpha}\hat{w} = -(z^l R_\alpha)_{(-)}\hat{w}, \quad R_\alpha = \hat{w}E_\alpha \hat{w}^{-1} \tag{18}$$

and equation (12) by

$$\partial_{l\alpha} w = B_{l\alpha} = (z^l R_a)_{(+)}.$$

It can be proven that the operators $\partial_{l\alpha}$ commute as well.

Proposition 1. *If $\hat{w}$ satisfies (18) then $\hat{v} = w_0^{-1}\hat{w}$ satisfies (11).*

Proof. Equation (18) implies

$$\partial_{l\alpha} w_0 = -(z^l \hat{w}E_\alpha \hat{w}^{-1})_0 w_0 =: -w_0(z^l \hat{v}E_\alpha \hat{v}^{-1})_0. \tag{19}$$

Then

$$
\begin{aligned}
\partial_{l\alpha}\hat{v} &= -w_0^{-1}\partial_{l\alpha} w_0 \cdot w_0^{-1}\hat{w} + w_0^{-1}\partial_{l\alpha}\hat{w} \\
&= (z^l \hat{v}E_\alpha \hat{v}^{-1})_0 \hat{v} - w_0^{-1}(z^l \hat{w}E_\alpha \hat{w}^{-1})_{(-)}\hat{w} \\
&= (z^l \hat{v}E_\alpha \hat{v}^{-1})_0 \hat{v} - (z^l \hat{v}E_\alpha \hat{v}^{-1})_{(-)}\hat{v} \\
&= -(z^l \hat{v}E_\alpha \hat{v}^{-1})_- \hat{v}.
\end{aligned}
$$

This is exactly equation (11). □

Proposition 1 allows us to express $\hat{v}$ in terms of a τ function. However, this is not what we need; w_0 remains indefinite. In order to determine it one has to solve the linear equation (19). We show further that the whole function $\hat{w}$ has an expression in terms of τ-functions

$$\hat{w}_{\alpha\beta}(\zeta) = \frac{\tau_{\alpha\beta}(..., t_{k\gamma} - \delta_{\beta\gamma}\cdot 1/(k\zeta^k), ...)}{\tau(t)} \tag{20}$$

for both $\alpha = \beta$ and $\alpha \neq \beta$.

First of all, one must write the bilinear identity. In this case, Eq.(17) holds in a stronger form: if $s > 0$ then (17) holds for $i \geq -1$, if $s = 0$ then it holds for $i \geq 0$ while $\mathrm{res}_z z^{-1} w \cdot w^{-1} = I$. The identity can be written in the dual form where all the derivatives act on w^{-1} rather than on w.

Proposition 2. *To every Baker function there exist functions τ and $\tau_{\alpha\beta}$ such that Eq.(20) holds up to an equivalence, i.e.,*

$$\hat{w}_{\alpha\beta}(\zeta) = c_\beta(\zeta)\frac{\tau_{\alpha\beta}(..., t_{k\gamma} - \delta_{\beta\gamma} \cdot 1/(k\zeta^k), ...)}{\tau(t)}$$

where $c_\beta(z)$ are constant series.

It follows from the bilinear identity (17) that

$$\mathrm{res}_z z^i w \cdot G_\beta(\zeta)w^{-1} = \left\{ \begin{array}{ll} 0, & \text{if } i \geq 0 \\ I, & \text{if } i = -1 \end{array} \right.$$

As in Section 4 this transforms to

$$\mathrm{res}_z z^i \hat{w}(z)(I - E_\beta + (1 - \tfrac{z}{\zeta})^{-1}E_\beta)G_\beta(\zeta)\hat{w}^{-1} = \left\{ \begin{array}{ll} 0, & \text{if } i \geq 0 \\ I, & \text{if } i = -1 \end{array} \right.$$

We have

$$\mathrm{res}_z z^i \hat{w}(z)(I - E_\beta)G_\beta(\zeta)\hat{w}^{-1}(z)$$

$$+\zeta(z^i\hat{w}(z)G_\beta(\zeta)E_\beta\hat{w}^{-1}(z))_-|_{z=\zeta} = \left\{ \begin{array}{ll} 0, & \text{if } i \geq 0 \\ I, & \text{if } i = -1 \end{array} \right.$$

Let $i = -1$. Then this equality becomes

$$w_0(I - E_\beta)G_\beta(\zeta)w_0^{-1} + \hat{w}(\zeta)G_\beta(\zeta)E_\beta\hat{w}^{-1}(\zeta) = I$$

or, multiplying by $G_\beta w_0$,

$$w_0(I - E_\beta) + \hat{w}(\zeta)G_\beta(\zeta)E_\beta\hat{w}^{-1}(\zeta)w_0 = G_\beta(\zeta)w_0.$$

For the (α, β)th element this is

$$\hat{w}_{\alpha\beta}G_\beta(\hat{v}^{-1})_{\beta\beta} = G_\beta(w_0)_{\alpha\beta}. \tag{21}$$

It is easy to see that the case $i = 0$ gives the same for $\hat{v}$ as in Section 4 for $\hat{w}$; in particular, the analogues of (2) and (3), and the possibility to express $\hat{v}$ in terms of a τ-function. Equation (2) becomes

$$G_\beta(\hat{v}^{-1})_{\beta\beta} = (\hat{v}_{\beta\beta})^{-1}. \tag{22}$$

We even do not need to prove this since we knew this in advance. We know also that there is a function $\tau(t)$ and constant series c_β such that $\hat{v}_{\beta\beta} = c_\beta G_\beta\tau \cdot \tau^{-1}$. With the help of (22), (21) transforms to

$$\hat{w}_{\alpha\beta}(\hat{v}_{\beta\beta})^{-1} = G_\beta(w_0)_{\alpha\beta}. \tag{23}$$

Now let

$$\tau_{\alpha\beta} = \tau(w_0)_{\alpha\beta}. \tag{24}$$

Using (23) and (22), we have

$$\frac{G_\beta \tau_{\alpha\beta}}{\tau} = \frac{G_\beta \tau}{\tau} \cdot G_\beta(w_0)_{\alpha\beta} = c_\beta^{-1} \hat{v}_{\beta\beta} \hat{w}_{\alpha\beta} (\hat{v}_{\beta\beta})^{-1} = c_\beta^{-1} \hat{w}_{\alpha\beta}$$

as required. $\qquad\qquad\square$

Notice that (22) and (23) look very nice being put together in the form

$$
\begin{aligned}
G_\beta(\hat{v}^{-1})_{\beta\beta} &= (\hat{v}_{\beta\beta})^{-1}, \\
G_\beta(\hat{w}\hat{v}^{-1})_{\alpha\beta} &= \hat{w}_{\alpha\beta}(\hat{v}_{\beta\beta})^{-1}.
\end{aligned}
$$

7. The General ZS Hierarchy

This hierarchy was introduced in [7]. Let a_k, $k = 1, ..., m$ be a given set of complex numbers. Let, for every k,

$$\hat{w}_k = \sum_0^\infty w_{ki}(z - a_k)^i,$$

be a formal series. The entries of $n \times n$ matrices w_{ki}, $w_{ki,\alpha\beta}$ are just letters. We consider the algebra $\mathcal{A}_w$ of polynomials of all this entries and $(\det w_{k0})^{-1}$. The formal series $\hat{w}_k$ can be inverted within this algebra. Let

$$R_{k\alpha} = \hat{w}_k E_\alpha \hat{w}_k^{-1}; \quad R_{k\alpha l} = R_{k\alpha}(z - a_k)^{-l}$$

where E_α is, as before, a matrix with only one nonvanishing element, equals 1 on the (α, α) place.

We have the following objects. Such quantities as $\hat{w}_k$ and $R_{k\alpha l}$ are formal series, or jets, at the points a_k. The algebra of all such jets will be called J_k and $J = \oplus J_k$. If $j_k \in J_k$ is a jet then j_k^- symbolizes its principal part, i.e., a sum of negative powers of $z - a_k$, and j_k^+ the rest of the series. Correspondingly, the jet algebras split into parts, $J_k = J_k^+ \oplus J_k^-$. If the principal part contains a finite number of terms (and we tacitly assume this unless the opposite is said or is evident from a context) it can be considered as a global meromorphic function; the algebra of global meromorphic functions is G. A global function gives rise to a jet at every a_k. In particular, j_k^- can be considered as a jet at a point a_{k_1}, different from a_k, more precisely, as an element of $J_{k_1}^+$. And finally, there will be formal products of jets or of global functions by expressions of the form $\exp \xi_k$ where

$$\xi_k = \sum_{\alpha=1}^n \sum_{l=0}^\infty t_{k\alpha l} E_\alpha (z - a_k)^{-l}.$$

Definitions. (i) *A hierarchy corresponding to a fixed set $\{a_k\}$ is the totality of equations*

$$\partial_{k\alpha l}\hat{w}_{k_1} = \begin{cases} -R^+_{k\alpha l}\hat{w}_{k_1}, & k = k_1 \\ R^-_{k\alpha l}\hat{w}_{k_1}, & \text{otherwise} \end{cases}, \quad \partial_{k\alpha l} = \partial/\partial t_{k\alpha l}. \quad (25)$$

In the second case $R^-_{k\alpha l}$ is considered as an element of $J^+_{k_1}$, see above; $t_{k\alpha l}$ are some variables.

(ii) *A ZS hierarchy is an inductive limit of hierarchies with fixed sets $\{a_k\}$, with respect to a natural embedding of a hierarchy corresponding to a subset into a hierarchy corresponding to a larger set, as a subhierarchy.*

In this article we deal with the hierarchy corresponding to a fixed set $\{a_k\}$. There was proven in [7] that all the equations of the hierarchy commute. The following proposition readily can be checked by a simple straightforward computation:

Proposition 1. *A dressing formula*

$$\hat{w}_{k_1}(\partial_{k\alpha l} - E_\alpha(z - a_k)^{-l}\delta_{kk_1})\hat{w}_{k_1}^{-1} = \partial_{k\alpha l} - B_{k\alpha l}, \quad B_{k\alpha l} = R^-_{k\alpha l} \quad (26)$$

is equivalent to (25).

The operator $\partial_{k\alpha l} - B_{k\alpha l}$ is assumed to act in J_{k_1}. However, it does not depend on k_1 at all and can be considered as a global function of z with the only pole of the lth order at a_k.

Let

$$w_k = \hat{w}_k \exp \xi_k.$$

Definition. *The collection $w = \{w_k\}$ is the formal Baker function of the hierarchy.*

Equation (25) can be written in terms of the Baker function as

$$\partial_{k\alpha l}w_{k_1} = B_{k\alpha l}w_{k_1} \quad (27)$$

and (26) as

$$w_{k_1}\partial_{k\alpha l}w_{k_1}^{-1} = \partial_{k\alpha l} - B_{k\alpha l}. \quad (28)$$

Proposition 2. *All the operators $\partial_{k\alpha l} - B_{k\alpha l}$ commute.*

Proof. This is a corollary of the fact that $\partial_{k\alpha l}$ commute and equation (28).
$\square$

One can consider arbitrary linear combinations of the above constructed operators,

$$L = \sum_{k,\alpha,l} \lambda_{k\alpha l}(\partial_{k\alpha l} - B_{k\alpha l}) = \partial + U$$

where $\partial = \sum_{k,\alpha,l} \lambda_{k\alpha l} \partial_{k\alpha l}$ and $U = -\sum_{k,\alpha,l} \lambda_{k\alpha l} B_{k\alpha l}$. Two such operators commute yielding equations of the Zakharov–Shabat type

$$\partial U_1 - \partial_1 U = [U_1, U].$$

Functions U and U_1 are rational functions of the parameter z.

Remark 1. A Baker function is determined up to equivalence. Two Baker functions $w^{(1)}$ and $w^{(2)}$ are equivalent if there are constant diagonal matrices $c_k(z) = \sum_0^\infty c_{ki} z^{-i} = \text{diag } (c_{k\beta}(z))$ such that $w_k^{(1)} = w_k^{(2)} c_k$, i.e., $w_{k,\alpha\beta}^{(1)} = c_{k,\beta} w_{k,\alpha\beta}^{(2)}$. Equivalent Baker functions generate the same Lax operator L.

Remark 2. Here we have a special case of ZS equation: the functions U and U_1 vanishing at infinity. If we make a gauge transformation $w_k \mapsto g(t) w_k$ then $\partial + U \mapsto g(\partial + U) g^{-1} = \partial + g U g^{-1} - (\partial g) g^{-1}$, the last term does not vanish at infinity. This yields the general case.

If we deal with only one component w_k of the Baker function, and consider its dependence solely on the variables $t_{k\alpha l}$ with the same k (local variables) ignoring all the others (alien variables), e.g., fixing their values as parameters, then we shall have a single-pole non-normalized hierarchy in the sense of the previous section. (One has to perform a transformation $(z - a_k)^{-1} = \zeta$.) This fact allows us to apply all the formulas obtained in that section to the present case. In particular, there are functions $\tau_k(t)$ and $\tau_{k\alpha\beta}(t)$ depending on local as well as on alien variables such that

$$w_{k,\alpha\beta}(t, z) = c_{k\beta}(z) \frac{G_{k\beta}(z) \tau_{k,\alpha\beta}(t)}{\tau_k(t)} e^{\xi_k} \qquad (29)$$

where operators of translation $G_{k\beta}(z)$ are defined by

$$G_{k\beta} f(t) = f(..., t_{k_1,\gamma,l} - \delta_{kk_1} \delta_{\beta\gamma} \frac{1}{l} (z - a_k)^l, ...)$$

and $c_{k\beta}(z)$ are constant series in $z - a_k$.

We have not yet used the equations of the hierarchy with respect to the alien variables. The rest of the section will be devoted to the proof that if those equations are taken into account, then, roughly speaking, all denominators τ_k in the previous formula are equal. More precisely, the following theorem holds:

Theorem. If $w = \{w_k\}$ is an arbitrary Baker function then there are functions $\tau(t)$ and $\tau_{k,\alpha\beta}(t)$ and constant series $c_{k\beta}(z)$ such that

$$w_{k,\alpha\beta}(t, z) = c_{k\beta}(z) \frac{G_{k\beta}(z) \tau_{k,\alpha\beta}(t)}{\tau(t)} e^{\sum_l \xi_l}.$$

.

Notice that the last factor is $\exp\sum_l \xi_l$ and not just $\exp\xi_k$. Therefore the expression in front of it is not $\hat{w}_k$. We call it $\hat{\hat{w}}_k$. Thus,

$$w_k = \hat{w}_k \exp\sum_l \xi_l, \quad \hat{w}_k = \hat{\hat{w}}_k \exp\sum_{l\neq k} \xi_l, \quad \hat{w}_{k0} = \hat{\hat{w}}_{k0} \exp\sum_{l\neq k} \xi_l(a_k).$$

Proof. We already have Baker functions w_k, $\hat{w}_k = w_k \exp(-\xi_k)$ and $\hat{\hat{w}}_k = \hat{w}_k \exp(-\sum_{l\neq k}\xi_l)$. Let us also introduce, as we did in Section 6,

$$v_k(z) = w_{k0}^{-1} w_k(z), \quad \hat{v}_k(z) = v_k(z)\exp(-\xi_k) = w_{k0}^{-1}\hat{w}_k(z)$$

and

$$\hat{\hat{v}}_k(z) = \hat{\hat{w}}_{k0}^{-1}\hat{\hat{w}}_k(z) = \exp\sum_{l\neq k}\xi_l(a_k)\hat{v}_k(z)\exp(-\sum_{l\neq k}\xi_l).$$

What is important, $\hat{v}_k$ and $\hat{\hat{v}}_k$ differ by two diagonal factors, on the left and on the right which do not depend on the variables with the same index k, the local variables. The series $\hat{w}_k$ and $\hat{\hat{w}}_k$ differ by a right factor of the same kind.

Considering w_k as a function of local variables, we have noticed that this is a Baker function of a single-pole not normalized hierarchy, and $v_k(z)$ that of the corresponding normalized hierarchy. Therefore, one can write for them (22), or in present notations,

$$G_{k\beta}(\zeta)(\hat{v}_k^{-1}(\zeta))_{\beta\beta} = (\hat{v}_{k,\beta\beta}(\zeta))^{-1} \tag{30}$$

and (23), or

$$\hat{w}_{k,\alpha\beta}(\zeta)(\hat{v}_{k,\beta\beta})^{-1} = G_{k\beta}(\zeta)w_{k0,\alpha\beta}. \tag{31}$$

The same equations can be written for $\hat{\hat{v}}_k$ and $\hat{\hat{w}}_{k0}$ since the diagonal factors we discussed above will cancel. They do not depend on local variables and the operators $G_{k\beta}$ do not act on them. Thus,

$$G_{k\beta}(\zeta)(\hat{\hat{v}}_k^{-1}(\zeta))_{\beta\beta} = (\hat{\hat{v}}_{k,\beta\beta}(\zeta))^{-1} \tag{30'}$$

and

$$\hat{\hat{w}}_{k,\alpha\beta}(\zeta)(\hat{\hat{v}}_{k,\beta\beta})^{-1} = G_{k\beta}\hat{\hat{w}}_{k0,\alpha\beta}. \tag{31'}$$

For the same reason we have

$$\frac{G_{k\beta_2}(\zeta_2)\hat{\hat{v}}_{k,\beta_1\beta_1}(\zeta_1)}{\hat{\hat{v}}_{k,\beta_1\beta_1}(\zeta_1)} = \frac{G_{k\beta_1}(\zeta_1)\hat{\hat{v}}_{k,\beta_2\beta_2}(\zeta_2)}{\hat{\hat{v}}_{k,\beta_2\beta_2}(\zeta_2)}. \tag{32}$$

It is correct for $\hat{v}_k$ since this is, virtually, equation (5). The additional diagonal factors cancel, so this is also correct for $\hat{\hat{v}}_k$.   □

Lemma. *The equality*

$$\frac{G_{k_2\beta}(\zeta_2)\hat{\hat{v}}_{k_1,\beta\beta}(\zeta_1)}{\hat{\hat{v}}_{k_1,\beta\beta}(\zeta_1)} = \frac{G_{k_1\beta}(\zeta_1)\hat{\hat{v}}_{k_2,\beta\beta}(\zeta_2)}{\hat{\hat{v}}_{k_2,\beta\beta}(\zeta_2)} \tag{33}$$

holds.

Proof of Lemma 1. Equation (27) implies that $\partial_{k_1\alpha l}w_k \cdot w_k^{-1} = B_{k_1\alpha l} = R_{k_1\alpha l}^-$ is a meromorphic function with a single pole at a_{k_1}, vanishing at infinity and *not depending on* k. Actually, it is easy to see that this is a characteristic property of the hierarchy which expresses its universality, but we do not use this fact below. More generally, $\partial_{k_1\alpha_1 l}...\partial_{k_s\alpha_s l}w_k \cdot w_k^{-1}$ does not depend on k and is a meromorphic function with the poles at $a_{k_1},...,a_{k_s}$ vanishing at infinity when $s > 0$. The same is also true for $w_k \cdot \partial_{k_1\alpha_1 l}...\partial_{k_s\alpha_s l}w_k^{-1}$. This implies that the expression $(z - a_{k_1})^{-1}(z - a_{k_2})^{-1}w_k G_{k_1\beta}(\zeta_1)G_{k_2\beta}(\zeta_2)w_k^{-1}$ is a meromorphic function having the only poles on the Riemann sphere at a_{k_1} and a_{k_2} and not depending on k. The sum of residues must vanish. Computing the residue at a_{k_1} we replace k by k_1 and doing this at a_{k_2} we replace k by k_2. For simplicity of writing, let res_{k_i} symbolize $\text{res}_{a_{k_i}}$. We have

$$\text{res}_{k_1}(z - a_{k_1})^{-1}(z - a_{k_2})^{-1}w_{k_1}(z)G_{k_1\beta}(\zeta_1)G_{k_2\beta}(\zeta_2)w_{k_1}^{-1}(z)$$
$$+\text{res}_{k_2}(z - a_{k_1})^{-1}(z - a_{k_2})^{-1}w_{k_2}(z)G_{k_1\beta}(\zeta_1)G_{k_2\beta}(\zeta_2)w_{k_2}^{-1}(z) = 0.$$

In terms of $\hat{w}_k$ this identity can be written as

$$\text{res}_{k_1}(z - a_{k_1})^{-1}(z - a_{k_2})^{-1}\hat{w}_{k_1}(z)(I - E_\beta + E_\beta(1 - \frac{\zeta_1 - a_{k_1}}{z - a_{k_1}})^{-1})$$
$$\cdot(I - E_\beta + E_\beta(1 - \frac{\zeta_2 - a_{k_2}}{z - a_{k_2}})^{-1})G_{k_1\beta}(\zeta_1)G_{k_2\beta}(\zeta_2)\hat{w}_{k_1}^{-1}(z)$$
$$+(k_1,\zeta_1 \Leftrightarrow k_2,\zeta_2) = 0,$$

i.e.,

$$\text{res}_{k_1}(z - a_{k_1})^{-1}(z - a_{k_2})^{-1}\hat{w}_{k_1}(z)(I - E_\beta + E_\beta(1 - \frac{\zeta_1 - a_{k_1}}{z - a_{k_1}})^{-1}$$
$$\cdot(1 - \frac{\zeta_2 - a_{k_2}}{z - a_{k_2}})^{-1})G_{k_1\beta}(\zeta_1)G_{k_2\beta}(\zeta_2)\hat{w}_{k_1}^{-1}(z)$$
$$+(k_1,\zeta_1 \Leftrightarrow k_2,\zeta_2) = 0, \tag{34}$$

Here $(k_1,\zeta_1 \Leftrightarrow k_2,\zeta_2)$ denotes a term obtained by switching k_1 and k_2, ζ_1 and ζ_2. In the previous sections we computed similar residues several times, so it does not need much explanation. The term with $I - E_\beta$ gives

$$(a_{k_1} - a_{k_2})\hat{w}_{k_1 0}(I - E_\beta)G_{k_1\beta}(\zeta_1)G_{k_2\beta}(\zeta_2)\hat{w}_{k_1 0}^{-1} + (k_1,\zeta_1 \Leftrightarrow k_2,\zeta_2).$$

Two others are

$$(\zeta_1 - a_{k_2})^{-1}\hat{w}_{k_1}(\zeta_1)E_\beta(1 - \frac{\zeta_2 - a_{k_2}}{\zeta_1 - a_{k_2}})^{-1}G_{k_1\beta}(\zeta_1)G_{k_2\beta}(\zeta_2)\hat{w}_{k_1}^{-1}(\zeta_1)$$
$$+(k_1,\zeta_1 \Leftrightarrow k_2,\zeta_2)$$
$$= (\zeta_1 - \zeta_2)^{-1}[\hat{w}_{k_1}(\zeta_1)E_\beta G_{k_1\beta}(\zeta_1)G_{k_2\beta}(\zeta_2)\hat{w}_{k_1}^{-1}(\zeta_1)$$
$$-(k_1,\zeta_1 \Leftrightarrow k_2,\zeta_2)].$$

Multiplying the transformed equation (34) by $\hat{w}_{k_10}^{-1}$ on the left and by $G_{k_1\beta}(\zeta_1)G_{k_2\beta}(\zeta_2)\hat{w}_{k_20}$ on the right, we obtain

$$(a_{k_1} - a_{k_2})[(I - E_\beta)G_{k_1\beta}(\zeta_1)G_{k_2\beta}(\zeta_2)\hat{w}_{k_10}^{-1}\hat{w}_{k_20} - \hat{w}_{k_10}^{-1}\hat{w}_{k_20}(I - E_\beta)]$$
$$+(\zeta_1 - \zeta_2)^{-1}[\hat{v}_{k_1}(\zeta_1)E_\beta G_{k_1\beta}(\zeta_1)G_{k_2\beta}(\zeta_2)\hat{v}_{k_1}^{-1}(\zeta_1)\hat{w}_{k_10}^{-1}\hat{w}_{k_20}$$
$$-\hat{w}_{k_10}^{-1}\hat{w}_{k_20}\hat{v}_{k_2}(\zeta_2)E_\beta G_{k_1\beta}(\zeta_1)G_{k_2\beta}(\zeta_2)\hat{v}_{k_2}^{-1}(\zeta_2)] = 0$$

where $\hat{v}_k(\zeta) = \hat{w}_{k0}^{-1}\hat{w}_k(\zeta)$. Now let us take the (β, β)th element of this identity. The first two terms are not involved by virtue of the factors $I - E_\beta$. Two others yield

$$\hat{v}_{k_1,\beta\beta}G_{k_1\beta}(\zeta_1)G_{k_2\beta}(\zeta_2)(\hat{v}_{k_1}^{-1}(\zeta)T_{k_1k_2})_{\beta\beta}$$

$$= (T_{k_1k_2}\hat{v}_{k_2}(\zeta_2))_{\beta\beta}G_{k_1\beta}(\zeta_1)G_{k_2\beta}(\zeta_2)(\hat{v}_{k_2}^{-1}(\zeta_2))_{\beta\beta} \qquad (35)$$

where

$$T_{k_1k_2} = \hat{w}_{k_10}^{-1}\hat{w}_{k_20},$$

the transition function. Using (30'), we can replace $G_{k_2\beta}(\zeta_2)$ $(\hat{v}_{k_2}^{-1}(\zeta_2))_{\beta\beta}$ by $(\hat{v}_{k_2,\beta\beta}(\zeta_2))^{-1}$. Equation (35) becomes

$$\hat{v}_{k,\beta\beta}G_{k_1\beta}(\zeta_1)G_{k_2\beta}(\zeta_2)(\hat{v}_{k_1}^{-1}(\zeta)T_{k_1k_2})_{\beta\beta}$$

$$= (T_{k_1k_2}\hat{v}_{k_2}(\zeta_2))_{\beta\beta}G_{k_1\beta}(\zeta_1)(\hat{v}_{k_2,\beta\beta}(\zeta_2))^{-1}. \qquad (36)$$

Let $\zeta_1 = a_{k_1}$. Then (36) becomes

$$G_{k_2\beta}(\zeta_2)T_{k_1k_2,\beta\beta} = (T_{k_1k_2}\hat{v}_{k_2}(\zeta_2))_{\beta\beta}(\hat{v}_{k_2,\beta\beta}(\zeta_2))^{-1}.$$

This enables us to rewrite the right-hand side of (36) as

$$\hat{v}_{k_2,\beta\beta}(\zeta_2)\frac{G_{k_2\beta}(\zeta_2)T_{k_1k_2,\beta\beta}}{G_{k_1\beta}(\zeta_1)\hat{v}_{k_2,\beta\beta}(\zeta_2)}.$$

Now, let $\zeta_2 = a_{k_2}$. Equation (36) transforms to

$$\hat{v}_{k_1,\beta\beta}(\zeta_1)G_{k_1\beta}(\zeta_1)(\hat{v}_{k_1}^{-1}(\zeta)T_{k_1k_2})_{\beta\beta} = T_{k_1k_2,\beta\beta}.$$

The left-hand side of (36) can be written as

$$\hat{v}_{k_1,\beta\beta}(\zeta_1)G_{k_2\beta}(\zeta_2)\frac{T_{k_1k_2,\beta\beta}}{\hat{v}_{k_1,\beta\beta}(\zeta_1)}.$$

Equating the left- and the right-hand sides and cancelling the common factor $G_{k_2\beta}(\zeta_2)T_{k_1k_2,\beta\beta}$, we obtain the required identity (33).   □

Lemma 2. *The equation*

$$\frac{G_{k_2\beta_2}(\zeta_2)\hat{v}_{k_1,\beta_1\beta_1}(\zeta_1)}{\hat{v}_{k_1,\beta_1\beta_1}(\zeta_1)} = \frac{G_{k_1\beta_1}(\zeta_1)\hat{v}_{k_2,\beta_2\beta_2}(\zeta_2)}{\hat{v}_{k_2,\beta_2\beta_2}(\zeta_2)} \tag{37}$$

holds for any k_1, k_2, β_1 and β_2.

Proof of Lemma 2. We already have two special cases of this lemma: Equation (32) for $k_1 = k_2$ and Lemma 1 for $\beta_1 = \beta_2$. Now suppose neither of these conditions holds. Similarly to what we did proving Lemma 1, we write a bilinear identity

$$\mathrm{res}_{k_1}(z - a_{k_1})^{-1}(z - a_{k_2})^{-1}w_{k_1}(z)G_{k_1\beta_1}(\zeta_1)G_{k_2\beta_2}(\zeta_2)w_{k_1}^{-1}(z)$$
$$+(k_1, \zeta_1, \beta_1 \Leftrightarrow k_2, \zeta_2, \beta_2) = 0.$$

In terms of $\hat{w}_k$ this identity can be written as

$$\mathrm{res}_{k_1}(z - a_{k_1})^{-1}(z - a_{k_2})^{-1}\hat{w}_{k_1}(z)$$
$$\cdot(I - E_{\beta_1} + E_{\beta_1}(1 - \frac{\zeta_1 - a_{k_1}}{z - a_{k_1}})^{-1})$$
$$\cdot(I - E_{\beta_2} + E_{\beta_2}(1 - \frac{\zeta_2 - a_{k_2}}{z - a_{k_2}})^{-1})$$
$$\cdot G_{k_1\beta_1}(\zeta_1)G_{k_2\beta_2}(\zeta_2)\hat{w}_{k_1}^{-1}(z)$$
$$+(k_1, \zeta_1, \beta_1 \Leftrightarrow k_2, \zeta_2, \beta_2) = 0,$$

i.e.,

$$\mathrm{res}_{k_1}(z - a_{k_1})^{-1}(z - a_{k_2})^{-1}\hat{w}_{k_1}(z)$$
$$\cdot(I - E_{\beta_1} - E_{\beta_2} + E_{\beta_1}(1 - \frac{\zeta_1 - a_{k_1}}{z - a_{k_1}})^{-1}$$
$$+E_{\beta_2}(1 - \frac{\zeta_2 - a_{k_2}}{z - a_{k_2}})^{-1})G_{k_1\beta_1}(\zeta_1)G_{k_2\beta_2}(\zeta_2)\hat{w}_{k_1}^{-1}(z)$$
$$+(k_1, \zeta_1, \beta_1 \Leftrightarrow k_2, \zeta_2, \beta_2) = 0.$$

Computing the residues, we have

$$(a_{k_1} - a_{k_2})^{-1}\hat{w}_{k_1 0}(I - E_{\beta_1} - E_{\beta_2})G_{k_1\beta_1}(\zeta_1)G_{k_2\beta_2}(\zeta_2)\hat{w}_{k_1 0}^{-1}$$
$$(\zeta_1 - a_{k_2})^{-1}\hat{w}_{k_1}(\zeta_1)E_{\beta_1}G_{k_1\beta_1}(\zeta_1)G_{k_2\beta_2}(\zeta_2)\hat{w}_{k_1}^{-1}(\zeta_1)$$
$$+(a_{k_1} - a_{k_2})^{-1}((1 - \frac{\zeta_2 - a_{k_2}}{a_{k_1} - a_{k_2}})^{-1}\hat{w}_{k_1 0}^{-1}$$
$$+(k_1, \zeta_1, \beta_1 \Leftrightarrow k_2, \zeta_2, \beta_2) = 0.$$

Dividing by $\hat{w}_{k_1 0}$ on the left, by $G_{k_1\beta_1}(\zeta_1)G_{k_2\beta_2}(\zeta_2)\hat{w}_{k_2 0}^{-1}$ on the right, we have

$$*(I - E_{\beta_1} - E_{\beta_2}) + (I - E_{\beta_1} - E_{\beta_2})*$$
$$+(\zeta_1 - a_{k_2})^{-1}\hat{v}_{k_1}(\zeta_1)E_{\beta_1}G_{k_1\beta_1}(\zeta_1)G_{k_2\beta_2}(\zeta_2)\hat{v}_{k_1}^{-1}(\zeta_1)T_{k_1 k_2}$$
$$+(\zeta_2 - a_{k_1}^{-1})T_{k_1 k_2}\hat{v}_{k_2}(\zeta_2)E_{\beta_2}G_{k_1\beta_1}(\zeta_1)G_{k_2\beta_2}(\zeta_2)\hat{v}_{k_2}^{-1}$$
$$-E_{\beta_2}*+*E_{\beta_1} = 0$$

where asterisks symbolize various factors which are not written in detail since they are not important below.

Take the (β_1, β_2)th element of this equality. The terms with asterisks vanish. The following terms remain:

$$(\zeta_1 - a_{k_2})^{-1}\hat{v}_{k_1\beta_1\beta_1}(\zeta_1)G_{k_1\beta_1}(\zeta_1)G_{k_2\beta_2}(\zeta_2)(\hat{v}_{k_1}^{-1}(\zeta_1)T_{k_1 k_2})_{\beta_1\beta_1}$$

$$+(\zeta_2 - a_{k_1})^{-1}(T_{k_1 k_2}\hat{v}_{k_2}(\zeta_2))_{\beta_1\beta_2}G_{k_1\beta_1}(\zeta_1)G_{k_2\beta_2}(\zeta_2)(\hat{v}_{k_2}^{-1}(\zeta_2))_{\beta_2\beta_2} = 0. \quad (38)$$

Now, let $\zeta_1 = a_{k_1}$:

$$(a_{k_1} - a_{k_2})^{-1}G_{k_2\beta_2}(\zeta_2)(T_{k_1 k_2})_{\beta_1\beta_2}$$

$$+(\zeta_2 - a_{k_1})^{-1}(T_{k_1 k_2}\hat{v}_{k_2}(\zeta_2))_{\beta_1\beta_2}G_{k_2\beta_2}(\zeta_2)(\hat{v}_{k_2}^{-1})_{\beta_2\beta_2} = 0. \quad (39)$$

Using this equality, transform the second term of (38):

$$-\frac{(a_{k_1} - a_{k_2})^{-1}G_{k_2\beta_2}(\zeta_2)(T_{k_1 k_2})_{\beta_1\beta_2}G_{k_1\beta_1}(\zeta_1)G_{k_2\beta_2}(\zeta_2)(\hat{v}_{k_2}^{-1}(\zeta_2))_{\beta_2\beta_2}}{G_{k_2\beta_2}(\zeta_2)(\hat{v}_{k_2}^{-1})_{\beta_2\beta_2}}$$

$$= -\frac{(a_{k_1} - a_{k_2})^{-1}G_{k_2\beta_2}(\zeta_2)(T_{k_1 k_2})_{\beta_1\beta_2}\hat{v}_{k_2,\beta_2\beta_2}(\zeta_2)}{G_{k_1\beta_1}(\zeta_1)\hat{v}_{k_2,\beta_2\beta_2}(\zeta_2)}.$$

(We have used (30') doing the last transformation).

Let $\zeta_2 = a_{k_2}$:

$$(\zeta_1 - a_{k_2})^{-1}\hat{v}_{k_1\beta_1\beta_1}(\zeta_1)G_{k_1\beta_1}(\zeta_1)(\hat{v}_{k_1}^{-1}(\zeta_1)T_{k_1 k_2})_{\beta_1\beta_1}$$

$$+(a_{k_1} - a_{k_2})^{-1}(T_{k_1 k_2})_{\beta_1\beta_2} = 0 \quad (40)$$

whence the first term can be written as

$$-(a_{k_1} - a_{k_2})^{-1}\hat{v}_{k_1\beta_1\beta_1}(\zeta_1)G_{k_2\beta_2}(\zeta_2) \cdot \frac{(T_{k_1 k_2})_{\beta_1\beta_2}}{\hat{v}_{k_1\beta_1\beta_1}(\zeta_1)}.$$

The identity (38) becomes, after a cancellation of the common factor,

$$\frac{\hat{v}_{k_2,\beta_2\beta_2}(\zeta_2)}{G_{k_1\beta_1}(\zeta_1)\hat{v}_{k_2,\beta_2\beta_2}(\zeta_2)} = \frac{\hat{v}_{k_1\beta_1\beta_1}(\zeta_1)}{G_{k_2\beta_2}(\zeta_2)\hat{v}_{k_1\beta_1\beta_1}(\zeta_1)}$$

which is the statement of the lemma.   □

Lemma 3. *The equation (37) implies that there is a function $\tau(t)$ and constant series $c_{k\beta}(\zeta)$ in powers of $\zeta - a_k$ such that $\hat{v}_{k,\beta\beta}(\zeta) = c_{k\beta}G_{k\beta}\tau \cdot \tau^{-1}$.*

Proof of Lemma 3. Taking the logarithm of (37) and denoting $\ln \hat{v}_{k,\beta\beta} = f_{k,\beta\beta}$, we have

$$(G_{k_1\beta_1}(\zeta_1) - 1)f_{k_2\beta_2\beta_2}(\zeta_2) = (G_{k_2\beta_2}(\zeta_2) - 1)f_{k_1\beta_1\beta_1}(\zeta_1),$$

and we just have to repeat the derivation of equation (8) from (5) in Section 4. $\qquad\qquad\qquad\square$

The end of the proof of the theorem. Put $\tau_{k,\alpha\beta} = \tau \cdot \hat{w}_{k0\alpha\beta}$. Taking into account (31′), we have

$$\frac{G_{k\beta}(\zeta)\tau_{k,\alpha\beta}}{\tau} = \frac{G_{k\beta}(\zeta)\tau}{\tau} \cdot G_{k\beta}(\zeta)\hat{w}_{k0,\alpha\beta} = c_{k\beta}^{-1}\hat{v}_{k,\beta\beta} \cdot \hat{w}_{k,\alpha\beta}\hat{v}_{k,\beta\beta}^{-1} = c_{k\beta}^{-1}\hat{w}_{k,\alpha\beta}$$

as required. $\qquad\qquad\qquad\square$

References

[1] Sato, M., Soliton equations as dynamical systems on infinite dimensional Grassmann manifolds, *RIMS Kokyuroku* **439** (1981), 30-46.

[2] Date, E., Jimbo, M., Kashiwara, M., and Miwa, T., Transformation groups for soliton equations, in: Jimbo and Miwa (ed.) *Non-linear integrable systems: classical theory and quantum theory*, Proc. RIMS symposium, Singapore, 1983.

[3] Date, E., Jimbo, M., Kashiwara, M., and Miwa, T., Operator approach to the Kadomtsev-Petviashvili equation - transformation groups for soliton equations III, *Journ. Phys. Soc. Japan* **50** (1981), 3806-3812.

[4] Ueno, K., and Takasaki, K., Toda lattice hierarchy, in: *Advanced Studies in Pure Mathematics* 4, World Scientific, 1-95, 1984.

[5] Dickey, L. A., On Segal-Wilson's definition of the τ-function and hierarchies AKNS-D and mcKP, in: *Integrable systems, The Verdier Memorial Conference*, Progress in Mathematics 115, 1993, Birkhäuser, 147-162.

[6] Dickey, L. A., On the τ-function of matrix hierarchies of integrable equations, *Journal Math. Physics* **32** (1991), 2996-3002.

[7] Dickey, L. A., Why the general Zakharov–Shabat equations form a hierarchy, *Com. Math. Phys.* **163** (1994), 509-521.

[8] Vassilev, S., Tau functions of algebraic geometrical solutions to the general Zakharov–Shabat hierarchy, preprint of the University of Oklahoma, 1994.

[9] Dickey, L. A., *Soliton Equations and Hamiltonian Systems*, Add Series in Mathematical Physics **12**, World Scientific, 1991.

University of Oklahoma, Norman, OK 73019
e-mail: ldickey@uoknor.edu

Received September 1995

On the Hamiltonian Representation of the Associativity Equations[1]

E.V. Ferapontov and O.I. Mokhov

Dedicated to the memory of Irina Dorfman

Abstract
We demonstrate that for an arbitrary number n of primary fields the equations of the associativity can be rewritten in the form of $(n-2)$ pairwise commuting systems of hydrodynamic type, which appear to be nondiagonalizable but integrable. We propose a natural Hamiltonian representation of the systems under study in the cases $n = 3$ and $n = 4$.

1. Introduction

Let us consider a function of n independent variables $F(t^1, \ldots, t^n)$ satisfying the following two conditions:

1. *The matrix*

$$\eta_{\alpha\beta} = \frac{\partial^3 F}{\partial t^1 \partial t^\alpha \partial t^\beta} \qquad (\alpha, \beta = 1, \ldots, n)$$

   is constant and nondegenerate.

Note that the matrix $\eta_{\alpha\beta}$ completely determines the dependence of the function F on the fixed variable t^1 (up to quadratic polynomials).

2. *For all $t = (t^1, \ldots, t^n)$ the functions*

$$c^\alpha_{\beta\gamma}(t) = \eta^{\alpha\mu} \frac{\partial^3 F}{\partial t^\mu \partial t^\beta \partial t^\gamma} \qquad (\text{here} \quad \eta^{\alpha\mu}\eta_{\mu\beta} = \delta^\alpha_\beta)$$

   are the structure constants of the associative algebra $A(t)$ of dimension n with the basis $e_1, \ldots, e_n$ and the multiplication

$$e_\beta \circ e_\gamma = c^\alpha_{\beta\gamma}(t) e_\alpha.$$

[1]This work was partially supported by the International Science Foundation (Grant No. RKR000), the Russian Foundation for Fundamental Research (Grant No. 94–01–01478 — O.I.M. and Grant No. 93–011–168 — E.V.F.) and the INTAS (Grant No. 93–0166 — O.I.M.).

Conditions 1 and 2 lead to a complicated overdetermined system of nonlinear partial differential equations of the third order on the function F. This system is known in two-dimensional topological field theory as the equations of the associativity or the Witten-Dijkgraaf-H.Verlinde-E.Verlinde (WDVV) system [10–13] (all necessary physical motivations and the theory of integrability of the equations of the associativity can be found in the survey of B. Dubrovin [1]).

For $n = 3$ Dubrovin considered two essentially different types of dependence of the function F on the fixed variable t^1:

$$F = \frac{1}{2}(t^1)^2 t^3 + \frac{1}{2}t^1(t^2)^2 + f(t^2, t^3)$$

and

$$F = \frac{1}{6}(t^1)^3 + t^1 t^2 t^3 + f(t^2, t^3).$$

In these cases the equations of the associativity reduce to the following two nonlinear equations of the third order on the function $f = f(x,t)$ of two independent variables ($x = t^2$, $t = t^3$):

$$f_{ttt} = f_{xxt}^2 - f_{xxx}f_{xtt} \tag{1.1}$$

and

$$f_{xxx}f_{ttt} - f_{xxt}f_{xtt} = 1, \tag{1.2}$$

respectively.

Following [2,3], we introduce the new variables

$$a = f_{xxx}, \quad b = f_{xxt}, \quad c = f_{xtt}.$$

As it was shown in the papers [2,3], in the new variables the equations (1.1) and (1.2) assume the form of the 3×3 systems of hydrodynamic type:

$$\begin{cases} a_t = b_x, \\ b_t = c_x, \\ c_t = (b^2 - ac)_x \end{cases} \tag{1.3}$$

and

$$\begin{cases} a_t = b_x, \\ b_t = c_x, \\ c_t = \left((1 + bc)/a\right)_x, \end{cases} \tag{1.4}$$

respectively.

Similar transformations reduce partial differential equations of the second order (equations of the Monge-Ampère type) to two-component systems of hydrodynamic type, which can be linearized, as it is known, by the hodograph transform. In particular, it was shown in [17] that the hyperbolic Monge-Ampère equation

$$u_{tt}u_{xx} - u_{xt}^2 = -1$$

transforms into the integrable system of Chaplygin gas

$$\begin{cases} U_t = UU_x + V^{-3}V_x, \\ V_t = VU_x + UV_x. \end{cases}$$

after the transformation

$$U = \frac{u_{xt}}{u_{xx}}, \quad V = u_{xx}$$

We recall that systems of hydrodynamic type are by definition systems of quasilinear equations of the form

$$u_t^i = v_j^i(u)u_x^j.$$

The main advantage of representation of the equations of the associativity in the form (1.3), (1.4) is the existence of the efficient and elaborate theory of integrability of systems of hydrodynamic type — see, for example, the surveys of Tsarev [4], Dubrovin and Novikov [5], and also the papers [6–9] devoted to systems of hydrodynamic type, which do not possess Riemann invariants.

In this paper we consider strictly hyperbolic systems only, i.e., the eigenvalues of the matrix v_j^i are assumed to be real and distinct. For example, both systems (1.3) and (1.4) are strictly hyperbolic.

For $n = 4$ the function F is of the form (see [1]):

$$F = \frac{1}{2}(t^1)^2 t^4 + t^1 t^2 t^3 + f(t^2, t^3, t^4),$$

and the corresponding equations of the associativity reduce to the following complicated overdetermined system [1]:

$$-2f_{xyz} - f_{xyy}f_{xxy} + f_{yyy}f_{xxx} = 0,$$

$$-f_{xzz} - f_{xyy}f_{xxz} + f_{yyz}f_{xxx} = 0,$$

$$-2f_{xyz}f_{xxz} + f_{xzz}f_{xxy} + f_{yzz}f_{xxx} = 0,$$

$$f_{zzz} - (f_{xyz})^2 + f_{xzz}f_{xyy} - f_{yyz}f_{xxz} + f_{yzz}f_{xxy} = 0,$$

$$f_{yyy}f_{xzz} - 2f_{yyz}f_{xyz} + f_{yzz}f_{xyy} = 0,$$

where $x = t^2$, $y = t^3$, $z = t^4$. Introducing the new dependent variables

$$f_{xxx} = a, \quad f_{xxy} = b, \quad f_{xxz} = c,$$

$$f_{xyy} = d, \quad f_{xyz} = e, \quad f_{xzz} = f,$$

one can rewrite these equations as a pair of 6×6 commuting systems of hydrodynamic type:

$$
\begin{pmatrix} a \\ b \\ c \\ d \\ e \\ f \end{pmatrix}_y = \begin{pmatrix} b \\ d \\ e \\ R \\ P \\ S \end{pmatrix}_x \tag{1.5}
$$

and

$$
\begin{pmatrix} a \\ b \\ c \\ d \\ e \\ f \end{pmatrix}_z = \begin{pmatrix} c \\ e \\ f \\ P \\ S \\ Q \end{pmatrix}_x , \tag{1.6}
$$

where we introduce the notation

$$
f_{yyz} = P = \frac{cd+f}{a}, \quad f_{yyy} = R = \frac{2e+bd}{a}, \quad f_{yzz} = S = \frac{2ec-bf}{a},
$$

$$
f_{zzz} = Q = e^2 - fd + \frac{c^2 d + cf - 2bec + b^2 f}{a}.
$$

In Section 2 the Hamiltonian property of the system (1.3) is established. For this system a local nondegenerate Hamiltonian structure of hydrodynamic type (a Poisson bracket of Dubrovin-Novikov type [5]) is found. In contrast to (1.3), the integrable system of hydrodynamic type (1.4) possesses only nonlocal Hamiltonian structure of hydrodynamic type (see [14–16]). Investigation [2,3] showed that both systems (1.3) and (1.4) are nondiagonalizable (i.e., do not possess Riemann invariants).

In Section 3 we consider the general theory of integrability of nondiagonalizable Hamiltonian 3×3 systems of hydrodynamic type following [6-9]. It turns out that any such system can be reduced to the integrable 3-wave system by some standard chain of transformations. We demonstrate this procedure for the system (1.3).

In Section 4 the explicit Bäcklund type transformation connecting solutions of the systems (1.3) and (1.4) is found.

In Section 5 we exhibit the Hamiltonian representation of the commuting systems (1.5) and (1.6). The corresponding Poisson bracket is again of Dubrovin-Novikov type.

2. Hamiltonian Representation of the System (1.3)

As it was noticed by Dubrovin [1], the equation (1.1) is connected with the spectral problem, which has the following form in the variables a, b, c:

$$\Psi_x = zA\Psi = z \begin{pmatrix} 0 & 1 & 0 \\ b & a & 1 \\ c & b & 0 \end{pmatrix} \Psi,$$

$$\Psi_t = zB\Psi = z \begin{pmatrix} 0 & 0 & 1 \\ c & b & 0 \\ b^2 - ac & c & 0 \end{pmatrix} \Psi. \tag{2.1}$$

The compatibility conditions of the spectral problem (2.1) are equivalent to the following two relations between the matrices A and B:

$$\begin{cases} A_t = B_x, \\ [A, B] = 0, \end{cases} \tag{2.2}$$

which are satisfied identically by virtue of the equations (1.3) (here $[\ ,\ ]$ denotes the commutator).

Lemma 1. *The eigenvalues of the matrix A are densities of conservation laws of the system (1.3).*

Proof. As far as the matrices A and B commute and have simple spectrum, they can be diagonalized simultaneously

$$A = PUP^{-1}, \quad B = PVP^{-1}.$$

Here $U = \text{diag}\,(u^1, u^2, u^3)$, $V = \text{diag}\,(v^1, v^2, v^3)$. Substitution in the equation (2.2) gives

$$[P^{-1}P_t, U] + U_t = [P^{-1}P_x, V] + V_x.$$

It remains to note that the matrices $[P^{-1}P_t, U]$ and $[P^{-1}P_x, V]$ are off-diagonal. Hence,

$$U_t = V_x.$$

Lemma 1 is proved.

Thus, besides the three evident conservation laws with the densities a, b, c, the system (1.3) has also three conservation laws with the densities u^1, u^2, u^3, which are the roots of the characteristic equation

$$\det(\lambda E - A) = \lambda^3 - a\lambda^2 - 2b\lambda - c = 0.$$

By virtue of the obvious linear relation $a = u^1 + u^2 + u^3$ only five conservation laws among them with the densities u^1, u^2, u^3, b, c are linearly independent. One can show that the system (1.3) has no other conservation laws of hydrodynamic type, i.e., with the densities of the form $h(a, b, c)$.

Let us change in the equations (1.3) from the variables a, b, c to the new field variables u^1, u^2, u^3, connected with a, b, c by the Viète formulas

$$a = u^1 + u^2 + u^3, \quad b = -\frac{1}{2}(u^1 u^2 + u^2 u^3 + u^3 u^1), \quad c = u^1 u^2 u^3.$$

To simplify the calculations we note that the matrices A and B are connected by the relation

$$B = A^2 - aA - bE.$$

Hence, the same relation is valid for the corresponding diagonal matrices U and V :

$$V = U^2 - aU - bE.$$

Substituting the expressions for a and b and using the equation (2.2), we obtain the following representation for the system (1.3)

$$U_t = \left(U^2 - (u^1 + u^2 + u^3)U + \frac{1}{2}(u^1 u^2 + u^2 u^3 + u^3 u^1)E\right)_x$$

or, in the components,

$$
\begin{pmatrix} u^1 \\ u^2 \\ u^3 \end{pmatrix}_t = \frac{1}{2} \begin{pmatrix} u^2 u^3 - u^1 u^2 - u^1 u^3 \\ u^1 u^3 - u^2 u^1 - u^2 u^3 \\ u^1 u^2 - u^3 u^1 - u^3 u^2 \end{pmatrix}_x
$$

$$
= \frac{1}{2} \begin{pmatrix} 1 & -1 & -1 \\ -1 & 1 & -1 \\ -1 & -1 & 1 \end{pmatrix} \frac{d}{dx} \begin{pmatrix} \partial h/\partial u^1 \\ \partial h/\partial u^2 \\ \partial h/\partial u^3 \end{pmatrix}, \quad (2.3)
$$

where $h = c = u^1 u^2 u^3$. Hence, the system under consideration is Hamiltonian with the Hamiltonian operator

$$M = \frac{1}{2} \begin{pmatrix} 1 & -1 & -1 \\ -1 & 1 & -1 \\ -1 & -1 & 1 \end{pmatrix} \frac{d}{dx} \quad (2.4)$$

and the Hamiltonian $H = \int c\, dx = \int u^1 u^2 u^3 dx$. The density of momentum and Casimirs of the corresponding Poisson bracket have the following form:

$$2b = -u^1 u^2 - u^2 u^3 - u^3 u^1 \quad \text{(the density of momentum)},$$

$$u^1, u^2, u^3 \quad \text{(Casimirs)}.$$

In the initial variables a, b, c the Hamiltonian operator (2.4) can be expressed in the form

$$
M = \begin{pmatrix} -\frac{3}{2} & \frac{1}{2}a & b \\ \frac{1}{2}a & b & \frac{3}{2}c \\ b & \frac{3}{2}c & 2(b^2 - ac) \end{pmatrix} \frac{d}{dx} + \begin{pmatrix} 0 & \frac{1}{2}a_x & b_x \\ 0 & \frac{1}{2}b_x & c_x \\ 0 & \frac{1}{2}c_x & (b^2 - ac)_x \end{pmatrix}.
$$

We recall that local nondegenerate Hamiltonian operators of hydrodynamic type were introduced and studied by Dubrovin and Novikov (see [5]). It was proved that the operator of the form

$$
P^{ij} = g^{ij}(u)\frac{d}{dx} + b^{ij}_k(u)u^k_x, \quad \det[g^{ij}(u)] \neq 0,
$$

is Hamiltonian if and only if

(1) $g^{ij}(u)$ is a metric of zero Riemannian curvature (i.e., simply a flat metric);

(2) $b^{ij}_k(u) = -g^{is}(u)\Gamma^j_{sk}(u)$, where $\Gamma^j_{sk}(u)$ are the coefficients of the differential geometric connection generated by the metric g^{ij}, i.e., the only symmetric connection compatible with the metric (the Levi-Civita connection).

Hamiltonian systems of hydrodynamic type, which were considered by Dubrovin and Novikov [5], have the form

$$
u^i_t = P^{ij}\frac{\partial H}{\partial u^j},
$$

where $H = \int h(u)dx$ is a functional of hydrodynamic type. Nonlocal generalization of the Hamiltonian theory of systems of hydrodynamic type was discovered in [14] (see also [15-16]). The efficient theory of integrability of diagonalizable Hamiltonian systems of hydrodynamic type, i.e., Hamiltonian systems of hydrodynamic type, which can be reduced to Riemann invariants:

$$
R^i_t = v^i(R)R^i_x,
$$

was proposed by Tsarev [4]. All such systems possess an infinite number of conservation laws and commuting flows of hydrodynamic type and can be integrated by generalized hodograph transform. However, it was shown in the papers [2,3] that the system (1.3) does not possess Riemann invariants. This explains, in particular, the fact that the system (1.3) has only finite number of hydrodynamic type integrals.

The general theory of integrability of nondiagonalizable (i.e., not possessing Riemann invariants) Hamiltonian systems of hydrodynamic type was developed in [6-9]. For three-component systems the following result was obtained:

Theorem [7,8]. *Nondiagonalizable Hamiltonian (with nondegenerate Poisson bracket of hydrodynamic type) 3×3 system of hydrodynamic type is integrable if and only if it is weakly nonlinear.*

We recall that a system of hydrodynamic type

$$u_t^i = v_j^i(u)u_x^j, \quad i,j = 1, \ldots, n, \tag{2.5}$$

is called weakly nonlinear if for eigenvalues $\lambda^i(u)$ of the matrix $v_j^i(u)$ the following relations are satisfied for any $i = 1, \ldots, n$:

$$L_{\vec{X}^i}(\lambda^i) = 0,$$

where $L_{\vec{X}^i}$ is the Lie derivative along the eigenvector $\vec{X}^i$ corresponding to the eigenvalue λ^i.

There exists a simple and efficient criterion of weak nonlinearity, which does not appeal to eigenvalues and eigenvectors.

Proposition 1 [7]. *A system of hydrodynamic type (2.5) is weakly nonlinear if and only if*

$$(\text{grad } f_1)v^{n-1} + (\text{grad } f_2)v^{n-2} + \ldots + (\text{grad } f_n)E = 0,$$

where f_i are the coefficients of the characteristic polynomial

$$\det(\lambda \delta_j^i - v_j^i(u)) = \lambda^n + f_1(u)\lambda^{n-1} + f_2(u)\lambda^{n-2} + \ldots + f_n(u),$$

and v^n denotes the n-th power of the matrix v_j^i.

As it was shown in [2,3], both systems (1.3) and (1.4) are weakly nonlinear.

3. Integrable Hamiltonian 3×3 Systems of Hydrodynamic Type Which do not Possess Riemann Invariants

Consider a system of hydrodynamic type

$$u_t^i = v_j^i(u)u_x^j. \tag{3.1}$$

Let $\lambda^i(u)$ be eigenvalues of the matrix v_j^i, i.e., the roots of characteristic equation $\det(v_j^i(u) - \lambda \delta_j^i) = 0$ (we assume that the system under consideration is strictly hyperbolic, i.e., all roots of the characteristic equation are real and distinct). Denote by $\vec{l}^i(u) = (l_1^i, \ldots, l_n^i)$ the left eigenvector of the matrix v_j^i, which corresponds to the eigenvalue λ^i, i.e., $l_k^i v_j^k = \lambda^i l_j^i$. Let us introduce the 1-forms $\omega^i = l_k^i du^k$ $(i = 1, \ldots, n)$. We emphasize that the

1-forms ω^i are defined up to normalization $\omega^i \mapsto p^i \omega^i$, $p^i \neq 0$. It is easy to verify that the equations (3.1) can be rewritten as the system of exterior equations

$$\omega^i \wedge (dx + \lambda^i dt) = 0, \quad i = 1, \ldots, n. \tag{3.2}$$

For the system (2.3) the eigenvalues λ^i and the corresponding left eigenvectors $\vec{l}^{\,i}$ have the form:

$$\lambda^1 = -u^1, \quad \vec{l}^{\,1} = (u^2 - u^3, u^1 - u^3, u^2 - u^1),$$

$$\lambda^2 = -u^2, \quad \vec{l}^{\,2} = (u^2 - u^3, u^1 - u^3, u^1 - u^2),$$

$$\lambda^3 = -u^3, \quad \vec{l}^{\,3} = (u^2 - u^3, u^3 - u^1, u^2 - u^1).$$

Hence, the equations (2.3) can be expressed as

$$\omega^i \wedge (dx - u^i dt) = 0, \quad i = 1, 2, 3,$$

where

$$
\begin{aligned}
\omega^1 &= (u^2 - u^3)du^1 + (u^1 - u^3)du^2 + (u^2 - u^1)du^3, \\
\omega^2 &= (u^2 - u^3)du^1 + (u^1 - u^3)du^2 + (u^1 - u^2)du^3, \\
\omega^3 &= (u^2 - u^3)du^1 + (u^3 - u^1)du^2 + (u^2 - u^1)du^3.
\end{aligned} \tag{3.3}
$$

Let $B(u)dx + A(u)dt$ and $N(u)dx + M(u)dt$ be two hydrodynamic type integrals of the system (3.1), i.e., the differential 1-forms closed on the solutions of the system. Changing from the variables x, t to the new independent variables $\tilde{x}, \tilde{t}$ by means of the following formulas

$$
\begin{aligned}
d\tilde{x} &= Bdx + Adt, \\
d\tilde{t} &= Ndx + Mdt,
\end{aligned} \tag{3.4}
$$

we arrive at the equations

$$u^i_{\tilde{t}} = \tilde{v}^i_j(u)u^j_{\tilde{x}},$$

where the matrix $\tilde{v}$ is related to v by the formula

$$\tilde{v} = (Bv - AE)(ME - Nv)^{-1}.$$

Using the language of exterior equations we can rewrite the transformed system in the following form

$$\omega^i \wedge (d\tilde{x} + \tilde{\lambda}^i d\tilde{t}) = 0, \quad i = 1, \ldots, n, \tag{3.5}$$

where

$$\tilde{\lambda}^i = \frac{\lambda^i B - A}{M - \lambda^i N}. \tag{3.6}$$

Hence, the 1-forms ω^i do not change under the transformations of the form (3.4) while the eigenvalues λ^i transform in accordance with the formula (3.6).

Theorem [7,8]. *If a 3×3 system of hydrodynamic type (3.1) is weakly nonlinear and Hamiltonian (with nondegenerate Poisson bracket of hydrodynamic type) then there exists a pair of integrals (3.4) of this system such that the transformed system has constant eigenvalues $\tilde{\lambda}^i$, which can be put equal to $1, -1, 0$ without loss of generality.*

For the system (2.3) the transformation (3.4), the existence of which is established by Theorem 2, has the following form:

$$
\begin{aligned}
d\tilde{x} &= B dx + A dt = (u^1 - u^2)dx + u^3(u^2 - u^1)dt, \\
d\tilde{t} &= N dx + M dt = (2u^3 - u^1 - u^2)dx + (2u^1u^2 - u^1u^3 - u^2u^3)dt.
\end{aligned}
\tag{3.7}
$$

According to the formula (3.6) the transformed eigenvalues will be $1, -1$ and 0, respectively. Hence, in new independent variables $\tilde{x}, \tilde{t}$ the system (2.3) can be rewritten in the form

$$
\omega^1 \wedge (d\tilde{x} + d\tilde{t}) = 0, \quad \omega^2 \wedge (d\tilde{x} - d\tilde{t}) = 0, \quad \omega^3 \wedge d\tilde{x} = 0.
$$

Theorem [7,8]. *If a 3×3 system of hydrodynamic type (3.1) is nondiagonalizable, weakly nonlinear and Hamiltonian (with nondegenerate Poisson bracket of hydrodynamic type) then the corresponding 1-forms $\omega^1, \omega^2, \omega^3$ can be normalized so that they will satisfy either the structure equations of the group $SO(3)$:*

$$
d\omega^1 = \omega^2 \wedge \omega^3, \quad d\omega^2 = \omega^3 \wedge \omega^1, \quad d\omega^3 = \omega^1 \wedge \omega^2,
\tag{3.8}
$$

(if the signature of the metric, determining the Poisson bracket of hydrodynamic type, is Euclidean), or the structure equations of the group $SO(2,1)$:

$$
d\omega^1 = \omega^2 \wedge \omega^3, \quad d\omega^2 = \omega^3 \wedge \omega^1, \quad d\omega^3 = -\omega^1 \wedge \omega^2
\tag{3.9}
$$

(in the case of Lorentzian signature of the metric).

For the system (2.3) the signature of the metric of the Poisson bracket (2.4) is Lorentzian. Hence, the forms (3.3) can be normalized so that they will satisfy the structure equations (3.9). The desired normalization has the form (we will not introduce a new notation for the normalized 1-forms ω^i):

$$
\begin{aligned}
\omega^1 &= \frac{(u^2 - u^3)du^1 + (u^1 - u^3)du^2 + (u^2 - u^1)du^3}{2(u^2 - u^3)\sqrt{(u^2 - u^1)(u^3 - u^1)}}, \\
\omega^2 &= \frac{(u^2 - u^3)du^1 + (u^1 - u^3)du^2 + (u^1 - u^2)du^3}{2(u^3 - u^1)\sqrt{(u^2 - u^1)(u^2 - u^3)}}, \\
\omega^3 &= \frac{(u^2 - u^3)du^1 + (u^3 - u^1)du^2 + (u^2 - u^1)du^3}{2(u^2 - u^1)\sqrt{(u^3 - u^1)(u^2 - u^3)}}
\end{aligned}
\tag{3.10}
$$

(for definiteness we consider $u^1 < u^3 < u^2$).

One can verify directly that the 1-forms (3.10) satisfy the structure equations (3.9). Thus, according to theorems 2 and 3 any nondiagonalizable weakly nonlinear Hamiltonian (with nondegenerate Poisson bracket of hydrodynamic type) 3×3 system of hydrodynamic type can be reduced to the canonical form

$$\omega^1 \wedge (d\tilde{x} + d\tilde{t}) = 0, \quad \omega^2 \wedge (d\tilde{x} - d\tilde{t}) = 0, \quad \omega^3 \wedge d\tilde{x} = 0 \qquad (3.11)$$

by suitable transformation (3.4). Moreover, for the forms ω^i the structure equations (3.8) or (3.9) are satisfied (note that the transformations of the form (3.4) do not change the structure equations). Introducing in the equations (3.11) the variables p^1, p^2, p^3 by the formulas (see [6,7])

$$\omega^1 = p^1(d\tilde{x} + d\tilde{t}), \quad \omega^2 = p^2(d\tilde{x} - d\tilde{t}), \quad \omega^3 = p^3 d\tilde{x}. \qquad (3.12)$$

and substituting (3.12) in the structure equations (for definiteness in (3.9)) we obtain the integrable 3-wave system

$$\begin{cases} p^1_{\tilde{t}} - p^1_{\tilde{x}} = -p^2 p^3, \\ p^2_{\tilde{t}} + p^2_{\tilde{x}} = -p^1 p^3, \\ p^3_{\tilde{t}} = -2p^1 p^2. \end{cases} \qquad (3.13)$$

Remark.   Using the explicit coordinate representation of the 1-forms $\omega^i = l^i_k(u)du^k$, we obtain for p^i the expressions of the form $p^i = l^i_k(u)u^k_{\tilde{x}}$. Hence, the change from u^i to p^i is a differential substitution of the first order.

Thus the transition from the equation (2.3) to the 3-wave system (3.13) can be decomposed in two steps:

1. The change from x, t to the new independent variables $\tilde{x}, \tilde{t}$ in accordance with the formulas (3.7).

2. The change of the field variables from u^1, u^2, u^3 to p^1, p^2, p^3 in accordance with the following formulas (compare with (3.10)):

$$\begin{aligned}
p^1 &= \frac{(u^2 - u^3)u^1_{\tilde{x}} + (u^1 - u^3)u^2_{\tilde{x}} + (u^2 - u^1)u^3_{\tilde{x}}}{2(u^2 - u^3)\sqrt{(u^2 - u^1)(u^3 - u^1)}}, \\
p^2 &= \frac{(u^2 - u^3)u^1_{\tilde{x}} + (u^1 - u^3)u^2_{\tilde{x}} + (u^1 - u^2)u^3_{\tilde{x}}}{2(u^3 - u^1)\sqrt{(u^2 - u^1)(u^2 - u^3)}}, \\
p^3 &= \frac{(u^2 - u^3)u^1_{\tilde{x}} + (u^3 - u^1)u^2_{\tilde{x}} + (u^2 - u^1)u^3_{\tilde{x}}}{2(u^2 - u^1)\sqrt{(u^3 - u^1)(u^2 - u^3)}}.
\end{aligned} \qquad (3.14)$$

Thus, any solution of the integrable 3-wave system (3.13) generates a three-parameter family of solutions of the equations of the associativity (1.1).

4. Relation between the systems (1.3) and (1.4)

The spectral problem corresponding to the system (1.4) has the form

$$\Psi_x = zA\Psi = z \begin{pmatrix} 0 & 1 & 0 \\ 0 & b & a \\ 1 & c & b \end{pmatrix} \Psi,$$

$$\Psi_t = zB\Psi = z \begin{pmatrix} 0 & 0 & 1 \\ 1 & c & b \\ 0 & (1+bc)/a & c \end{pmatrix} \Psi. \qquad (4.1)$$

It is easy to verify that the matrix B is related to A by

$$B = \frac{1}{a}(A^2 - bA).$$

The compatibility condition of the spectral problem (4.1)

$$A_t = B_x,$$

rewritten in terms of the eigenvalues of matrices A and B (see Lemma 1, Section 2), has the form

$$w_t^i = \left(\frac{1}{a}((w^i)^2 - w^i b)\right)_x, \qquad (4.2)$$

where w^i are eigenvalues of the matrix A, i.e., the roots of the characteristic equation

$$\det(\lambda E - A) = \lambda^3 - 2b\lambda^2 + (b^2 - ac)\lambda - a = 0.$$

Expressing a and b by the Viète formulas

$$b = \frac{1}{2}(w^1 + w^2 + w^3), \qquad a = w^1 w^2 w^3$$

and substituting these expressions in (4.2) we obtain the explicit representation of the equations (1.4) in the coordinates w^i:

$$\begin{pmatrix} w^1 \\ w^2 \\ w^3 \end{pmatrix}_t = \frac{1}{2} \begin{pmatrix} (w^1 - w^2 - w^3)/w^2 w^3 \\ (w^2 - w^1 - w^3)/w^1 w^3 \\ (w^3 - w^1 - w^2)/w^1 w^2 \end{pmatrix}_x. \qquad (4.3)$$

Note that the integrable systems of hydrodynamic type (1.4) and (4.3) do not possess local Hamiltonian structures of hydrodynamic type (the Poisson brackets of Dubrovin-Novikov type [5]). Hamiltonian structures of hydrodynamic type corresponding to them are strictly nonlocal ([14–16]).

We exhibit now the explicit relation between the systems (2.3) and (4.3). For this reason we change in the equations (2.3) from x, t to the new independent variables $\tilde{x}$, $\tilde{t}$ according to the formulas

$$d\tilde{x} = -\frac{1}{2}(u^1u^2 + u^1u^3 + u^2u^3)dx + u^1u^2u^3 dt, \quad d\tilde{t} = dx. \tag{4.4}$$

After the transformation (4.4) the system (2.3) assumes the form

$$\begin{pmatrix} 1/u^1 \\ 1/u^2 \\ 1/u^3 \end{pmatrix}_{\tilde{t}} = \frac{1}{2} \begin{pmatrix} (u^2u^3)/u^1 - u^2 - u^3 \\ (u^1u^3)/u^2 - u^1 - u^3 \\ (u^1u^2)/u^3 - u^1 - u^2 \end{pmatrix}_{\tilde{x}},$$

which coincides with (4.3) after the transformation

$$w^i = \frac{1}{u^i}. \tag{4.5}$$

In terms of the initial equations (1.1) and (1.2) we can represent the transformations (4.4) and (4.5) in the following way: the equation

$$f_{ttt} = f_{xxt}^2 - f_{xxx}f_{xtt}$$

transforms into the equation

$$\tilde{f}_{\tilde{x}\tilde{x}\tilde{x}}\tilde{f}_{\tilde{t}\tilde{t}\tilde{t}} - \tilde{f}_{\tilde{x}\tilde{x}\tilde{t}}\tilde{f}_{\tilde{x}\tilde{t}\tilde{t}} = 1$$

after the transformation

$$\begin{aligned} \tilde{x} &= f_{xt}, \quad \tilde{t} = x, \\ \tilde{f}_{\tilde{x}\tilde{x}} &= t, \quad \tilde{f}_{\tilde{x}\tilde{t}} = -f_{xx}, \quad \tilde{f}_{\tilde{t}\tilde{t}} = f_{tt}. \end{aligned} \tag{4.6}$$

Note that transformation (4.6), connecting solutions of the associativity equations (1.1) and (1.2), is not contact.

5. Hamiltonian Representation of the Commuting

Systems (1.5) and (1.6)

As it was pointed out by Dubrovin [1], equations (1.5) and (1.6) are connected with the following spectral problem:

$$\Psi_x = \lambda A\Psi, \quad \Psi_y = \lambda B\Psi, \quad \Psi_z = \lambda C\Psi, \tag{5.1}$$

with the matrices A, B, C given by

$$A = \begin{pmatrix} 0 & 1 & 0 & 0 \\ c & b & a & 0 \\ e & d & b & 1 \\ f & e & c & 0 \end{pmatrix}, \quad B = \begin{pmatrix} 0 & 0 & 1 & 0 \\ e & d & b & 1 \\ P & R & d & 0 \\ S & P & e & 0 \end{pmatrix},$$

$$C = \begin{pmatrix} 0 & 0 & 0 & 1 \\ f & e & c & 0 \\ S & P & e & 0 \\ Q & S & f & 0 \end{pmatrix},$$

where we adopt the notation

$$P = \frac{cd+f}{a}, \quad R = \frac{2e+bd}{a}, \quad S = \frac{2ec-bf}{a},$$

$$Q = \frac{ae^2 + c^2d + cf - 2bce + b^2f - adf}{a}.$$

The compatibility conditions of the spectral problem (5.1) imply the commutativity of the matrices A, B, C

$$[A, B] = [A, C] = [B, C] = 0,$$

and the equations

$$A_y = B_x, \quad A_z = C_x, \quad B_z = C_x,$$

which are satisfied identically by virtue of equations (1.5) and (1.6).

Following the approach of Section 2, we introduce the eigenvalues u^i of the matrix A, being the roots of the 4th order characteristic polynomial

$$\det(A - \lambda E) = \lambda^4 - 2b\lambda^3 + (b^2 - ad - 2c)\lambda^2 + 2(bc - ae)\lambda + c^2 - af = 0.$$

The variables b, c, e, f are related to u^1, u^2, u^3, u^4, a, d through the Viète formulas:

$$b = \frac{1}{2}(u^1 + u^2 + u^3 + u^4),$$

$$c = \frac{1}{4}((u^1)^2 + (u^2)^2 + (u^3)^2 + (u^4)^2) - \frac{1}{8}(u^1 + u^2 + u^3 + u^4)^2 - \frac{1}{2}ad,$$

$$e = \frac{bc}{a} + \frac{1}{2a}(u^1 u^2 u^3 + u^1 u^2 u^4 + u^1 u^3 u^4 + u^2 u^3 u^4),$$

$$f = \frac{c^2 - u^1 u^2 u^3 u^4}{a}.$$

In the new variables a, u^1, u^2, u^3, u^4, d both systems (1.5) and (1.6) assume the Hamiltonian form

$$\begin{pmatrix} a \\ u^1 \\ u^2 \\ u^3 \\ u^4 \\ d \end{pmatrix}_y = \begin{pmatrix} 0 & 0 & 0 & 0 & 0 & -2 \\ 0 & 1 & -1 & -1 & -1 & 0 \\ 0 & -1 & 1 & -1 & -1 & 0 \\ 0 & -1 & -1 & 1 & -1 & 0 \\ 0 & -1 & -1 & -1 & 1 & 0 \\ -2 & 0 & 0 & 0 & 0 & 0 \end{pmatrix} \frac{d}{dx} \begin{pmatrix} \partial e / \partial a \\ \partial e / \partial u^1 \\ \partial e / \partial u^2 \\ \partial e / \partial u^3 \\ \partial e / \partial u^4 \\ \partial e / \partial d \end{pmatrix}, \quad (5.2)$$

with the Hamiltonian $H = \int e\,dx$, and

$$
\begin{pmatrix} a \\ u^1 \\ u^2 \\ u^3 \\ u^4 \\ d \end{pmatrix}_z = \frac{1}{2} \begin{pmatrix} 0 & 0 & 0 & 0 & 0 & -2 \\ 0 & 1 & -1 & -1 & -1 & 0 \\ 0 & -1 & 1 & -1 & -1 & 0 \\ 0 & -1 & -1 & 1 & -1 & 0 \\ 0 & -1 & -1 & -1 & 1 & 0 \\ -2 & 0 & 0 & 0 & 0 & 0 \end{pmatrix} \frac{d}{dx} \begin{pmatrix} \partial f/\partial a \\ \partial f/\partial u^1 \\ \partial f/\partial u^2 \\ \partial f/\partial u^3 \\ \partial f/\partial u^4 \\ \partial f/\partial d \end{pmatrix},
$$

$$(5.3)$$

with the Hamiltonian $H = \int \frac{1}{2} f\,dx$, respectively.

The characteristic velocities of the systems (5.2) and (5.3) are given by

$$
\lambda^1 = \frac{u^1 + u^2 - u^3 - u^4}{2a}, \quad \lambda^2 = -\lambda^1,
$$

$$
\lambda^3 = \frac{u^1 - u^2 + u^3 - u^4}{2a}, \quad \lambda^4 = -\lambda^3,
$$

$$
\lambda^5 = \frac{-u^1 + u^2 + u^3 - u^4}{2a}, \quad \lambda^6 = -\lambda^5
$$

and

$$
\mu^1 = \mu + \frac{u^3 u^4}{a}, \quad \mu^2 = \mu + \frac{u^1 u^2}{a},
$$

$$
\mu^3 = \mu + \frac{u^2 u^4}{a}, \quad \mu^4 = \mu + \frac{u^1 u^3}{a},
$$

$$
\mu^5 = \mu + \frac{u^1 u^4}{a}, \quad \mu^6 = \mu + \frac{u^2 u^3}{a},
$$

where

$$
\mu = -\frac{1}{2} d + \frac{(u^1)^2 + (u^2)^2 + (u^3)^2 + (u^4)^2}{8a}
$$
$$
- \frac{u^1 u^2 + u^1 u^3 + u^1 u^4 + u^2 u^3 + u^2 u^4 + u^3 u^4}{4a},
$$

respectively.

We have calculated also the explicit form of our Hamiltonian operator

$$
M = \begin{pmatrix} 0 & 0 & 0 & 0 & 0 & -2 \\ 0 & 1 & -1 & -1 & -1 & 0 \\ 0 & -1 & 1 & -1 & -1 & 0 \\ 0 & -1 & -1 & 1 & -1 & 0 \\ 0 & -1 & -1 & -1 & 1 & 0 \\ -2 & 0 & 0 & 0 & 0 & 0 \end{pmatrix} \frac{d}{dx}
$$

in the initial variables a, b, c, d, e, f:

$$
M = \begin{pmatrix}
0 & 0 & a & -2 & b & 2c \\
0 & -2 & b & 0 & d & 2e \\
a & b & 2c & d & 2e & 3f \\
-2 & 0 & d & 0 & R & 2P \\
b & d & 2e & R & 2P & 3S \\
2c & 2e & 3f & 2P & 3S & 4Q
\end{pmatrix} \frac{d}{dx}
$$

$$
+ \begin{pmatrix}
0 & 0 & a_x & 0 & b_x & 2c_x \\
0 & 0 & b_x & 0 & d_x & 2e_x \\
0 & 0 & c_x & 0 & e_x & 2f_x \\
0 & 0 & d_x & 0 & R_x & 2P_x \\
0 & 0 & e_x & 0 & P_x & 2S_x \\
0 & 0 & f_x & 0 & S_x & 2Q_x
\end{pmatrix}.
$$

The approach similar to that adopted in Section 2 can be applied to transform both systems (1.5) and (1.6) into the integrable 6-wave systems. However, in this case all the formulas become more complicated.

Acknowledgement. We would like to thank B.A. Dubrovin for drawing our attention to the equations of the associativity, I.M. Krichever and S.P. Novikov for helpful discussions.

References

[1] Dubrovin, B., Geometry of 2D topological field theories, preprint SISSA–89/94/FM, SISSA, Trieste (1994).

[2] Mokhov, O.I., Differential equations of associativity in 2D topological field theories and geometry of nondiagonalizable systems of hydrodynamic type, In: *Abstracts of Internat. Conference on Integrable Systems "Nonlinearity and Integrability: from Mathematics to Physics", February 21-24, 1995 , Montpellier, France* (1995).

[3] Mokhov, O.I., Symplectic and Poisson geometry on loop spaces of manifolds and nonlinear equations, *AMS Translations*, series 2 **Vol. 170** *Topics in topology and mathematical physics*, S.P. Novikov, ed., (1995); hep-th/9503076.

[4] Tsarev, S.P., Geometry of Hamiltonian systems of hydrodynamic type, The generalized hodograph method, *Izvestiya Akad. Nauk SSSR, Ser. mat.* **54** (5) (1990),1048–1068.

[5] Dubrovin, B.A. and Novikov, S.P., Hydrodynamics of weakly deformed soliton lattices. Differential geometry and Hamiltonian theory, *Uspekhi Mat. Nauk* **44** (6) (1989), 29–98.

[6] Ferapontov, E.V., On integrability of 3×3 semi-Hamiltonian hydrodynamic type systems which do not possess Riemann invariants, *Physica D* **63** (1993), 50–70.

[7] Ferapontov, E.V., On the matrix Hopf equation and integrable Hamiltonian systems of hydrodynamic type, which do not possess Riemann invariants, *Phys. Lett. A* **179** (1993), 391–397.

[8] Ferapontov, E.V., Dupin hypersurfaces and integrable Hamiltonian systems of hydrodynamic type which do not possess Riemann invariants, *Diff. Geometry and its Appl.* **5** (1995), 121–152.

[9] Ferapontov, E.V., Several conjectures and results in the theory of integrable Hamiltonian systems of hydrodynamic type, which do not possess Riemann invariants, *Teor. and Mat. Physics* **99** (2) (1994), 257–262.

[10] Witten, E., On the structure of the topological phase of two-dimensional gravity, *Nucl. Physics B* **340** (1990), 281–332.

[11] Dijkgraaf, R., Verlinde, H. and Verlinde, E., Topological strings in $d < 1$, *Nucl. Physics B* **352** (1991), 59–86.

[12] Witten, E., Two-dimensional gravity and intersection theory on moduli space, *Surveys in Diff. Geometry* **1** (1991), 243–310.

[13] Dubrovin, B., Integrable systems in topological field theory, *Nucl. Physics B* **379** (1992), 627–689.

[14] Mokhov, O.I. and Ferapontov, E.V., On the nonlocal Hamiltonian operators of hydrodynamic type, connected with constant curvature metrics, *Uspekhi Mat. Nauk* **45** (3) (1990), 191–192.

[15] Ferapontov, E.V., Differential geometry of nonlocal Hamiltonian operators of hydrodynamic type, *Funkts. Analiz i ego Prilozh.* **25** (3) (1991), 37–49.

[16] Mokhov, O.I., Hamiltonian systems of hydrodynamic type and constant curvature metrics, *Phys. Letters A* **166** (3–4) (1992), 215–216.

[17] Mokhov, O.I. and Nutku, Y., Bianchi transformation between the real hyperbolic Monge-Ampère equation and the Born-Infeld equation, *Letters in Math. Phys.* **32** (2) (1994), 121–123.

E.V. Ferapontov
Inst. for Math. Modelling
Miusskaya, 4
Moscow, 125047, Russia
e-mail: fer#9@imamod.msk.su

O.I. Mokhov
Steklov Mathematical Institute
ul. Vavilova, 42
Moscow, GSP-1, 117966, Russia
e-mail: mokhov@class.mian.su

Received December 1995

A Plethora of Integrable
Bi-Hamiltonian Equations

A.S. Fokas, P.J. Olver, P. Rosenau

1. Introduction

This paper discusses several algorithmic ways of constructing integrable evolution equations based on the use of multi-Hamiltonian structures. The recognition that integrable soliton equations, such as the Korteweg-deVries (KdV) and nonlinear Schrödinger (NLS) equations, can be constructed using a biHamiltonian method dates back to the late 1970's. An extension of the method was proposed by the first author and Fuchssteiner in the early 1980's and was used to derive integrable generalizations of the KdV and of the modified KdV. However it was not until these models reappeared in physical problems, and their novel solutions such as compactons and peakons were discovered, that the method achieved recognition. In this paper, we describe the basic approach to constructing a wide variety of integrable bi-Hamiltonian equations. In addition to usual soliton equations, these new hierarchies include equations with nonlinear dispersion which support novel types of solitonic solutions.

Let us start with the simple case of a scalar evolution equation

$$u_t = K[u], \tag{1.1}$$

where $K[u]$ is a smooth function depending on the scalar spatial variable x, the dependent variable u, and its x derivatives u_x, u_{xx}, etc. Later, we shall generalize K by allowing it to be either non-local, or depend on additional spatial variables. A second evolution equation

$$u_s = Q[u]$$

is called a **symmetry** of equation (1.1) iff the two flows commute. Cross-differentiation produces the basic symmetry condition

$$K'Q - Q'K = 0, \tag{1.2}$$

where the prime denotes Fréchet differentiation, i.e.

$$K'[u]Q = \frac{\partial}{\partial \varepsilon} K[u + \varepsilon Q]\Big|_{\varepsilon=0}.$$

More explicitly,

$$K' = \frac{\partial K}{\partial u} + \frac{\partial K}{\partial u_x}\partial + \frac{\partial K}{\partial u_{xx}}\partial^2 + \cdots, \qquad \text{where} \qquad \partial = \frac{\partial}{\partial x}. \tag{1.3}$$

For example, if equation (1.1) does not depend explicitly on x then it is invariant under the group of x-translations, which gives rise to the symmetry $Q = u_x$. See [1] for details.

According to the general symmetry approach to integrability, [1], [2], [3], the existence of an infinite hierarchy of higher order symmetries is a manifestation of the integrability of the evolution equation (1.1). The basic method for constructing such hierarchies is through the use of a **recursion operator** which is defined as an operator $\Phi[u]$ which maps symmetries to symmetries. It was shown in [4] that if the operator $\Phi[u]$ satisfies the operator equation

$$\Phi'[K] = [K', \Phi], \tag{1.4}$$

where $[\cdot, \cdot]$ denotes the usual commutator, then Φ is a recursion operator for (1.1). If Φ is a recursion operator for (1.1), then there is an associated hierarchy of commuting flows (symmetries) taking the form

$$u_t = \Phi^n K. \tag{1.5}$$

A large class of recursion operators satisfy a remarkable property introduced by Fuchssteiner [5], namely they are **hereditary** (or Nijenhuis) operators. The operator $\Phi[u]$ is called hereditary iff $\Phi'[\Phi v]w - \Phi\Phi'[v]w$ defines a symmetric bilinear form of the functions v and w. It can be shown that if Φ is hereditary and if Φ is a recursion operator for the seed equation $u_t = K_0$, then Φ is also a recursion operator for any constant coefficient linear combination

$$u_t = c_0 K_0 + c_1 \Phi K_0 + \ldots + c_n \Phi^n K_0, \tag{1.6}$$

of the associated hierarchy of symmetries, or, even more generally,

$$u_t = B(\Phi)K_0, \tag{1.7}$$

where $B(z)$ is any rational, or even analytic function of z; see [1] for details. (Indeed, one can even regard the Bäcklund transformation as an exponential series in the higher order symmetries [6] and hence in the recursion operator Φ.) Thus the question of constructing integrable equations reduces to the question of constructing hereditary operators and "starting symmetries" K_0, i.e. functions K_0 which satisfy equation (1.4). Usually, the hereditary operator is independent of the spatial variable x, in which case the starting symmetry is $K_0 = u_x$, and the seed equation is the linear wave equation $u_t = u_x$. Indeed, $\partial(\Phi v) = \Phi'[u_x]v + \Phi \partial v$, or $\Phi'[u_x] = [\partial, \Phi]$, and equation (1.4) is satisfied with $K_0 = u_x$.

An operator $\theta[u]$ is called **Hamiltonian** if it is skew symmetric (with respect to a suitable inner product — usually a variant of the L^2 inner product), and such that the associated Poisson bracket $\{F, H\} = \int \delta F \theta \delta H \, dx$ on the space of functionals satisfies the Jacobi identity. Here δH denotes the variational derivative of the Hamiltonian functional H. An evolution equation is called **Hamiltonian** if it can be written in the form

$$u_t = \theta \delta H, \tag{1.8}$$

where θ is a Hamiltonian operator.

Following the fundamental discovery of Magri [7] that Hamiltonian integrable equations are actually bi-Hamiltonian systems, an algorithmic way of constructing hereditary operators was proposed independently in [8] and [9]. Two operators θ_1 and θ_2 are said to form a **Hamiltonian pair** if every linear combination $\alpha\theta_1 + \beta\theta_2$ for α, β constant, is a Hamiltonian operator. This requires that θ_1 and θ_2 are Hamiltonian, and, moreover, that they satisfy a certain bilinear compatibility condition [7]. Given a Hamiltonian pair θ_1, θ_2, it can be shown [8], [9] that the operator $\Phi = \theta_2\theta_1^{-1}$ is a hereditary operator. Furthermore, if θ_1 and θ_2 do not depend explicitly on x, then $K_0 = u_x$ is a seed symmetry for Φ.

If a hereditary operator Φ is derived from a pair of compatible Hamiltonian operators θ_1 and θ_2, then there exists an algorithmic way of constructing additional starting symmetries. Let $C[u]$ be a function which is annihilated by the Hamiltonian operator θ_1, i.e. $\theta_1 C = 0$. In most cases, C is the variational derivative of a **Casimir** (or distinguished) functional B for the Hamiltonian operator θ_1, i.e. $C = \delta B$ [1], [10]. By abuse of terminology, we shall call all such functions $C[u]$ Casimirs. Then

$$K_0 = \theta_2 C; \qquad \text{where} \qquad \theta_1 C = 0, \tag{1.9}$$

is a starting symmetry of the hereditary operator $\Phi = \theta_2\theta_1^{-1}$. Indeed, we can write $K_0 = \Phi \cdot 0$ as the image of the trivial symmetry $u_t = 0$ under the recursion operator Φ since (formally) $C = \theta_1^{-1}0$. Clearly $K = 0$ satisfies the recursion operator criterion (1.4), which implies that K_0 does also.

Combining the above discussion with the general form (1.7) of the hierarchy generated by a recursion operator, suggests the following algorithmic construction of integrable evolution equations:

Let θ_1, θ_2 be a Hamiltonian pair which do not depend on x. Let $\Phi = \theta_2\theta_1^{-1}$. Then the following equations are integrable,

$$u_t = \Phi u_x, \tag{1.10}$$

$$\Phi u_t = u_x, \tag{1.11}$$

$$u_t = \Phi u_y, \tag{1.12}$$

$$u_t = \theta_2 C, \quad \theta_1 C = 0, \tag{1.13}$$

$$\sum_{\kappa=1}^{i} a_\kappa \Phi^\kappa u_t = \sum_{\kappa=1}^{j} b_\kappa \Phi^\kappa u_x + \sum_{\kappa=1}^{k} d_\kappa \Phi^\kappa u_y + \sum_{\kappa=1}^{l} e_\kappa \Phi^\kappa \theta_2 C, \tag{1.14}$$

where $a_\kappa, b_\kappa, d_\kappa, e_\kappa$ are constants.

The integrability of equation (1.11) is a consequence of the fact that if Φ is hereditary then Φ^{-1} is also hereditary. In (1.12), y is an additional arbitrary independent variable that does not occur in the operator Φ, and thus u_y is a starting symmetry for Φ. The integrability of equations (1.13) has been commented upon in [11]. The integrability of equations (1.14) follows from the fact that on can take arbitrary constant coefficient linear combinations of all the equations in the hierarchies associated with (1.10–13).

In many cases, one of the operators in the Hamiltonian pair is itself a linear combination of two Hamiltonian operators, so that the linear combinations $\alpha\theta_1 + \beta\theta_2$ are actually members of a three parameter family $\widetilde{\alpha}\widetilde{\theta}_1 + \widetilde{\beta}\widetilde{\theta}_2 + \widetilde{\gamma}\widetilde{\theta}_3$, where $\widetilde{\theta}_1, \widetilde{\theta}_2, \widetilde{\theta}_3$ form a Hamiltonian triple. In such cases, there are several distinct choices of the Hamiltonian pair θ_1, θ_2, that can be used to generate a hereditary operator $\Phi = \theta_2\theta_1^{-1}$. Usually this happens because there is a scaling transformation which decouples one of the operators into two components having different scaling properties, and hence decomposing into a sum of two Hamiltonian operators. In such cases, a single Hamiltonian triple can lead to a plethora of associated integrable equations. This fact was first exploited in [12], where certain generalizations of both KdV and modified KdV were presented. Analogous generalizations of the nonlinear Schrödinger equation and of the sine-Gordon equations are given in [13].

It is interesting that although equations (1.10–14) are derived from the same basic mathematical structure, namely a Hamiltonian pair or triple, they admit very different types of solutions. These include, solitons, peakons (peaked solitons) [14], compactons (solitons with compact support) [15], 2-hump solitons [13], infinitely many solitons [16], and twisted solitons [17]. It is also remarkable that many of these equations appear in physical applications. For example the generalized KdV and the generalized modified KdV appear in the modeling of unidirectional idealized water waves [13], [14], [18].

2. Integrable Generalizations of the KdV Equation

The prototypical example, leading to the Korteweg-deVries equation and its generalizations, starts with the operator

$$\theta = \alpha\partial + \beta\partial^3 + \gamma(u\partial + \partial u). \tag{2.1}$$

It can be shown that θ is a Hamiltonian operator for all values of the constant parameters α, β, γ, and hence its component parts ∂, ∂^3 and $u\partial + \partial u = 2u\partial + u_x$ form a compatible Hamiltonian triple. In what follows we use this operator to illustrate the constructions (1.10–13).

(i) Let $\theta_2 = \theta$, $\theta_1 = \partial + \nu\partial^3$. Then equation (1.10) becomes

$$u_t = (\alpha\partial + \beta\partial^3 + \gamma(u\partial + \partial u))(\partial + \nu\partial^3)^{-1}u_x. \tag{2.2}$$

If $\nu = 0$, this equation becomes the celebrated KdV equation

$$u_t = \alpha u_x + \beta u_{xxx} + 3\gamma u u_x. \tag{2.3}$$

If $\nu \neq 0$, equation (2.2) can be written in a local form by letting $(\partial + \nu\partial^3)^{-1}u_x = q$, or $u = q + \nu q_{xx}$. Then equation (2.2) becomes

$$(q + \nu q_{xx})_t = \left(\alpha\partial + \beta\partial^3 + 2\gamma(q + \nu q_{xx})\partial + \gamma(q_x + \nu q_{xxx})\right)q,$$

or

$$q_t + \nu q_{xxt} = \alpha q_x + \beta q_{xxx} + 3\gamma q q_x + \gamma\nu(qq_{xxx} + 2q_x q_{xx}). \tag{2.4}$$

Equation (2.2) is equation (26.e) of [12], while equation (2.4) is equation (5.3) of [19]. Equation (2.4) with $\nu < 0$ was derived from physical considerations in [14] where also its Lax pair as well as its peakon solutions were given. For $\nu > 0$ (2.4) admits compacton solutions [15]. The inverse spectral method for equation (2.4) is discussed in [16]; other interesting aspects of this equation are discussed in [13], [20], [21].

Letting $\alpha = \beta = 0$, $\nu = \varepsilon^{-1}$, $\varepsilon = 0$, in equation (2.4), and integrating once, one finds

$$q_{xt} = \gamma(qq_{xx} + \frac{1}{2}q_x^2). \tag{2.5}$$

Equation (2.5) was first shown to be integrable by Calogero [22]; its relation to the generalized KdV was pointed out in [11].

(ii) Let $\theta_2 = \theta$, $\theta_1 = \partial + \nu\partial^3$, $u = q_x + \nu q_{xxx}$, then equation (1.11) becomes

$$\alpha q_{xt} + \beta q_{xxxt} + \gamma(2q_x q_{xt} + q_{xx}q_t) + \gamma\nu(q_{xxxx}q_t + 2q_{xxx}q_{xt}) = q_{xx} + \nu q_{xxxx}. \tag{2.6}$$

The particular case when $\nu = 0$ is given in [12].

(iii) Let $\theta_2 = \theta$, $\theta_1 = \partial + \nu\partial^3$, $u = q_x + \nu q_{xxx}$, then equation (1.12) becomes

$$q_{xt} + \nu q_{xxxt} = \alpha q_{xy} + \beta q_{xxxy} + \gamma(q_{xx}q_y + 2q_xq_{xy}) + \gamma\nu(q_{xxxx}q_y + 2q_{xxx}q_{xy}).$$
(2.7)

The particular case of $\nu = 0$ has been discussed earlier, and it has been shown that in this case equation (2.7) supports breaking solitons.

(iv) Let $\theta_2 = \theta$, $\theta_1 = u\partial + \partial u$. Then, since equation $\theta_1 C = 0$ implies $C = u^{-1/2}$, equation (1.13) becomes

$$u_t = (\alpha\partial + \beta\partial^3)u^{-1/2}.$$
(2.8)

This equation, which admits the hereditary operator $\Phi = \theta(u\partial + \partial u)^{-1}$, where θ is defined by equation (2.1), was derived in [11]. If $\alpha = 0$, equation (2.8) reduces to the Harry–Dym equation.

3. Integrable Generalizations of the mKdV Equation

A second collection of integrable Hamiltonian systems starts with the nonlocal Hamiltonian operator

$$\theta = \alpha\partial + \beta\partial^3 + \gamma\partial u\partial^{-1}u\partial,$$
(3.1)

which is a Hamiltonian operator for all values of the constant parameters α, β, γ. This operator leads to a similar set of integrable equations associated with the modified Korteweg-deVries equation.

(i) Let $\theta_2 = \theta$, $\theta_1 = \partial + \nu\partial^3$. Then equation (1.10) becomes

$$u_t = (\alpha\partial + \beta\partial^3 + \gamma\partial u\partial^{-1}u\partial)(\partial + \nu\partial^3)^{-1}u_x.$$
(3.2)

If $\nu = 0$, this equation reduces to the mKdV equation

$$u_t = \alpha u_x + \beta u_{xxx} + \frac{3}{2}\gamma u^2 u_x.$$
(3.3)

If $\nu \neq 0$, equation (3.2) can be written in a local form by letting $u = q + \nu q_{xx}$. Then equation (3.2) becomes

$$(q + \nu q_{xx})_t = \alpha q_x + \beta q_{xxx} + \frac{1}{2}\gamma\partial[(q + \nu q_{xx})(q^2 + \nu q_x^2)].$$
(3.4)

(ii) Let $\theta_2 = \theta$, $\theta_1 = \partial + \nu\partial^3$, then equation (1.11) becomes the nonlocal equation

$$(\alpha + \beta\partial^2 + \gamma\partial u\partial^{-1}u)(1 + \nu\partial^2)^{-1}u_t = u_x.$$
(3.5)

Setting $u = q + \nu q_{xx}$ removes the second non-locality, but there appear to be no direct way to remove the first ∂^{-1}.

(iii) Similarly, setting $\theta_2 = \theta$, $\theta_1 = \partial + \nu \partial^3$, then equation (1.12) leads to a non-local $2 + 1$ dimensional equation

$$u_t = (\alpha + \beta \partial^2 + \gamma \partial u \partial^{-1} u)(1 + \nu \partial^2)^{-1} u_y. \tag{3.6}$$

Again, as in (3.5) the non-locality appears essential.

(iv) Let $\theta_2 = \partial + \lambda \partial^3$, $\theta_1 = \partial^3 + \sigma^2 \partial u \partial^{-1} u \partial$. If $\sigma \neq 0$, then the Casimir functional for θ_1 is found to be $\int \sigma^{-1} \cos(\sigma \partial^{-1} u) \, dx$ with variational derivative $C = \partial^{-1} \sin(\sigma \partial^{-1} u)$. Then equation (1.13) becomes

$$u_t = (\partial + \lambda \partial^3) \sin(\sigma \partial^{-1} u). \tag{3.7}$$

Setting $u = q_x + \lambda q_{xxx}$, we see that (3.7) can be rewritten in the local form

$$q_{xt} = \sin \sigma (q + \lambda q_{xx}) \tag{3.8}$$

an equation whose integrability was first noted in [13]. In particular, for $\lambda = 0$, (3.8) reduces to the well-known sine–Gordon equation.

On the other hand, if we consider the "singular limit" $\sigma \to \infty$, then the Casimir for $\theta_1 = \partial u \partial^{-1} u \partial$ has variational derivative $C = u^{-2}$, leading to the equation

$$u_t = (\partial + \lambda \partial^3) \, u^{-2}. \tag{3.9}$$

According to the formal symmetry approach of Shabat the two Casimir equations (2.8), (3.9), are the only two integrable cases of the general class of equations $u_t = \partial(1 + \lambda \partial^2) u^k$. As with (2.8), equation (3.9) admits solitons, whereas replacing u by $r = 1/u$ in (3.9), we obtain

$$r_t = r^2 (\partial + \partial^3) r^2, \tag{3.10}$$

which admits both traveling and stationary compactons [23].

4. The Nonlinear Schrödinger Equation

As our final example, we consider the integrable Hamiltonian systems that are associated with the nonlinear Schrödinger equation. In this case, the function u is complex-valued, as are the associated evolution equations. We use bars to denote complex conjugates, and $i = \sqrt{-1}$. Consider the non-local Hamiltonian operator

$$\theta F = \alpha i F + \beta \partial F + \gamma u \partial^{-1} (\bar{u} F - u \bar{F}), \tag{4.1}$$

which is Hamiltonian for all values of the constant parameters α, β, γ.

Let $\theta_2 = \theta$, $\theta_1 = i + \nu \partial$. Then equation (1.7) becomes

$$u_t = (\alpha i + \beta \partial)(i + \nu \partial)^{-1} u_x + \gamma u \partial^{-1} \left[\bar{u}(i + \nu \partial)^{-1} u_x - u(-i + \nu \partial)^{-1} \bar{u}_x \right]. \tag{4.2}$$

Setting $u = (i + \nu \partial)q$, equation (4.2) becomes the local equation

$$iq_t + \nu q_{xt} = \alpha i q_x + \beta q_{xx} - i\gamma(iq + \nu q_x)|q|^2, \qquad (4.3)$$

first noted in [13] (see also [11]). If $\alpha = \nu = 0$, $\beta = \gamma = 1$, this equation reduces to the nonlinear Schrödinger equation

$$u_t = i(u_{xx} + |u|^2 u). \qquad (4.3)$$

On the other hand, if $\alpha = \beta = 0$, and $v = qe^{-ix}$, it follows that

$$-iv_{xt} = |v|^2 v_x. \qquad (4.4)$$

This equation has a first integral $|v_x|^2$. Here, in contrast to the KdV and mKdV cases, the dispersion remains linear due to the fact that the Hamiltonian operator θ_2 is a pure integral operator. The construction of an associated hierarchy is more problematic in this case due to nonlocalities.

Acknowledgement. A.S. Fokas was partially supported by an NSF grant DMS 9111611 and by an AFOSR grant F49620-93. P.J. Olver was partially supported by an NSF grant DMS 92-04192. P. Rosenau was partially supported by an AFOSR grant F49620-95. P.J. Olver and P. Rosenau were partially supported by the US-Israel Binational Science Foundation, BSF grant 94-00283.

References

[1] P.J. Olver, *Applications of Lie Groups to Differential Equations*, Second Ed., Vol. 107, Springer-Verlag, NY, 1993.

[2] A.S. Fokas, *J. Math. Phys.* **2** (1980), 1318; *Stud. Appl. Math.* **77** (1987), 253.

[3] A.V. Mikhailov, A.B. Shabat and V.V. Sokolov, in *What is Integrability?* V.E. Zakharov, ed., Springer Verlag, New York (1990), 115–184.

[4] P.J. Olver, *J. Math. Phys.* **18** (1977), 1212.

[5] B. Fuchssteiner, *Non. Anal. Theory Meth. Appl.*, **3** (1979), 849.

[6] S. Kumei, *J. Math. Phys.* **16** (1975), 2461.

[7] F. Magri, *J. Math. Phys.* **19** (1978), 1156.

[8] A.S. Fokas and B. Fuchssteiner, *Lett. Nuovo Cimento* **28** (1980), 299.

[9] I.M. Gel'fand, and I.Y. Dorfman, *Funct. Anal. Appl.* **3** (1980), 14.

[10] I.Y. Dorfman, *Dirac Structures and Integrabilities of Nonlinear Evolution Equations*, J. Wiley, NY (1993).

[11] P.J. Olver and P. Rosenau, Tri-Hamiltonian Duality Between Solitons and Compactons, *Phys. Rev. E* **53** (1996), 1900.

[12] B. Fuchssteiner and A.S. Fokas, *Physica* **4D** (1981), 47.

[13] A.S. Fokas, *Acta Appl. Math.* **34** (1995), 295; *Physica* **87D** (1995), 145.

[14] D. Holm and R. Camassa, *Phys. Rev. Lett.* **71** (1993), 1671.

[15] P. Rosenau, *Phys. Rev. Lett.* **73** (1994), 1737.

[16] A.S. Fokas and P.M. Santini, preprint (1995).

[17] M. Wadati, K. Konno, and Y.H. Ichikawa, *J. Phys. Soc. Japan* **46** (1979), 1965.

[18] A.S. Fokas and Q. Liu, Asymptotic Integrability of Water Waves, 1995.

[19] B. Fuchssteiner, *Prog. Theor. Phys.* **65** (1981), 3.

[20] M.S. Alber, R. Camassa, D.D. Holm, J.E. Marsden, *Proc. Roy. Soc.* (1995).

[21] M.S. Alber, R. Camassa, D.D. Holm, J.E. Marsden, *Lett. Math. Phys.* **32** (1994), 137.

[22] F. Calogero, Stud. Appl. Math. **70**, 189 (1984).

[23] P. Rosenau, *Phys. Rev. Lett.*, to appear.

A.S. Fokas
Department of Mathematics
Imperial College
180 Queen's Gate
London, SW7 2BZ, UK
a.fokas@ic.ac.uk

P.J. Olver
School of Mathematics
University of Minnesota
Minneapolis, MN 55455
olver@ima.umn.edu

P. Rosenau
School of Mathematical Sciences
Tel Aviv University
69978 Tel Aviv, Israel
rosenau@math.tau.ac.il

Submitted November 1995

Hamiltonian Structures in Stationary Manifold Coordinates

Allan P. Fordy and Simon D. Harris

Dedicated to the memory of Irina Dorfman

Abstract

We consider the restriction of isospectral flows to stationary manifolds. Specifically, we present a systematic construction of Hamiltonian structures written in stationary manifold coordinates, which demonstrates the close relationship between the Hamiltonian formulations of nonlinear evolution equation (PDE) and its stationary reduction. We illustrate these ideas in the context of the KdV and 5th order KdV equations.

We then apply these ideas to the Boussinesq hierarchy, associated with the (trace free) 3rd order Lax operator, together with the Sawada-Kotera and Kaup-Kupershmidt reductions.

We use our results to study the integrable cases of the Hénon Heiles equation.

1. Introduction

Most of Irene Dorfman's work was concerned with the algebraic properties of Hamiltonian operators and the integrability of nonlinear evolution equations (see, for instance, [14, 7]). This culminated in the publication of her book [8]. The present article is very much in the spirit of her work.

We consider the restriction of an integrable nonlinear evolution equation to its stationary manifold. It is well known [6] that the stationary flows of many integrable nonlinear evolution equations (themselves completely integrable, infinite dimensional Hamiltonian systems) constitute completely integrable, *finite* dimensional Hamiltonian systems. In some cases we can even find *bi-Hamiltonian* formulations of these finite dimensional systems [3]. However, the relationship between the Poisson brackets of the infinite dimensional and corresponding finite dimensional systems is rather obscure. The approach presented in this paper sheds light on this relationship.

We start, in Section 2, by constructing the Poisson brackets and Hamiltonians associated with the isospectral flows of a given spectral problem. By choosing the correct coordinates (and reversing the roles of x and t) these can be adapted to the stationary manifolds. These ideas are illustrated through the stationary KdV equation.

2. Stationary Manifolds: Poisson Brackets

In this section we explain two ways of constructing the Poisson brackets for stationary flows of integrable nonlinear evolution equations. These are illustrated through the KdV hierarchy. The mth flow in the KdV hierarchy is defined by

$$u_{t_m} = \left(\frac{1}{2}\partial^3 + 2u\partial + u_x \right) \delta_u \mathcal{H}_m = \frac{1}{2}\partial \delta_u \mathcal{H}_{m+1}, \quad m \geq 1,$$

the first few Hamiltonian densities being given by

$$\mathcal{H}_0 = u, \quad \mathcal{H}_1 = u^2, \quad \mathcal{H}_2 = 2u^3 - u_x^2, \quad \mathcal{H}_3 = u_{xx}^2 - 10uu_x^2 + 5u^4.$$

The stationary flow $u_{t_m} = 0$ is in the form of a generalised Lagrangian equation:

$$\delta_u \mathcal{L}_{m+1} = 0, \quad \mathcal{L}_{m+1} = \mathcal{H}_{m+1} - c_m u. \tag{1}$$

It is possible to write these equations as an m-degrees of freedom canonical Hamiltonian system (see Appendix A and [17]). If we take c_m as a dynamical variable then the Poisson bracket is degenerate, with c_m as Casimir:

$$\{f, g\} = \frac{\partial f}{\partial x_i} J_{ij} \frac{\partial g}{\partial x_j} = (\nabla f)^T J (\nabla g), \quad J = \begin{pmatrix} 0 & I_m & 0 \\ -I_m & 0 & 0 \\ 0 & 0 & 0 \end{pmatrix}, \tag{2}$$

where I_m is the $m \times m$ unit matrix. Bogoyavlenskii and Novikov [6] showed that this Hamiltonian system is completely integrable with first integrals given by the fluxes of the Hamiltonians $\mathcal{H}_1, \cdots, \mathcal{H}_m$:

$$
\begin{aligned}
0 &= \mathcal{H}_{0t_m} = \mathcal{F}_{0mx} \quad \text{the equation,} \\
0 &= \mathcal{H}_{kt_m} = \mathcal{F}_{kmx} \quad k = 1, \cdots, m.
\end{aligned}
$$

The equation $\mathcal{F}_{0m} = c_m$ is an ODE of order $2m$. This is used to eliminate higher derivatives from the fluxes $\mathcal{F}_{km}$, which can then be written as functions of the variables (q_i, p_i, c_m): $\mathcal{F}_{km} = h_k(q_i, p_i, c_m)$, $k = 1, \cdots, m$. The later fluxes become functionally dependent upon $h_1, \cdots, h_m$. Since $\{\mathcal{H}_i, \mathcal{H}_m\} = 0$ for all elements of the KdV hierarchy, the stationary manifold is invariant under the action of the flows. The first component of each of the commuting Hamiltonian flows of the stationary flow are precisely

$$u_{t_k} = \frac{1}{2}\partial \delta \mathcal{H}_{k+1}, \quad k = 1, \cdots, m-1,$$

when written in terms of the stationary manifold coordinates, since $q_1 = u$.

Example 2.1 (The stationary KdV and MKdV equations). The stationary KdV equation (which defines c_1) is

$$2u_{xx} + 6u^2 = c_1, \tag{3}$$

with Lagrangian $\mathcal{L}$ given by $\mathcal{L} = 2u^3 - u_x^2 - c_1 u$ and canonical coordinates: $q = u$, $p = -2u_x$, c_1 (the Casimir of the degenerate canonical bracket). The Legendre transformation gives $h_{\mathcal{L}} = c_1 q - 2q^3 - \frac{1}{4}p^2$, which is the flux of $\mathcal{H}_1 = u^2$.

The Miura map $u = -v_x - v^2 + \frac{1}{4}\lambda$ induces the diffeomorphism M given by

$$q = \frac{1}{4}\lambda - \frac{1}{2}\tilde{p} - \tilde{q}^2, \quad p = -\tilde{c}_1 + 4\tilde{q}^3 + 2\tilde{q}\tilde{p} - 3\lambda\tilde{q},$$

$$c_1 = \frac{1}{2}\tilde{p}^2 - 2\tilde{q}^4 + 2\tilde{q}\tilde{c}_1 + 3\lambda\tilde{q}^2 + \frac{3}{8}\lambda^2, \tag{4}$$

relating (3) to the stationary MKdV equation:

$$-2v_{xx} + 4v^3 - 3\lambda v = \tilde{c}_1, \tag{5}$$

which has Lagrangian $\mathcal{L} = v_x^2 + v^4 - \tilde{c}_1 v - \frac{3}{2}\lambda v^2$ and canonical variables: $\tilde{q} = v$, $\tilde{p} = 2v_x$, $\tilde{c}_1$. The canonical Poisson bracket between the modified variables: $\{\tilde{q}, \tilde{p}\} = 1$, $\{\tilde{q}, \tilde{c}_1\} = \{\tilde{p}, \tilde{c}_1\} = 0$, induces:

$$\{q, p\} = -2q - \lambda, \quad \{q, c_1\} = -p, \quad \{p, c_1\} = -2c_1 + 12q^2. \tag{6}$$

This is a non-canonical Poisson bracket for the stationary KdV equation. In fact it is a one parameter family of Poisson brackets: the coefficients of λ is just the canonical bracket, while the remaining part is the *second* bracket [3]. This proves not only that each coefficient is independently a Poisson bracket, but also that they are compatible.

The variable c_1 is a Casimir for the degenerate canonical bracket with its level surfaces being symplectic leaves. The symplectic leaves of the second Poisson bracket are different, being the level surfaces of $h_{\mathcal{L}}$, while c_1 now generates the stationary KdV equation.

A similar calculation for system (1) gives rise to a $(2m + 1) \times (2m + 1)$ non-canonical Poisson bracket, which is compatible with (2).

2.1. Reversing the Role of x and t

Each member of the KdV hierarchy is bi-Hamiltonian, in both the PDE and stationary flow cases. However, the latter case required a Legendre transformation and a Miura map, so showed no clear relationship between the Hamiltonian structures of the PDE and those of the stationary flow. This deficit was overcome in [1] where they took t as the 'spatial' variable and x as the evolution parameter. In so doing, the KdV equation becomes a system of three equations in x, with a 3×3 matrix Poisson bracket, *which reduces to the stationary case when the potential functions are independent of t*

With the same canonical variables as before, the KdV equation can be written *as a PDE* in the form

$$
\begin{pmatrix} q \\ p \\ c_1 \end{pmatrix}_x = \begin{pmatrix} -\frac{1}{2}p \\ 6q^2 - c_1 \\ 2q_t \end{pmatrix} = \begin{pmatrix} 0 & 1 & 0 \\ -1 & 0 & 0 \\ 0 & 0 & 2\partial_t \end{pmatrix} \begin{pmatrix} c_1 - 6q^2 \\ -\frac{1}{2}p \\ q \end{pmatrix},
$$

where this last vector is just $\delta_q(c_1 q - 2q^3 - \frac{1}{4}p^2)$. Similarly the MKdV equation can be written:

$$
\begin{pmatrix} \tilde{q} \\ \tilde{p} \\ \tilde{c}_1 \end{pmatrix}_x = \begin{pmatrix} 0 & 1 & 0 \\ -1 & 0 & 0 \\ 0 & 0 & -2\partial_t \end{pmatrix} \begin{pmatrix} \tilde{c}_1 - 4\tilde{q}^3 + 3\lambda\tilde{q} \\ \frac{1}{2}\tilde{p} \\ \tilde{q} \end{pmatrix} = \hat{B}\delta_{\tilde{q}}\tilde{h}, \qquad (7)
$$

where $\tilde{h} = \tilde{q}\tilde{c}_1 - \tilde{q}^4 + \frac{1}{4}\tilde{p}^2 + \frac{3}{2}\lambda\tilde{q}^2$.

The Miura map (4) is only changed by $-2\tilde{q}_t$ in the definition of c_1. The above bracket for the MKdV equation is then transformed to

$$
\begin{pmatrix} 0 & -2q & \partial_t - p \\ 2q & -2\partial_t & 12q^2 - 2c_1 \\ \partial_t + p & 2c_1 - 12q^2 & 4(q\partial_t + \partial_t q) \end{pmatrix} - \lambda \begin{pmatrix} 0 & 1 & 0 \\ -1 & 0 & 0 \\ 0 & 0 & 2\partial_t \end{pmatrix}. \qquad (8)
$$

These define Poisson brackets for the *full* KdV equation written in terms of the stationary manifold coordinates and clearly reduce to those of the stationary equations when the potential functions are independent of t.

The $x - t$ reversal flows and Poisson brackets have also been considered in [13, 16].

2.2. The Spectral Problem

The above Poisson brackets can be systematically constructed by considering the corresponding spectral problem and its isospectral flows.

In order to consider the t_m stationary flow we just reverse the roles of U (the original 'x−part' of the zero-curvature representation) and $V_{(m)}$ (the original 't_m−part'). The latter now plays the role of the spectral problem. We write the zero curvature equations as

$$
V_{(m)x} = U_\xi - [V_{(m)}, U],
$$

where $\xi = t_m$. It should be noted that $V_{(m)}$ depends upon $u, u_x, \ldots, u_{x \cdots x}$, where the last term has $2m$ derivatives, so can be written in terms of any set of coordinates on the stationary manifold. For comparison of results we choose the canonical coordinates.

The KdV equation. The spectral problem is

$$\Psi_\xi = \mathcal{U}\Psi, \quad \text{where} \quad \mathcal{U} \equiv V_{(1)} = \begin{pmatrix} \frac{1}{2}p & \lambda + 2q \\ \frac{1}{4}\lambda^2 - \frac{1}{2}\lambda q + q^2 - \frac{1}{2}c_1 & -\frac{1}{2}p \end{pmatrix}.$$

We are interested in the hierarchy rather than just the KdV flow, so consider the time evolution

$$\Psi_\tau = \mathcal{V}\Psi \equiv \begin{pmatrix} A & B \\ C & -A \end{pmatrix} \begin{pmatrix} \psi_1 \\ \psi_2 \end{pmatrix}, \tag{9}$$

where A, B and C are functions of q, p, c_1, their ξ−derivatives and λ.

In terms of $\mathcal{U}$ and $\mathcal{V}$ the integrability conditions are just

$$\mathcal{U}_\tau = \mathcal{V}_\xi - [\mathcal{U}, \mathcal{V}]. \tag{10}$$

We are interested in the equation for $\mathbf{q} = (q, p, c_1)^T$ and have $\mathcal{U}_\tau = \frac{D\mathcal{U}}{D\mathbf{q}} \cdot \mathbf{q}_\tau$, where $\frac{D\mathcal{U}}{D\mathbf{q}}$ is the Fréchet derivative. If we represent $\mathcal{U} = \begin{pmatrix} a & b \\ c & -a \end{pmatrix}$ by the vector $\mathbf{b} = (b, c, a)^T$, and $\mathcal{J}_\varphi$ is the Fréchet derivative of the map $\varphi : \mathbf{q} \mapsto \mathbf{b}$, then

$$\mathbf{b}_\tau = \mathcal{J}_\varphi \mathbf{q}_\tau, \quad \text{where} \quad \mathcal{J}_\varphi = \begin{pmatrix} 2 & 0 & 0 \\ 2q - \frac{1}{2}\lambda & 0 & -\frac{1}{2} \\ 0 & \frac{1}{2} & 0 \end{pmatrix},$$

whilst from the standard Poisson bracket associated with the general 2×2 matrix spectral problem, we have

$$\mathbf{b}_\tau \equiv \begin{pmatrix} b \\ c \\ a \end{pmatrix}_\tau = \begin{pmatrix} 0 & -\partial + 2a & -b \\ -\partial - 2a & 0 & c \\ b & -c & -\frac{1}{2}\partial \end{pmatrix} \begin{pmatrix} C \\ B \\ 2A \end{pmatrix} \equiv \tilde{\mathbf{B}}\delta_b\tilde{h}, \tag{11}$$

where $\partial \equiv \frac{\partial}{\partial \xi}$ and $(C, B, 2A) = -\left(\frac{\delta \tilde{h}}{\delta b}, \frac{\delta \tilde{h}}{\delta c}, \frac{\delta \tilde{h}}{\delta a} \right)$, $h(q, p, c_1) = \tilde{h} \circ \varphi(\mathbf{q})$. We pull this gradient back to the $\mathbf{q}$ space

$$\delta_q h = \mathcal{J}_\varphi^\dagger \cdot \delta_b \tilde{h} \quad \Rightarrow \quad \begin{pmatrix} \delta_q h \\ \delta_p h \\ \delta_{c_1} h \end{pmatrix} = \begin{pmatrix} -2 & \frac{1}{2}\lambda - 2q & 0 \\ 0 & 0 & -\frac{1}{2} \\ 0 & \frac{1}{2} & 0 \end{pmatrix} \begin{pmatrix} C \\ B \\ 2A \end{pmatrix}.$$

In this case it is possible to invert $\mathcal{J}_\varphi$ to obtain

$$\mathbf{q}_\tau = \mathcal{J}_\varphi^{-1} \tilde{\mathbf{B}} \left(\mathcal{J}_\varphi^{-1} \right)^\dagger \delta_q h = \mathbf{B}\delta_q h, \tag{12}$$

where

$$\mathbf{B} = \begin{pmatrix} 0 & -2q - \lambda & -p + \partial \\ 2q + \lambda & -2\partial & 12q^2 - 2c_1 \\ p + \partial & -12q^2 + 2c_1 & (4q - \lambda)\partial + \partial(4q - \lambda) \end{pmatrix}. \quad (13)$$

As previously mentioned, this is related to the simple Hamiltonian structure of (7) through the Miura map. This is an alternative way of proving that $\mathbf{B}$ is indeed Hamiltonian.

To calculate the hierarchy of Hamiltonians we repeat the standard argument used with the KdV hierarchy in the usual coordinates. Since $\tilde{c}_1$ is a *flux* in the usual MKdV language, it is now a conserved density under τ-evolutions. Furthermore, it is a Casimir function of the Poisson bracket given in (7), so generates the bi-Hamiltonian ladder (8):

$$\hat{\mathbf{B}}\delta_{\tilde{q}}\tilde{c}_1 = 0 \quad \Rightarrow \quad \mathbf{B}\delta_q\tilde{c}_1 = M'\hat{\mathbf{B}}(M')^\dagger\delta_q\tilde{c}_1 = M'\hat{\mathbf{B}}\delta_{\tilde{q}}\tilde{c}_1 = 0,$$

where $\mathbf{B}$ is given by (13). The asymptotic series for $\tilde{c}_1$ can be derived from the *known* series in the usual KdV coordinates, replacing x-derivatives of u by q, p, c_1 and their ξ-derivatives. The resulting Hamiltonians:

$$h_{-1} = \frac{1}{2}c_1, \quad h_0 = -\frac{1}{4}p^2 + qc_1 - 2q^3, \quad h_1 = \frac{1}{4}c_1^2 + q_\xi p,$$

and second Hamiltonian structure, generate

$$\begin{pmatrix} q \\ p \\ c_1 \end{pmatrix}_{\tau_{-1}} = \begin{pmatrix} -\frac{1}{2}p \\ 6q^2 - c_1 \\ 2q_\xi \end{pmatrix}, \quad \begin{pmatrix} q \\ p \\ c_1 \end{pmatrix}_{\tau_0} = \begin{pmatrix} q \\ p \\ c_1 \end{pmatrix}_\xi,$$

$$\begin{pmatrix} q \\ p \\ c_1 \end{pmatrix}_{\tau_1} = \begin{pmatrix} \frac{1}{2}c_{1\xi} - 2qq_\xi - \frac{1}{2}pc_1 \\ -2q_{\xi\xi} - 2qp_\xi + 6q^2c_1 - c_1^2 \\ (4c_1q - 4q^3 - \frac{1}{2}p^2 - p_\xi)_\xi \end{pmatrix}.$$

In general we have

$$q_{\tau_r} = (\mathbf{B}_1 - \lambda\mathbf{B}_0)\delta_q h_{(r)} = \mathbf{B}_1\delta_q h_r, \quad h_{(r)} = \lambda^{r+1}h_{-1} + \cdots + h_r.$$

The time evolution Ψ_{τ_r} is given in terms of $\delta_q h_{(r)}$

$$A = -\frac{\delta h_{(r)}}{\delta p}, \quad B = 2\frac{\delta h_{(r)}}{\delta c_1}, \quad C = -\frac{1}{2}\frac{\delta h_{(r)}}{\delta q} + (\frac{1}{2}\lambda - 2q)\frac{\delta h_{(r)}}{\delta c_1}. \quad (14)$$

Thus h_{-1} generates the KdV equation and h_0 the 'translational flow'. We can use the τ_{-1} flow to write p, c_1 and q_ξ in terms of q and its τ_{-1} derivatives after which

$$q_{\tau_1} = q_{xxxxx} + 10qq_{xxx} + 20q_xq_{xx} + 30q^2q_x,$$

where we have written $\tau_{-1} = x$.

Thus, the hierarchy is just the KdV hierarchy in disguise. However, without the τ_{-1} flow, this hierarchy shows no hint of being reducible from the (q, p, c_1) space to the q space. In these coordinates the reduction to the stationary flow $q_\xi = 0$ is very natural, with (13) reducing to (6) and the bi-Hamiltonian ladder becoming finite.

The 5th order KdV equation. The KdV equation itself is deceptively simple, since the map $\varphi : \mathbf{q} \mapsto \mathbf{b}$ is invertible, so the systems (11) and (12) are equivalent. However, when $m \geq 2$, the stationary manifold of the mth member of the KdV hierarchy is of dimension $2m + 1$. Thus, while $\mathbf{b}$ is a 3–vector, $\mathbf{q}$ is a $(2m+1)$–vector and $\mathcal{J}_\varphi$ projects $\mathbf{q}_\tau$ onto $\mathbf{b}_\tau$: $\mathbf{b}_\tau = \mathcal{J}_\varphi \mathbf{q}_\tau$, since $\mathcal{J}_\varphi$ is a $3 \times (2m + 1)$ matrix. Thus, the simple formula (12) is no longer possible. Nevertheless, it *is* possible to modify the procedure to find $\mathbf{B}_0$ and $\mathbf{B}_1$ so that

$$\mathbf{q}_{\tau_r} = (\mathbf{B}_1 - \lambda \mathbf{B}_0)\, \delta_q h_{(r)}, \tag{15}$$

where $h_{(r)} = \sum_{i=0}^{r+m} \lambda^{r+m-i} h_{i-m}$, with h_k being easily calculated from the relevant fluxes.

Corresponding to the t_2–flow of the KdV hierarchy, we have the time evolution matrix $V_{(2)}$:

$$V_{(2)} = \begin{pmatrix} -\lambda u_x - u_{xxx} - 6uu_x & \lambda^2 + 2\lambda u + 2u_{xx} + 6u^2 \\ C_{(2)} & \lambda u_x + u_{xxx} + 6uu_x \end{pmatrix},$$

where $C_{(2)} = \frac{1}{4}\lambda^3 - \frac{1}{2}\lambda^2 u - \frac{1}{2}\lambda\left(u_{xx} + u^2\right) - u_{xxxx} - 8uu_{xx} - 6u_x^2 - 6u^3$. When written in terms of *any* coordinates on the stationary manifold, $V_{(2)}$ gives an appropriate spectral problem. Here we use the canonical coordinates (q_i, p_i, c_2), defined by

$$\begin{aligned} q_1 &= u, \quad q_2 = u_x, \quad p_2 = 2u_{xx}, \quad p_1 = -2u_{xxx} - 20uu_x, \\ c_2 &= 2u_{xxxx} + 20uu_{xx} + 10u_x^2 + 20u^3, \end{aligned} \tag{16}$$

(where c_2 will be a Casimir function of the resulting degenerate canonical bracket).

When written in terms of these coordinates, $V_{(2)}$ takes the form of $\mathcal{U}$ below:

$$\Psi_\xi = \mathcal{U}\Psi, \quad \text{where} \quad \mathcal{U} = \begin{pmatrix} a & b \\ c & -a \end{pmatrix}, \tag{17}$$

with a, b, c given by

$$a = -\lambda q_2 + \frac{1}{2}p_1 + 4q_1 q_2, \quad b = \lambda^2 + 2\lambda q_1 + p_2 + 6q_1^2,$$

$$c = \frac{1}{4}\lambda^3 - \frac{1}{2}\lambda^2 q_1 - \frac{1}{4}\lambda(p_2 + 2q_1^2) - \frac{1}{2}c_2 + q_1 p_2 + 4q_1^3 - q_2^2.$$

We denote this mapping by $\varphi : \mathbf{q} \mapsto \mathbf{b}$. As before, we consider the hierarchy (9), where A, B and C are functions of (q_i, p_i, c_2), their ξ-derivatives and λ. The integrability conditions are again (10).

The Jacobian $\mathcal{J}_\varphi$ is a 3×5 matrix:

$$\mathcal{J}_\varphi = \begin{pmatrix} 12q_1 + 2\lambda & 0 & 0 & 1 & 0 \\ 12q_1^2 + p_2 - \lambda q_1 - \frac{1}{2}\lambda^2 & -2q_2 & 0 & q_1 - \frac{1}{4}\lambda & -\frac{1}{2} \\ 4q_2 & 4q_1 - \lambda & \frac{1}{2} & 0 & 0 \end{pmatrix},$$

which does not, therefore, have a unique inverse. Thus the equation

$$\mathbf{b}_\tau = \mathcal{J}_\varphi \, \mathbf{q}_\tau \qquad (18)$$

is underdetermined for the components of $\mathbf{q}_\tau$, whilst the equation

$$\delta_q h = \mathcal{J}_\varphi^\dagger \, \delta_b \tilde{h} \qquad (19)$$

gives an overdetermined system of equations for the components of $\delta_b \tilde{h}$, which we would like to solve in terms of $\delta_q h$. Thus $\delta_q h$ must be orthogonal to $\mathbf{w} \in Ker \, \mathcal{J}_\varphi$:

$$\mathbf{w}^\dagger \, \delta_q \, h = (\mathcal{J}_\varphi \mathbf{w})^\dagger \delta_b \tilde{h} \Rightarrow \mathbf{w}^\dagger \delta_q h = 0,$$

apparently giving rise to some constraints on $\delta_q h$. In fact these 'constraints' can be identified with part of the recursion relation of the bi-Hamiltonian ladder, so do not constitute a restriction. The (right) kernel of $\mathcal{J}_\varphi$ is important in what follows, and is spanned by

$$\mathbf{v}_1 = (1, 0, -8q_2, -12q_1 - 2\lambda, 2p_2)^T, \quad \mathbf{v}_2 = (0, 1, -8q_1 + 2\lambda, 0, -4q_2)^T.$$

The left kernel of $\mathcal{J}_\varphi$ is $\mathbf{0}$. We can then solve (uniquely) for $\delta_b \tilde{h}$:

$$\delta_a \tilde{h} = 2\delta_{p_1} h, \quad \delta_b \tilde{h} = \delta_{p_2} h + (2q_1 - \frac{1}{2}\lambda)\delta_{c_2} h, \quad \delta_c \tilde{h} = -2\delta_{c_2} h, \qquad (20)$$

with 'constraints':

$$\begin{aligned} \lambda \delta_{p_1} h &= -\frac{1}{2}\delta_{q_2} h + 4q_1 \delta_{p_1} h + 2q_2 \delta_{c_2} h, \\ \lambda \delta_{p_2} h &= \frac{1}{2}\delta_{q_1} h - 4q_2 \delta_{p_1} h - 6q_1 \delta_{p_2} h + p_2 \delta_{c_2} h. \end{aligned} \qquad (21)$$

Since we do not have a left inverse for $\mathcal{J}_\varphi$, we cannot use (12). We write (10) as:

$$\mathcal{J}_\varphi \, \mathbf{q}_\tau = \tilde{\mathbf{B}} \delta_b \tilde{h}, \qquad (22)$$

where $\tilde{\mathbf{B}}$ is the 3×3 Hamiltonian structure given in (11). We write (20) as $\delta_b \tilde{h} = D \delta_q h$, where

$$D = \begin{pmatrix} 0 & 0 & 0 & 1 & 2q_1 - \frac{1}{2}\lambda \\ 0 & 0 & 0 & 0 & -2 \\ 0 & 0 & 2 & 0 & 0 \end{pmatrix},$$

but are free to add the transpose of an element of the right kernel of $\mathcal{J}_\varphi$:

$$\delta_b \tilde{h} = \left(D + \sum_{i=1}^{2} \mathbf{f}_i \mathbf{v}_i^\dagger\right)\delta_q h \equiv \hat{D}\delta_q h. \tag{23}$$

Indeed,

$$\mathcal{J}_\varphi^\dagger \delta_b \tilde{h} = \delta_q h \quad \Rightarrow \quad \hat{D}\mathcal{J}_\varphi^\dagger \delta_b \tilde{h} = \hat{D}\delta_q h = \delta_b \tilde{h} \quad \Rightarrow \quad \hat{D}\mathcal{J}_\varphi^\dagger = I_3,$$

where I_3 is the 3×3 identity matrix. Thus, for *any* choice of $\mathbf{f}_i$, $\hat{D}$ is a left inverse of $\mathcal{J}_\varphi^\dagger$.

While the $\mathbf{f}_i$ make no difference to the formula (20) for $\delta_b \tilde{h}$ they *can* change the right hand side of

$$\mathcal{J}_\varphi \mathbf{q}_{\tau_r} = \tilde{\mathbf{B}}\left(D + \sum_{i=1}^{2} \mathbf{f}_i \mathbf{v}_i^\dagger\right)\delta_q h_{(r)}, \tag{24}$$

which involves the *truncated* Hamiltonian $h_{(r)}$, which no longer satisfies the infinite recursion relation represented by the 'constraint' (21).

We now use our knowledge of the standard KdV hierarchy. The Hamiltonian $h_{(r)} = \left(\lambda^{r+2}h\right)_+$ is given in terms of the fluxes previously calculated,

$$h_{-2} = \frac{1}{2}c_2, \quad h_{-1} = -5q_1^4 + 10q_1 q_2^2 + q_1 c_2 + q_2 p_1 + \frac{1}{4}p_2^2.$$

We have the following correspondence between the usual KdV t_k and τ_m: $\tau_{-2} = t_0 = x$, $\tau_{-1} = t_1$, $\tau_0 = t_2 = \xi$, $\tau_1 = t_3, \cdots$. The τ_{-2} flow is just the $x - t_2$ reversed fifth order KdV equation:

$$\mathbf{q}_{\tau_{-2}} = \left(q_2, \frac{1}{2}p_2, 20q_1^3 - 10q_2^2 - c_2, -20q_1 q_2 - p_1, 2q_{1\xi}\right)^T,$$

where $\mathbf{q} = (q_1, q_2, p_1, p_2, c_2)^T$, whilst the τ_{-1} flow is the KdV equation for q_1 and its prolongation to the other variables (but written in these coordinates):

$$\begin{pmatrix} q_1 \\ q_2 \\ p_1 \\ p_2 \\ c_2 \end{pmatrix}_{\tau_{-1}} = \begin{pmatrix} -4q_1 q_2 - \frac{1}{2}p_1 \\ -10q_1^3 - 2q_1 p_2 + q_2^2 + \frac{1}{2}c_2 \\ 120q_1^4 + 30q_1^2 p_2 + 20q_1 q_2^2 - 6q_1 c_2 + 4q_2 p_1 + p_2^2 - 2q_{2\xi} \\ 20q_1^2 q_2 + 4q_1 p_1 - 2q_2 p_2 + 2q_{1\xi} \\ 12q_1 q_{1\xi} + p_{2\xi} \end{pmatrix}.$$

Substituting these equation into (24) determines $\mathbf{f}_i$ to be zero. Thus, for this example the correct choice of $\hat{D}$ is D.

Remark 2.1. Whilst for the KdV hierarchy and several other examples, we find $\mathbf{f}_i = 0$, this is *not* always true. In this paper we mention the Boussinesq t_5−flow. Other examples, such as Ito's equation, can be found in [15].

We wish to find a 5×5 matrix $\mathbf{B}$ such that

$$\tilde{\mathbf{B}} = \mathcal{J}_\varphi \mathbf{B} \mathcal{J}_\varphi^\dagger. \tag{25}$$

Using $\hat{D} (= D)$ we can find a particular solution $\bar{\mathbf{B}}$ of (25) satisfying:

$$\bar{\mathbf{B}} = D^\dagger \tilde{\mathbf{B}} D \quad \text{so that} \quad \mathcal{J}_\varphi \bar{\mathbf{B}} \mathcal{J}_\varphi^\dagger = (\mathcal{J}_\varphi D^\dagger) \tilde{\mathbf{B}} (D \mathcal{J}_\varphi^\dagger) = \tilde{\mathbf{B}}.$$

This $\bar{\mathbf{B}}$ explicitly takes the form:

$$\bar{\mathbf{B}} = \begin{pmatrix} 0 & 0 & 0 & 0 & 0 \\ 0 & 0 & 0 & 0 & 0 \\ 0 & 0 & -2\partial & \Gamma_3 & \Gamma_1 \\ 0 & 0 & -\Gamma_3 & 0 & 2\partial + \Gamma_2 \\ 0 & 0 & -\Gamma_1 & 2\partial - \Gamma_2 & 4(\partial q_1 + q_1 \partial) - 2\lambda\partial \end{pmatrix},$$

where

$$\Gamma_1 = 40q_1^3 + 8q_1 p_2 - 4q_2^2 - 2c_2 - 2\lambda p_2,$$
$$\Gamma_2 = -16q_1 q_2 - 2p_1 + 4\lambda q_2, \quad \Gamma_3 = 12q_1^2 + 2p_2 + 4\lambda q_1 + 2\lambda^2.$$

To this we can add any skew symmetric matrix which is in the kernel of $M \mapsto \mathcal{J}_\varphi M \mathcal{J}_\varphi^\dagger$:

$$\mathbf{B} = \bar{\mathbf{B}} + \sum_{i,j=1}^{2} \left(\mathbf{a}_i \mathbf{v}_i^\dagger - \mathbf{v}_i \mathbf{a}_i^\dagger \right).$$

We wish to write the evolution equations for $\mathbf{q}$ as

$$\mathbf{q}_{\tau_r} = \mathbf{B} \delta_q h_{(r)} = \left(\bar{\mathbf{B}} + \sum_{i,j=1}^{2} \left(\mathbf{a}_i \mathbf{v}_i^\dagger - \mathbf{v}_i \mathbf{a}_i^\dagger \right) \right) \delta_q h_{(r)}, \tag{26}$$

but also have (22) after replacing $\tilde{\mathbf{B}}$

$$\mathcal{J}_\varphi \left(\mathbf{q}_{\tau_r} - \bar{\mathbf{B}} \delta_q h_{(r)} \right) = 0. \tag{27}$$

To be consistent we require: $\mathcal{J}_\varphi \sum_{i,j=1}^{2} \left(\mathbf{a}_i \mathbf{v}_i^\dagger - \mathbf{v}_i \mathbf{a}_i^\dagger \right) \delta_q h_{(r)} = 0$. Since we do *not* have $\mathbf{v}_i^\dagger \delta_q h_{(r)} = 0$, we need $\mathcal{J}_\varphi \mathbf{a}_i = 0$, so that $\mathbf{a}_i = \sum_{j=1}^{2} \mathbf{v}_j d_{ji}$, and

$$\mathbf{B} = \bar{\mathbf{B}} + \sum_{i,j=1}^{2} \mathbf{v}_i b_{ij} \mathbf{v}_j^\dagger,$$

where $b_{ij} = (\mathbf{b})_{ij}$ with $\mathbf{b}$ a skew adjoint matrix differential operator (in the present case, a simple skew symmetric matrix). Equation (27) implies $\mathbf{q}_{\tau_r} - \bar{\mathbf{B}}\delta_q h_{(r)} \in Ker \, \mathcal{J}_\varphi$. From the forms of $\bar{\mathbf{B}}$ and $\mathbf{v}_i$, we have explicitly

$$\mathbf{q}_{\tau_r} = \bar{\mathbf{B}}\delta_q h_{(r)} + \mathbf{v}_1 q_{1\tau_r} + \mathbf{v}_2 q_{2\tau_r}. \tag{28}$$

Matching (26) and (28) we have

$$q_{i\tau_r} = \sum_{j=1}^{2} b_{ij}\mathbf{v}_j^\dagger \delta_q h_{(r)}, \quad i = 1, 2.$$

The coordinates (16) define the fifth order KdV equation as an $x-t$ reversed flow, with $\tau_{-2} = x$, from which we find: $q_{1\tau_{-2}} = q_2$, $q_{2\tau_{-2}} = \frac{1}{2}p_2$. In our Hamiltonian formulation, this should be generated by the function $h_{-2} = \frac{1}{2}c_2$. Substituting these into (28) gives: $b_{11} = b_{22} = 0$, $b_{12} = -b_{21} = -\frac{1}{2}$. The Hamiltonian operator $\mathbf{B}$ then takes the form: $\mathbf{B} = \mathbf{B}_1 - \lambda \mathbf{B}_0$, where

$$\mathbf{B}_0 = \begin{pmatrix} 0 & 0 & 1 & 0 & 0 \\ 0 & 0 & 0 & 1 & 0 \\ -1 & 0 & 0 & 0 & 0 \\ 0 & -1 & 0 & 0 & 0 \\ 0 & 0 & 0 & 0 & 2\partial \end{pmatrix},$$

$$\mathbf{B}_1 = \begin{pmatrix} 0 & -\frac{1}{2} & 4q_1 & 0 & 2q_2 \\ \frac{1}{2} & 0 & -4q_2 & -6q_1 & p_2 \\ -4q_1 & 4q_2 & -2\partial & 60q_1^2 + 2p_2 & \Omega_1 \\ 0 & 6q_1 & -60q_1^2 - 2p_2 & 0 & 2\partial + \Omega_2 \\ -2q_2 & -p_2 & -\Omega_1 & 2\partial - \Omega_2 & 4(\partial q_1 + q_1 \partial) \end{pmatrix},$$

and $\Omega_1 = 40q_1^3 - 20q_2^2 - 2c_2$, $\Omega_2 = -40q_1 q_2 - 2p_1$.

The modified 5th order hierarchy has a *single* local Hamiltonian structure (almost identical to $\mathbf{B}_0$ above). This simple operator is transformed onto $\mathbf{B}_1 - \lambda \mathbf{B}_0$ by the Miura map, when extended to this 5 dimensional space. Once again, this is the simplest proof of the fact that $\mathbf{B}_0$ and $\mathbf{B}_1$ are both Hamiltonian *and* compatible. As usual, we can use the Miura map (with parameter λ) to generate the infinite sequence of Hamiltonians.

The time evolution (9) is calculated from $h_{(r)}$ by (23) (compare with (14)):

$$A = \delta_{p_1} h_{(r)}, \quad B = -2\delta_{c_2} h_{(r)}, \quad C = \delta_{p_2} h_{(r)} + (2q_1 - \frac{1}{2}\lambda)\delta_{c_2} h_{(r)}.$$

When ξ−derivatives of the phase space variables are zero, then $\mathbf{B}_0$ and $\mathbf{B}_1$ reduce to the previously known [3] Poisson matrices of the stationary 5th order KdV equation. We then have just three independent Hamiltonians:

c_2, h_{-2}, h_{-1}. The integral c_2 is the Casimir of $\mathbf{B}_0$ and h_{-1} the Casimir of $\mathbf{B}_1$ in this reduction.

3. The Boussinesq Hierarchy

The isospectral flows of: $L = \partial^3 + u\partial + v$, with $\partial = \partial/\partial x$, form two bi-Hamiltonian ladders, labelled by t_{3m-2} and t_{3m-1}, $m \geq 1$;

$$\begin{pmatrix} u \\ v \end{pmatrix}_{t_{3m-i}} = \mathbf{B}_1 \delta \mathcal{H}_m^i = \mathbf{B}_0 \delta \mathcal{H}_{m+1}^i, \quad i = 1, 2,$$

where

$$\mathbf{B}_0 = \begin{pmatrix} 0 & -3\partial \\ -3\partial & 0 \end{pmatrix}, \quad \mathbf{B}_1 = \begin{pmatrix} X & Y \\ -Y^\dagger & Z \end{pmatrix}, \tag{29}$$

with: $X = 2\partial^3 + u\partial + \partial u$, $Y = -\partial^4 - \partial^2 u + 2\partial v + v\partial$ and $Z = -\frac{2}{3}\left(\partial^5 + \partial^3 u + u\partial^3\right) + \partial^2 v - v\partial^2 - \frac{1}{3}\left(\partial u^2 + u^2\partial\right)$. The first few Hamiltonians in these series are

$$\mathcal{H}_1^2 = u, \quad \mathcal{H}_2^2 = \frac{1}{3}uv, \quad \mathcal{H}_1^1 = -v,$$

$$\mathcal{H}_2^1 = \frac{1}{9}\left(-u_x^2 + 3u_x v + \frac{1}{3}u^3 - 3v^2\right),$$

$$\mathcal{H}_3^2 = \frac{1}{9}\left(-\frac{1}{3}u_{xx}^2 + u_{xx}v_x + uu_x^2 - v_x^2 + u^2 v_x - \frac{1}{9}u^4 + 2uv^2\right),$$

$$\mathcal{H}_3^1 = \frac{1}{27}\left(u_{xx}v_{xx} - 5u_x^2 v - 5uu_x v_x - 15uvv_x + \frac{5}{3}u^3 v - 5v^3\right).$$

When $m = i = 1$, the t_2–flow is a system which can be written in scalar form as the Boussinesq equation:

$$u_{tt} = -\frac{1}{3}\left(u_{xx} + 2u^2\right)_{xx}.$$

In this paper we only consider the t_2, t_4 and t_5 flows. The latter has two reductions to single component equations: the Sawada-Kotera and Kaup-Kupershmidt equations. We consider the $x-t$ reversed flows corresponding to each of these.

It is evident from the simplicity of the first Hamiltonian structure $\mathbf{B}_0$ that the stationary flows take Lagrangian form, with

$$\mathcal{L}_m^i = \mathcal{H}_{m+1}^i - \frac{1}{3}uc_1 - \frac{1}{3}vc_2,$$

for two constants c_i. Sometimes this Lagrangian is degenerate, but we can eliminate some variables to obtain a related, non-degenerate Lagrangian system.

The Boussinesq hierarchy can be written in zero curvature form

$$U_{t_m} = V_{(m)x} - [U, V_{(m)}],$$

where

$$U = \begin{pmatrix} 0 & 1 & 0 \\ 0 & 0 & 1 \\ \lambda - v & -u & 0 \end{pmatrix}$$

and $V_{(m)}$, while more complicated, can be systematically derived for each m (see [15] and the examples below).

3.1. The t_2-Flow

The t_2-flow is generated by $\mathcal{H}_2^1$ when using the Hamiltonian structure $\mathbf{B}_0$, giving the degenerate Lagrangian:

$$\mathcal{L}_1^1 = \mathcal{H}_2^1 - \frac{1}{3}uc_1 - \frac{1}{3}vc_2$$

for the stationary flow, where: $c_1 = \frac{2}{3}u_{xx} - v_x + \frac{1}{3}u^2$, $c_2 = u_x - 2v$. We can eliminate v to obtain the non-degenerate Lagrangian and canonical coordinates:

$$\tilde{\mathcal{L}}_1^1 = -\frac{1}{36}u_x^2 - \frac{1}{6}u_xc_2 + \frac{1}{27}u^3 - \frac{1}{3}uc_1, \quad q = u, \quad p = -\frac{1}{18}u_x - \frac{1}{6}c_2.$$

With these coordinates, the time evolution $V_{(2)}$ gives the spectral problem:

$$\Psi_\xi = \mathcal{U}\Psi, \quad \text{where} \quad \mathcal{U} = \begin{pmatrix} a & b & c \\ d & e & f \\ g & h & -a-e \end{pmatrix}, \tag{30}$$

where:

$$a = -2e = -\frac{2}{3}q, \ b = f = c - 1 = 0, \ d = 3p - \lambda,$$

$$g = \frac{1}{3}q^2 - c_1, \ h = -3p - c_2 - \lambda.$$

In this case we reduce equation (39) from 8 to 4 components. The Hamiltonian structure $\bar{\mathbf{B}}_1$ reduces in a straightforward way to $\mathbf{B}_1 - \lambda\mathbf{B}_0$, where

$$\mathbf{B}_0 = \begin{pmatrix} 0 & 1 & 0 & 0 \\ -1 & 0 & 0 & 0 \\ 0 & 0 & 0 & 3\partial \\ 0 & 0 & 3\partial & 0 \end{pmatrix}, \quad \mathbf{B}_1 = \begin{pmatrix} X & Y \\ -Y^\dagger & Z \end{pmatrix},$$

with

$$X = \begin{pmatrix} -6\partial & 3p \\ -3p & 0 \end{pmatrix}, \quad Y = \begin{pmatrix} 3\partial^2 - \partial q & -18p - 3c_2 \\ p\partial - \partial p & \frac{1}{3}(\partial^2 - q\partial - c_1 + \frac{1}{3}q^2) \end{pmatrix},$$

$$Z = \begin{pmatrix} 2\partial^3 - c_1\partial - \partial c_1 & -9\partial p - 2\partial c_2 - c_2\partial \\ -9p\partial - \partial c_2 - 2c_2\partial & q\partial + \partial q \end{pmatrix},$$

where $\partial = \partial_\xi$. The operators $\mathbf{B}_0$ and $\mathbf{B}_1$ form a Hamiltonian pair (most easily seen by using Miura map techniques) and the isospectral flows of (30) are bi-Hamiltonian, forming 2 ladders:

$$\mathbf{q}_{\tau_{2r-j+3}} = \mathbf{B}_1 \, \delta_q h_r^j = \mathbf{B}_0 \, \delta_q h_{r+1}^j, \quad j = 1, 2,$$

where $\mathbf{q} = (q, p, c_1, c_2)^T$ and the first few Hamiltonian are: $h_{-1}^1 = -c_1$, $h_{-1}^2 = c_2$ and

$$h_0^1 = \frac{1}{3}c_1 c_2 + q_\xi p, \quad h_0^2 = -\frac{1}{27}q^3 + \frac{1}{3}qc_1 - 9p^2 - 3pc_2 - \frac{1}{3}c_2^2.$$

The τ_{-1} flow is just the $x - t$ reversed Boussinesq equation:

$$\mathbf{q}_{\tau_{-1}} = \begin{pmatrix} -18p - 3c_2 \\ \frac{1}{9}q^2 - \frac{1}{3}c_1 \\ -9p_\xi - c_{2\xi} \\ q_\xi \end{pmatrix}, \tag{31}$$

while the τ_0 flow is just the ξ-translation: $\mathbf{q}_{\tau_0} = \mathbf{q}_\xi$. The first component of the τ_1 flow is:

$$q_{\tau_1} = 12pc_2 + 2c_2^2 + q_{\xi\xi} + \frac{2}{3}qq_\xi - 2c_{1\xi}. \tag{32}$$

With $\tau_{-1} = x$ and substituting ξ-derivatives from (31), equation (32) becomes the u-component of the fourth order element of the usual Boussinesq hierarchy.

3.2. The t_4 Flow

The t_4-flow is generated by $\mathcal{H}_3^2$ when using the Hamiltonian structure $\mathbf{B}_0$, giving the degenerate Lagrangian:

$$\mathcal{L}_3^2 = \mathcal{H}_3^2 - \frac{1}{3}uc_1 - \frac{1}{3}vc_2.$$

The degeneracy of this follows from the non-occurrence of v_{xx} in $\mathcal{L}_3^2$. However, if we perform a generalised Legendre transformation (see Appendix

A) with respect to u, but only a standard Legendre transformation with respect to v, we obtain:

$$q_1 = u, \quad q_2 = u_x, \quad p_2 = -\tfrac{2}{27}u_{xx} + \tfrac{1}{9}v_x, \quad p_1 = \tfrac{2}{27}u_{xxx} - \tfrac{1}{9}v_{xx} + \tfrac{2}{9}uu_x,$$
$$q_3 = v, \quad p_3 = \tfrac{1}{9}u_{xx} - \tfrac{2}{9}v_x + \tfrac{1}{9}u^2.$$

$$(33)$$

Remark 3.1. This apparent difficulty only arises because we chose to use the Legendre transformation. A less 'tricky' approach would be to use coordinates:

$$r_1 = u, \quad r_2 = u_x, \quad r_3 = u_{xx}, \quad r_4 = v, \quad r_5 = v_x, \quad r_6 = v_{xx},$$

together with the previously defined c_1, c_2, which would give rise to a slightly more complicated Hamiltonian pair. Reducing the first of these to 'canonical form' recovers our coordinates (q_i, p_i).

Using coordinates (q_i, p_i, c_1, c_2), the time evolution matrix $V_{(2)}$ leads to the spectral problem:

$$\Psi_\xi = \mathcal{U}\Psi, \quad \text{where} \quad \mathcal{U} = \begin{pmatrix} a & b & c \\ d & e & f \\ g & h & -a-e \end{pmatrix}, \quad (34)$$

with

$$a = \frac{2}{9}q_1^2 - 3p_2, \quad b = -\frac{1}{3}q_2 + \frac{1}{3}q_3 + \lambda, \quad c = \frac{1}{3}q_1,$$

$$d = -\frac{2}{9}q_1 q_2 - \frac{1}{3}q_1 q_3 + 3p_1 + \frac{1}{3}\lambda q_1,$$

$$e = -\frac{4}{9}q_1^2 + 6p_2 + 3p_3, \quad f = \frac{1}{3}q_3 + \lambda,$$

$$g = -\frac{4}{27}q_1^3 + 3q_1 p_2 + \frac{1}{9}q_2^2 - \frac{1}{3}q_2 q_3 + \frac{1}{3}q_3^2 - c_1 + \frac{1}{3}\lambda q_2 - \frac{2}{3}\lambda q_3 + \lambda^2,$$

$$h = \frac{2}{9}q_1 q_2 + \frac{2}{3}q_1 q_3 - 3p_1 - c_2 - \frac{2}{3}\lambda q_1.$$

In this case the map $\varphi : \mathbf{q} \mapsto \mathbf{u}$ is invertible (compare with the KdV_3 case), so we can invert the Jacobian matrix $\mathcal{J}_\varphi$ to directly construct the Hamiltonian pair in these coordinates:

$$\mathcal{J}_\varphi^{-1} \bar{\mathbf{B}} \left(\mathcal{J}_\varphi^{-1} \right)^\dagger = \mathbf{B}_1 - \lambda \mathbf{B}_0,$$

where

$$\mathbf{B}_0 = \begin{pmatrix} 0 & I_3 & 0 & 0 \\ -I_3 & 0 & 0 & 0 \\ 0 & 0 & 0 & 3\partial \\ 0 & 0 & 3\partial & 0 \end{pmatrix}, \quad \mathbf{B}_1 = \begin{pmatrix} X & Y \\ -Y^\dagger & Z \end{pmatrix},$$

with I_3 denoting the 3×3 identity matrix, 0 representing appropriately sized blocks of zeros, and:

$$X = \begin{pmatrix} 0 & 0 & 0 & -\frac{1}{3}q_3 \\ 0 & 0 & -3q_1 & \Omega_{10} \\ 0 & 3q_1 & 0 & \frac{2}{9}q_1^2 \\ \frac{1}{3}q_3 & -\Omega_{10}^\dagger & -\frac{2}{9}q_1^2 & \Omega_{11} \end{pmatrix},$$

$$Y = \begin{pmatrix} \frac{1}{3}q_1 & -\frac{2}{3}q_1 & \Omega_4 & q_2 \\ \frac{1}{3}q_2 - \frac{1}{3}q_3 & -\frac{1}{3}q_2 & \Gamma_4 & \Gamma_1 \\ 0 & -\frac{1}{3}q_3 & \Gamma_5 & \Omega_1 \\ \Gamma_7 & \Gamma_6 & \Omega_5 & \Omega_2 \end{pmatrix},$$

$$Z = \begin{pmatrix} -\frac{2}{27}\partial & \Omega_9 & \Omega_6 & \Gamma_2 \\ -\Omega_9^\dagger & -\frac{2}{9}\partial & \Omega_7 & \Gamma_3 \\ -\Omega_6^\dagger & -\Omega_7^\dagger & \Omega_8 & \Omega_3 \\ -\Gamma_2 & -\Gamma_3 & -\Omega_3^\dagger & \partial q_1 + q_1\partial \end{pmatrix},$$

where:

$$\Omega_1 = 2q_1^2 - 27p_2 - 18p_3 + 3\partial,$$

$$\Omega_2 = \frac{32}{81}q_1^3 - 6q_1p_2 - 4q_1p_3 + \frac{1}{9}q_2^2 + \frac{2}{9}q_3^2 - \frac{1}{3}c_1 + \frac{1}{3}q_1\partial,$$

$$\Omega_3 = -q_2\partial + \partial q_3 + 2q_3\partial, \quad \Omega_4 = q_1^2 - 9p_3 + 3\partial,$$

$$\Omega_5 = \frac{8}{27}\left(2q_1^2q_2 - q_1^2q_3\right) - 2q_1p_1 + \frac{2}{9}q_1c_2 - 2q_2p_2 - \frac{4}{3}q_2p_3 + 4q_3p_2$$
$$+ \frac{2}{9}\left(\partial q_2 + q_2\partial\right) + \frac{1}{3}\left(q_3\partial - \partial q_3\right),$$

$$\Omega_6 = \frac{16}{81}q_1^3 - 2q_1p_2$$
$$- \frac{4}{3}q_1p_3 - \frac{1}{9}q_2^2 + \frac{8}{27}q_2q_3 - \frac{2}{9}q_3^2 + \frac{1}{3}c_1 + \frac{2}{9}\left(2q_1\partial - \partial q_1\right),$$

$$\Omega_7 = -\frac{8}{81}q_1^3 + 4q_1p_2 + \frac{4}{27}q_2^2 - \frac{4}{9}q_2q_3 + \frac{4}{9}q_3^2 - \frac{2}{3}c_1 + \frac{1}{3}\partial q_1,$$

$$\Omega_8 = \partial(-\frac{1}{3}q_1^2 + 9p_2) + (-\frac{1}{3}q_1^2 + 9p_2)\partial, \quad \Omega_9 = -\frac{8}{81}q_1^2 + \frac{1}{9}\partial,$$

$$\Omega_{10} = -q_1^2 + 9p_2 + 3p_3 + \partial, \quad \Omega_{11} = \frac{2}{27}(\partial q_1 + q_1\partial),$$

$$\Gamma_1 = 3q_1^2 - 54p_2 - 27p_3, \quad \Gamma_2 = \frac{2}{9}q_1q_2 - p_1,$$

$$\Gamma_3 = \frac{4}{9}q_1q_3 - \frac{1}{3}c_2, \quad \Gamma_4 = 2q_1q_2 - 4q_1q_3 + 3c_2,$$

$$\Gamma_5 = \frac{4}{3}q_1q_2 - 9p_1, \quad \Gamma_6 = -\frac{8}{81}q_1q_2 - \frac{4}{27}q_1q_3 + \frac{1}{3}p_1,$$

$$\Gamma_7 = \frac{2}{27}q_1q_2 + \frac{8}{81}q_1q_3 - \frac{1}{3}p_1.$$

As before, we obtain two bi-Hamiltonian ladders:

$$\mathbf{q}_{\tau_{2r-j+3}} = \mathbf{B}_1 \delta_q h_r^j = \mathbf{B}_0 \delta_q h_{r+1}^j, \quad j = 1, 2,$$

where $\mathbf{q} = (q_i, p_i, c_1, c_2)^T$. The sequence of Hamiltonians can be generated directly by Miura map techniques or by calculating the fluxes of the standard Boussinesq hierarchies. The first few are: $h_{-2}^1 = c_2$, $h_{-1}^2 = -c_1$ and

$$h_{-1}^1 = -9(3p_2^2 + 3p_2 p_3 + p_3^2) + q_2 p_1 + q_1^2(3p_2 + 2p_3)$$
$$- \frac{1}{9}(q_1 q_2^2 + 2q_1 q_3^2 + \frac{8}{9}q_1^4) + \frac{1}{3}(q_1 c_1 + q_3 c_2).$$

The τ_{-2} flow:

$$\mathbf{q}_{\tau_{-2}} = \begin{pmatrix} q_2 \\ 3q_1^2 - 54p_2 - 27p_3 \\ 2q_1^2 - 27p_2 - 18p_3 \\ \frac{32}{81}q_1^3 - 6q_1 p_2 - 4q_1 p_3 + \frac{1}{9}q_2^2 + \frac{2}{9}q_3^2 - \frac{1}{3}c_1 \\ \frac{2}{9}q_1 q_2 - p_1 \\ \frac{4}{9}q_1 q_3 - \frac{1}{3}c_2 \\ q_3 \xi \\ q_1 \xi \end{pmatrix},$$

is just the $x - t$ reversed Boussinesq t_4-flow. The (q_1, q_3) $(= (u, v))$ components of the τ_{-1} flow are

$$q_{1\tau_{-1}} = -q_1^2 + 9p_3, \quad q_{3\tau_{-1}} = -\frac{4}{3}q_1 q_2 + 9p_1,$$

which, after substituting from (33), is just the Boussinesq system.

Remark 3.3. Starting with the spectral problem (34) the τ_{-2} flow can be considered to *define* the coordinates (q_i, p_i, c_i) in terms of u, v and their $x (= \tau_{-2})$ derivatives. Without this flow, it would not be apparent that we are generating a "disguised" form of the Boussinesq hierarchy.

The τ_{-1} flow gives the Boussinesq t_2 flow for q_1 and q_3 (after using the τ_{-2} flow to replace q_2, p_i, etc). The τ_0 flow just defines τ_0 as $\xi (= t_4)$. The τ_1 flow has not been presented, but gives the t_5 flow for q_1 and q_3.

3.3. The t_5-Flow

The Lagrangian $\mathcal{L}_3^1 = \mathcal{H}_3^1 - \frac{1}{3}uc_1 - \frac{1}{3}vc_2$ is non-degenerate so can be used in the standard way to generate canonical coordinates (q_i, p_i, c_1, c_2), $i = 1, \cdots, 4$, and the corresponding Hamiltonian. We thus have 10 coordinates, whilst the zero-curvature conditions only generate 8 equations. We must proceed in the way described above for the KdV$_5$ equation. For this example the vectors $\mathbf{f}_i$ are non-trivial. The details, leading to a pair of 10×10 matrix Hamiltonian structures, can be found in [15]. Here we just consider the Sawada-Kotera and Kaup-Kupershmidt reductions.

The Sawada-Kotera reduction. The 2 component Boussinesq hierarchy is reduced to that of the single component Sawada-Kotera (SK) by setting $v = 0$. We are thus considering the isospectral flows of: $L_{sk} = \partial^3 + u\partial$. Only the odd-order members of the full Boussinesq hierarchy survive this reduction (the recursion operator has to be squared [11, 12]). The first Hamiltonian structure is no longer applicable, while the second reduces to $\mathbf{B}^{sk}_{u1} = \frac{1}{2}\partial^3 + 2u\partial + u_x$. The Hamiltonians $\mathcal{H}^2_1$, $\mathcal{H}^1_2$ and $\mathcal{H}^2_3$, reduce respectively to

$$
\mathcal{H}^{sk}_0 = u, \quad \mathcal{H}^{sk}_1 = \frac{1}{27}(u^3 - 3u_x^2), \quad \mathcal{H}^{sk}_2 = -\frac{1}{27}(u_{xx}^2 - 3uu_x^2 + \frac{1}{3}u^4),
$$

and $\mathcal{H}^{sk}_1$ generates the SK equation:

$$
u_{t_1} = \frac{1}{9}(u_{xxxxx} + 5uu_{xxx} + 5u_xu_{xx} + 5u^2u_x).
$$

Corresponding to $v = 0$, we set 5 of the previous (but not given) 10 coordinates to zero ($q_3 = q_4 = p_1 = p_2 = c_1 = 0$), while $\mathbf{q}^{sk} = (q_1, q_2, p_3, p_4, c_2)^T$ with p_3 and p_4 respectively conjugate to q_1 and q_2. We re-label: $\mathbf{q}^{sk} = (q_1, q_2, p_1, p_2, c_1)^T$. The previous formulae [15] reduce to:

$$
q_1 = u, \quad q_2 = u_x, \quad p_1 = -\frac{1}{27}(u_{xxx} + 5uu_x), \quad p_2 = \frac{1}{27}u_{xx},
$$

$$
c_1 = \frac{1}{9}(u_{xxxx} + 5uu_{xx} + \frac{5}{3}u^3).
$$

With these coordinates, the second 10×10 matrix Hamiltonian structure of the full Boussinesq hierarchy reduces to:

$$
\mathbf{B}^{sk}_1 = \begin{pmatrix}
0 & -\frac{9}{2} & \frac{2}{3}q_1 & 0 & q_2 \\
\frac{9}{2} & 0 & -\frac{2}{3}q_2 & -\frac{1}{6}q_1 & 27p_2 \\
-\frac{2}{3}q_1 & \frac{2}{3}q_2 & -\frac{1}{18}\partial & \Gamma_2 & \Gamma_1 \\
0 & \frac{1}{6}q_1 & -\Gamma_2 & 0 & \Omega_1 \\
-q_2 & -27p_2 & -\Gamma_1 & -\Omega_1^\dagger & \partial q_1 + q_1\partial
\end{pmatrix},
$$

where

$$
\Gamma_1 = \frac{5}{81}q_1^3 - \frac{5}{27}q_2^2 - \frac{1}{3}c_1, \quad \Gamma_2 = \frac{5}{162}q_1^2 + \frac{1}{3}p_2, \quad \Omega_1 = -\frac{5}{27}q_1q_2 - p_1 + \frac{1}{6}\partial.
$$

The equations take the form: $\mathbf{q}_{\tau_r}^{sk} = \mathbf{B}_1^{sk}\delta h_r$, with the first few h_r being given by $h_{-1} = c_1$ and:

$$
\begin{aligned}
h_0 &= -\frac{2}{243}q_1^5 - \frac{10}{9}q_1^3 p_2 - \frac{5}{27}q_1^2 q_2^2 + \frac{1}{9}q_1^2 c_1 - \frac{8}{3}q_1 q_2 p_1 \\
&\quad -36q_1 p_2^2 + \frac{2}{3}q_2^2 p_2 - 9p_1^2 + 6p_2 c_1 - \frac{2}{9}q_{1\xi}q_2, \\
h_1 &= -\frac{5}{729}q_1^6 - \frac{20}{27}q_1^4 p_2 - \frac{20}{243}q_1^3 q_2^2 + \frac{2}{27}q_1^3 c_1 - \frac{14}{9}q_1^2 q_2 p_1 \\
&\quad -21q_1^2 p_2^2 + \frac{2}{9}q_1 q_2^2 p_2 - 6q_1 p_1^2 + 4q_1 p_2 c_1 - \frac{1}{81}q_2^4 - \frac{1}{9}q_2^2 c_1 \\
&\quad -6q_2 p_1 p_2 - 54p_2^3 - \frac{1}{3}c_1^2 - \frac{4}{27}q_1 q_2 q_{1\xi} - 2q_{1\xi}p_1 - 2q_{2\xi}p_2,
\end{aligned}
$$

and flows:

$$
\mathbf{q}_{\tau_{-1}}^{sk} = \left(q_2, 27p_2, \frac{5}{81}q_1^3 - \frac{5}{27}q_2^2 - \frac{1}{3}c_1, -\frac{5}{27}q_1 q_2 - p_1, q_{1\xi}\right)^T, \qquad \mathbf{q}_{\tau_0}^{sk} = \mathbf{q}_\xi^{sk},
$$

with $q_{\tau_1}^{sk}$ giving the SK$_7$ equation for q_1.

The Kaup-Kupershmidt reduction. The 2 component Boussinesq hierarchy is reduced to that of the single component Kaup-Kupershmidt (KK) by setting $v = \frac{1}{2}u_x$. We are thus considering the isospectral flows of (putting $u = 2w$): $L_{kk} = \partial^3 + 2w\partial + w_x$. Once again, only the odd-order members of the Boussinesq hierarchy survive this reduction. The first Hamiltonian operator is, again, no longer applicable while the second reduces to: $\mathbf{B}_{w1}^{kk} = \frac{1}{2}\partial^3 + w\partial + \frac{1}{2}w_x$. The Hamiltonians $\mathcal{H}_1^2, \mathcal{H}_2^1$ and $\mathcal{H}_3^2$ reduces respectively to:

$$
\mathcal{H}_0 = 2w, \quad \mathcal{H}_1 = \frac{1}{27}(8w^3 - 3w_x^2), \quad \mathcal{H}_2 = -\frac{1}{27}\left(w_{xx}^2 - 12ww_x^2 + \frac{16}{3}w^4\right),
$$

with $\mathcal{H}_1$ generating the KK equation:

$$
w_{t_1} = \frac{1}{9}(w_{xxxxx} + 10ww_{xxx} + 25w_x w_{xx} + 20w^2 w_x).
$$

Once again, the stationary manifold reduces from 10 to 5 dimensions and, after a relabelling, we have:

$$
\begin{aligned}
q_1 &= w, \quad q_2 = w_x, \quad p_1 = -\frac{1}{27}(w_{xxx} + 25ww_x), \\
p_2 &= \frac{1}{27}w_{xx}, \quad c_1 = \frac{1}{9}\left(w_{xxxx} + 10ww_{xx} + \frac{15}{2}w_x^2 + \frac{20}{3}w^3\right).
\end{aligned}
$$

The second Hamiltonian structure (a 10×10 matrix) then reduces to a 5×5 matrix on our stationary KK manifold:

$$
\mathbf{B}_1^{kk} = \begin{pmatrix}
0 & -\frac{9}{2} & \frac{17}{6}q_1 & 0 & \frac{1}{2}q_2 \\
\frac{9}{2} & 0 & -\frac{17}{6}q_2 & -\frac{4}{3}q_1 & \frac{27}{2}p_2 \\
-\frac{17}{6}q_1 & \frac{17}{6}q_2 & -\frac{1}{18}\partial & \Gamma_2 & \Gamma_1 \\
0 & \frac{4}{3}q_1 & -\Gamma_2 & 0 & \Omega_1 \\
-\frac{1}{2}q_2 & -\frac{27}{2}p_2 & -\Gamma_1 & -\Omega_1^\dagger & \frac{1}{2}(\partial q_1 + q_1 \partial)
\end{pmatrix},
$$

where

$$
\Gamma_1 = \frac{10}{81}q_1^3 - \frac{15}{2}q_1 p_2 - \frac{35}{108}q_2^2 - \frac{1}{6}c_1, \quad \Gamma_2 = \frac{70}{81}q_1^2 + \frac{1}{6}p_2,
$$
$$
\Omega_1 = -\frac{25}{54}q_1 q_2 - \frac{1}{2}p_1 + \frac{1}{6}\partial.
$$

The equations take the form: $\mathbf{q}_{\tau_r} = \mathbf{B}_1^{kk}\delta h_r$, with the first h_r being given by $h_{-1} = 2c_1$ and

$$
\begin{aligned}
h_0 = \ & -\frac{64}{243}q_1^5 - \frac{40}{9}q_1^3 p_2 - \frac{95}{27}q_1^2 q_2^2 + \frac{8}{9}q_1^2 c_1 - \frac{34}{3}q_1 q_2 p_1 \\
& -18 q_1 p_2^2 + \frac{1}{3}q_2^2 p_2 - 9 p_1^2 + 6 p_2 c_1 - \frac{2}{9}q_{1\xi}q_2,
\end{aligned}
$$

corresponding to flows

$$
\mathbf{q}_{\tau_{-1}} = \begin{pmatrix}
q_2 \\
27 p_2 \\
\frac{20}{81}q_1^3 - 15 q_1 p_2 - \frac{35}{54}q_2^2 - \frac{1}{3}c_1 \\
-\frac{25}{27}q_1 q_2 - p_1 \\
q_{1\xi}
\end{pmatrix}, \quad \mathbf{q}_{\tau_0} = \mathbf{q}_\xi.
$$

4. The Hénon-Heiles System

The Hénon-Heiles Hamiltonian

$$
H = \frac{1}{2}(p_1^2 + p_2^2 + c_1 q_1^2 + c_2 q_2^2) + a q_1 q_2^2 - \frac{1}{3}b q_1^3
$$

is known to be integrable for the following parameter values:

(i) $a/b = -1$, $\quad c_1 = c_2$,

(ii) $a/b = -\frac{1}{6}$, $\quad c_1, c_2$ arbitrary,

(iii) $a/b = -\frac{1}{16}$, $\quad c_2 = 16 c_1$.

These values were originally determined by the Painlevé method and by a direct search for a second integral. However, *our* starting point is the connection noted in [10] between these integrable cases of the Hénon-Heiles system and the stationary flows of 3 known 5th order nonlinear evolution equations. Specifically, cases (i)-(iii) above are related respectively to the Sawada-Kotera, 5th order KdV and Kaup-Kupershmidt equations. For this paper we take $c_1 = c_2 = 0$, since the formulae are simplified.

Each of the above mentioned 5th order nonlinear evolution equations can be represented in Hamiltonian form: $u_t = \mathbf{B}_u \delta \mathcal{H}$, where $\mathbf{B}_u$ and $\mathcal{H}$ take the form:

$$\mathbf{B}_u = \frac{1}{2}\partial^3 + 4au\partial + 2au_x, \quad \mathcal{H} = -u_x^2 - \frac{2}{3}bu^3.$$

The 3 cases correspond respectively to: $a = -b = \frac{1}{2}$, $6a = -b = 3$ and $16a = -b = 4$. The stationary solutions satisfy:

$$\frac{1}{2}\Delta_{xxx} + 4au\Delta_x + 2au_x\Delta = 0, \tag{35}$$

where $\Delta = \delta \mathcal{H} = 2u_{xx} - 2bu^2$. Multiplying (35) by Δ and integrating, we obtain

$$\frac{1}{2}\Delta\Delta_{xx} - \frac{1}{4}\Delta_x^2 + 2au\Delta^2 = 4a^2\alpha,$$

where α is a constant of integration. Defining

$$Q_1 = u, \quad Q_2 = (-\frac{1}{2}a^{-1}\Delta)^{1/2},$$

leads to a system which can be written in canonical Hamiltonian form, with:

$$H_\alpha = \frac{1}{2}(P_1^2 + P_2^2) + aQ_1Q_2^2 - \frac{1}{3}bQ_1^3 + \frac{1}{2}\alpha Q_2^{-2},$$

which reduces to H (with $c_i = 0$) when $\alpha = 0$. Choosing a and b to match the 3 integrable cases of the 5th order nonlinear evolution equation, gives rise to the 3 cases of integrable Hénon-Heiles sytem, listed above.

Remark 4.1. A different approach is adopted in [5, 4], with applications to the Hénon-Heiles system and to quartic and higher degree potentials. In particular, the Hirota-Satsuma coupled KdV equation is shown to be related to one of the known integrable quartic potentials.

Comparing the definitions of the 2 coordinate systems (q_i, p_i, c) and (Q_i, P_i, K_1) we can transform the Poisson tensors for the *PDE hierarchies*, which reduce to the integrable Hénon-Heiles systems as stationary flows.

The Sawada-Kotera equation.   Writing the $x - t$ reversed Sawada-Kotera equation in Hénon-Heiles coordinates, leads to the Hamiltonian hierarchy:

$$\mathbf{Q}_{\tau_r} = \mathbf{B}_1^{sk} \delta h_r, \qquad (36)$$

where $\mathbf{Q} = (Q_1, Q_2, P_1, P_2, K_1)^T$ and

$$\mathbf{B}_1^{sk} = \begin{pmatrix} 0 & 0 & 1 & 0 & 0 \\ 0 & 0 & 0 & 1 & \partial Q_2 - \frac{1}{2} Q_2 \partial \\ -1 & 0 & 0 & 0 & 0 \\ 0 & -1 & 0 & -2Q_2^{-1}\partial Q_2^{-1} & \partial P_2 - \frac{1}{2} P_2 \partial \\ 0 & -\frac{1}{2}\partial Q_2 + Q_2\partial & 0 & -\frac{1}{2}\partial P_2 + P_2\partial & \partial K_1 + K_1 \partial \end{pmatrix},$$

with $\partial \equiv \partial_\xi$, ξ being the original Sawada-Kotera time. The first two Hamiltonians are:

$$h_{-1} = -\frac{9}{2}\left(\frac{1}{2}(P_1^2 + P_2^2) + \frac{2}{81}(Q_1 Q_2^2 + \frac{1}{3}Q_1^3) + 2K_1 Q_2^{-2}\right),$$
$$h_0 = K_1 + Q_{1\xi} P_1.$$

The Hamiltonian h_{-1} generates the $x - t$ reversed Sawada-Kotera equation:

$$\mathbf{Q}_{\tau_{-1}} = \left(-\frac{9}{2}P_1, -\frac{9}{2}P_2, \frac{1}{9}Q_1^2 + \frac{1}{9}Q_2^2, \frac{2}{9}Q_1 Q_2 - 18K_1 Q_2^{-3}, -\frac{1}{9}Q_2^2 Q_{1\xi}\right)^T.$$

In these coordinates the equation is rational in $Q_{1xx} + \frac{1}{2}Q_1^2$, where $x = \tau_{-1}$. However, the stationary flow, with $Q_{1\xi} = 0$, is just case (i) of the Hénon-Heiles system.

The 5th order KdV equation.   Writing the $x - t$ reversed 5th order KdV equation in Hénon-Heiles coordinates leads to the bi-Hamiltonian hierarchy:

$$\mathbf{Q}_{\tau_r} = \mathbf{B}_1^{kdv}\delta h_r = \mathbf{B}_0^{kdv}\delta h_{r+1},$$

where $\mathbf{Q} = (Q_1, Q_2, P_1, P_2, K_1)^T$ and

$$\mathbf{B}_0^{kdv} = \begin{pmatrix} 0 & 0 & 0 & -Q_2^{-1} & \frac{1}{2}Q_2 P_2 \\ 0 & 0 & -Q_2^{-1} & 2Q_1 Q_2^{-2} & -Q_1 P_2 + \frac{1}{2}Q_2 P_1 \\ 0 & Q_2^{-1} & 0 & -P_2 Q_2^{-2} & \Gamma_1 \\ Q_2^{-1} & -2Q_1 Q_2^{-2} & P_2 Q_2^{-2} & 0 & \Gamma_2 \\ -\frac{1}{2}Q_2 P_2 & Q_1 P_2 - \frac{1}{2}Q_2 P_1 & -\Gamma_1 & -\Gamma_2 & \frac{1}{2}Q_2^2 \partial Q_2^2 \end{pmatrix},$$

$$\mathbf{B}_1^{kdv} = \begin{pmatrix} 0 & 0 & 1 & 0 & 0 \\ 0 & 0 & 0 & 1 & \partial Q_2 - \frac{1}{2}Q_2\partial \\ -1 & 0 & 0 & 0 & 0 \\ 0 & -1 & 0 & -2Q_2^{-1}\partial Q_2^{-1} & \partial P_2 - \frac{1}{2}P_2\partial \\ 0 & -\frac{1}{2}\partial Q_2 + Q_2\partial & 0 & -\frac{1}{2}\partial P_2 + P_2\partial & \partial K_1 + K_1\partial \end{pmatrix},$$

with $\partial \equiv \partial_\xi$, ξ being the original KdV_5 time and

$$\Gamma_1 = -2Q_1 Q_2^2 + \frac{1}{2}P_2^2 + 2K_1 Q_2^{-2},$$

$$\Gamma_2 = -2Q_1^2 Q_2 - Q_2^3 - \frac{1}{2}P_1 P_2 - 4Q_1 K_1 Q_2^{-3}.$$

The first 3 Hamiltonians are

$$
\begin{aligned}
h_{-2} &= -2Q_1^3 - Q_1 Q_2^2 - \frac{1}{4}P_1^2 - \frac{1}{4}P_2^2 - K_1 Q_2^{-2} \\
h_{-1} &= Q_1^2 Q_2^2 - \frac{1}{2}Q_1 P_2^2 + \frac{1}{4}Q_2^4 + \frac{1}{2}Q_2 P_1 P_2 - 2Q_1 K_1 Q_2^{-2},
\end{aligned}
$$

and $h_0 = K_1 + Q_{1\xi}P_1$. The τ_{-2} flow is the $x - t$ reversed KdV$_5$ equation:

$$\mathbf{Q}_{\tau_{-2}} = \left(-\frac{1}{2}P_1, -\frac{1}{2}P_2, 6Q_1^2 + Q_2^2, 2Q_1 Q_2 - 2K_1 Q_2^{-3}, -Q_2^2 Q_{1\xi}\right)^T.$$

This reduces to the Hénon-Heiles equation, case (ii), when $\mathbf{Q}_\xi = 0$. The τ_{-1} and τ_1 flows are respectively the KdV and KdV$_7$ equations in these coordinates.

Remark 4.2. In the original Bogoyavlenskii-Novikov coordinates, $\mathbf{B}_0^{kdv}$ took simple canonical form, while $\mathbf{B}_1^{kdv}$ was more complicated. In the Hénon-Heiles coordinates it is $\mathbf{B}_1^{kdv}$ which reduces to (degenerate) canonical form for the stationary flows. There are, in fact, 2 classes of special coordinates, which respectively reduce one or other of the finite dimensional (stationary case) Poisson brackets to canonical form.

The Kaup-Kupershmidt equation. Here, a similar analysis leads to this hiearchy taking Hamiltonian form (36) with

$$
\mathbf{B}_1^{kk} = \begin{pmatrix}
0 & 0 & 1 & 0 & 0 \\
0 & 0 & 0 & 1 & \partial Q_2 - \frac{1}{2}Q_2 \partial \\
-1 & 0 & 0 & 0 & 0 \\
0 & -1 & 0 & -8Q_2^{-1}\partial Q_2^{-1} & \partial P_2 - \frac{1}{2}P_2 \partial \\
0 & -\frac{1}{2}\partial Q_2 + Q_2 \partial & 0 & -\frac{1}{2}\partial P_2 + P_2 \partial & \partial K_1 + K_1 \partial
\end{pmatrix}.
$$

The first two Hamiltonians are:

$$
\begin{aligned}
h_{-1} &= -\frac{9}{2}\left(\frac{1}{2}(P_1^2 + P_2^2) + \frac{1}{81}Q_1 Q_2^2 + \frac{16}{243}Q_1^3 + 8K_1 Q_2^{-2}\right), \\
h_0 &= K_1 + Q_{1\xi}P_1.
\end{aligned}
$$

The $x - t$ reversed Kaup-Kupershmidt equation is generated by h_{-1}:

$$\mathbf{Q}_{\tau_{-1}} = \left(-\frac{9}{2}P_1, -\frac{9}{2}P_2, \frac{8}{9}Q_1^2 + \frac{1}{18}Q_2^2, \frac{1}{9}Q_1 Q_2 - 72K_1 Q_2^{-3}, -\frac{1}{18}Q_2^2 Q_{1\xi}\right)^T.$$

This reduces to the Hénon-Heiles equation, case (iii) when $\mathbf{Q}_\xi = 0$.

5. Conclusions

In this paper we have been interested in the relationship between an integrable nonlinear evolution equation (PDE) and its stationary flow. We were particularly interested in the reduction of the infinite dimensional Hamiltonian structures to their finite dimensional counterparts. This reduction is most transparent when the PDE is written as a flow in a larger space whose coordinates are those of the stationary manifold, together with their t-derivatives.

Starting from a zero-curvature representation (reversing the roles of U and V) we gave a systematic construction of the isospectral flows and their Hamiltonian structures. We have adopted the approach presented in [2], which *simultaneously* constructs the isospectral flows, time evolutions of the wave functions, the Hamiltonians and Hamiltonian structures. This close relationship between the zero curvature representation and the Hamiltonian formulation of the equations is perhaps best seen by the formula (14).

While the importance of our results is mainly in the realm of the stationary reductions, we have, in passing, answered a number of the questions raised in [13], where they believed they had found a *new* hierarchy of equations. We have seen that, in fact, this hierarchy is just the KdV hierarchy in disguise.

Appendix A: Generalised Lagrangians

and Legendre's Transformation

In this paper we consider many examples of (generalised) Lagrangian:

$$\mathcal{L}(q^{(0)}, q^{(1)} \cdots, q^{(n)}), \quad q^{(i)} = \frac{d^i q}{dx^i}, \, n \geq 1.$$

The corresponding Euler-Lagrange equations are

$$\sum_{i=0}^{n} (-\partial)^i \frac{\partial \mathcal{L}}{\partial q^{(i)}} = 0. \tag{37}$$

When $\mathcal{L}$ is non-degenerate ($\frac{\partial^2 \mathcal{L}}{\partial q^{(n)2}} \neq 0$) we can perform a (generalised) Legendre transformation to obtain canonical coordinates, defined by:

$$q_i = q^{(i-1)}, \quad i = 1, \cdots, n,$$

$$p_n = \frac{\partial \mathcal{L}}{\partial q^{(n)}}, \quad p_i = \frac{\partial \mathcal{L}}{\partial q^{(i)}} - \dot{p}_{i+1}, \quad i = 1, \cdots, n-1.$$

The Euler-Lagrange equations (37) then take canonical Hamiltonian form with

$$h = \sum_{i=1}^{n-1} q_{i+1}p_i + q_{nx}p_n - \mathcal{L}(q_1, \cdots, q_n, q_{nx}),$$

where q_{nx} is removed by inverting the non-degenerate Legendre transformation [17]. The generalisation to more than one component q_i is straightforward.

Appendix B: Zero Curvature Equations in Hamiltonian Form

Consider the pair of matrix equations:

$$\Psi_x = U\Psi, \quad \Psi_t = V\Psi,$$

with

$$U = \begin{pmatrix} w & q \\ r & -w \end{pmatrix}, \quad V = \begin{pmatrix} \gamma & \alpha \\ \beta & -\gamma \end{pmatrix},$$

where q, r, w are potential functions (dependent on λ in some specified way) and V is to be determined algorithmically for some classes of λ–dependence. The integrability conditions: $U_t = (\partial - adU)V$ are written

$$\sum_i u_{it}\, e_i = -\left(\sum_{i,j} B_{ij}e_i \otimes e_j \right) \left(\sum_k v_k^* e_k^* \right) = -\sum_{i,j} \left(B_{ij}v_j^* \right) e_i,$$

with e_i^* the dual basis of e_i , so that $B_{ij}e_i \otimes e_j$ is a $(2,0)$ tensor. In components,

$$u_{it} = -B_{ij}v_j^*, \quad v_i^* = g_{ij}v_j, \quad g_{ij} = tr(e_ie_j).$$

The simplest way of calculating this is to check the trace form $tr(U_tV)$ since

$$tr(U_tV) = \mathbf{u}_t \cdot \mathbf{v}^*.$$

For our example,

$$\mathbf{u} = (q, r, w)^T, \, \mathbf{v} = (\alpha, \beta, \gamma)^T, \, \mathbf{v}^* = (\beta, \alpha, 2\gamma)^T = -\delta h$$

and

$$B_{ij} = \begin{pmatrix} 0 & -\partial + 2w & -q \\ -\partial - 2w & 0 & r \\ q & -r & -\tfrac{1}{2}\partial \end{pmatrix}. \tag{38}$$

The general 3×3 zero-curvature formulation. The Hamiltonian structures (29) for the Boussinesq hierarchy can be obtained through a reduction of the general 3×3 zero-curvature formulation where $\mathbf{U}$, $\mathbf{V}_{(r)}$ are given by:

$$
\mathbf{U} = \begin{pmatrix} a & b & c \\ d & e & f \\ g & h & -a-e \end{pmatrix}, \quad \mathbf{V}_{(r)} = \begin{pmatrix} A_{(r)} & B_{(r)} & C_{(r)} \\ D_{(r)} & E_{(r)} & F_{(r)} \\ G_{(r)} & H_{(r)} & -A_{(r)}-E_{(r)} \end{pmatrix}.
$$

Defining the vectors

$$
\bar{\mathbf{u}} = (b,c,d,f,g,h,a,e)^T,
$$
$$
\mathbf{v}_{(r)} = (D_{(r)}, G_{(r)}, B_{(r)}, H_{(r)}, C_{(r)}, F_{(r)}, 2A_{(r)}+E_{(r)}, A_{(r)}+2E_{(r)})^T,
$$

the zero curvature condition can be written in the form

$$
\bar{\mathbf{u}}_{t_r} = \bar{\mathbf{B}}_1 \delta_{\bar{u}} \bar{\mathcal{H}}_r[\bar{u}], \quad \text{where} \quad \bar{\mathbf{B}}_1 = \begin{pmatrix} \bar{X} & \bar{Y} \\ -\bar{Y}^\dagger & \bar{Z} \end{pmatrix} \tag{39}
$$

with

$$
\bar{X} = \begin{pmatrix} 0 & 0 & -\partial+a-e & c \\ 0 & 0 & -f & 0 \\ -\partial-a+e & f & 0 & 0 \\ -c & 0 & 0 & 0 \end{pmatrix},
$$

$$
\bar{Y} = \begin{pmatrix} -h & 0 & -b & b \\ -\partial+2a+e & b & -c & 0 \\ 0 & -g & d & -d \\ d & -\partial+a+2e & 0 & -f \end{pmatrix},
$$

$$
\bar{Z} = \begin{pmatrix} 0 & 0 & g & 0 \\ 0 & 0 & 0 & h \\ -g & 0 & -\frac{2}{3}\partial & \frac{1}{3}\partial \\ 0 & -h & \frac{1}{3}\partial & -\frac{2}{3}\partial \end{pmatrix},
$$

and $\mathbf{v}_{(r)} = -\delta_{\bar{u}} \bar{\mathcal{H}}_r[\bar{u}]$, for some sequence of functionals $\bar{\mathcal{H}}_0, \bar{\mathcal{H}}_1, \cdots$.

The Boussinesq hierarchy in the usual coordinates corresponds to the reduction: $a = c = d = e = 0$, $b = f = 1$, $g = -v+\lambda$, $h = -u$. The Boussinesq hierarchy in various stationary manifold coordinates are derived from (39) in section 3.

Acknowledgement. SDH was supported by an SERC Studentship.

References

[1] M. Antonowicz and M. Blaszak, On a non-standard Hamiltonian description of NLEE, in: it Nonlinear Evolution Equations and Dynamical Systems, S. Carillo and O. Ragnisco, eds., Springer, Berlin (1990), 152–156.

[2] M. Antonowicz and A.P. Fordy, Hamiltonian structures of nonlinear evolution equations, in: *Soliton Theory: A Survey of Results*, A.P. Fordy, ed., MUP, Manchester (1990), 273–312.

[3] M. Antonowicz, A.P. Fordy, and S. Wojciechowski, Integrable stationary flows: Miura maps and bi-Hamiltonian structures, *Phys. Letts. A* **124** (1987), 143–50.

[4] S. Baker, Integrable nonlinear evolution equations and related finite dimensional systems, PhD thesis, University of Leeds, 1995.

[5] S. Baker, V.Z. Enolskii, and A.P. Fordy, Integrable quartic potentials and coupled KdV equations, *Phys. Letts. A* **201** (1995), 167–74.

[6] O.I. Bogoyavlenskii and S.P. Novikov, The relationship between Hamiltonian formalisms of stationary and nonstationary problems, *Func. Anal & Apps.* **10** (1976), 8–11.

[7] I.Ya. Dorfman, Dirac structures of integrable evolution equations, *Phys. Letts. A* **125** (1987), 240–6.

[8] I.Ya. Dorfman, *Dirac Structures and Integrability of Nonlinear Evolution Equations*, Wiley, Chichester (1993).

[9] A.P. Fordy, ed., *Soliton Theory: A Survey of Results*, MUP, Manchester (1990).

[10] A.P. Fordy, The Hénon-Heiles system revisited, *Physica* **52D** (1991), 201–210.

[11] A.P. Fordy and J. Gibbons, Factorization of operators I: Miura transformations, *J. Math. Phys.* **21** (1980), 2508–10.

[12] A.P. Fordy and J. Gibbons, Factorization of operators II, *J. Math. Phys.* **22** (1981), 1170–75.

[13] B. Fuchssteiner and W. Oevel, New hierarchies of nonlinear completely integrable systems related to a change of variables for evolution parameters, *Physica* **145A** (1987), 67–95.

[14] I.M. Gelfand and I.Ya. Dorfman, Hamiltonian operators and algebraic structures related to them, *Func. Anal & Apps.* **13** (1979), 13–30.

[15] S.D. Harris, Integrable nonlinear evolution equations and their Stationary Flows, PhD thesis, University of Leeds, 1994.

[16] O.I. Mokhov, On the Hamiltonian property of an arbitrary evolution system on the set of stationary points of its integral, *Math. USSR Izvestiya* **31** (1988), 657–664.

[17] E.T. Whittaker, *A Treatise on the Analytical Dynamics of Particles and Rigid Bodies*, CUP, Cambridge, UK (1988).

Department of Applied Mathematical Studies and
Centre for Nonlinear Studies,
University of Leeds,
Leeds LS2 9JT, UK.
e-mail: allan@amsta.leeds.ac.uk

Received October 11, 1995

Compatibility in Abstract Algebraic Structures

Benno Fuchssteiner

Abstract

Compatible Hamiltonian pairs play a crucial role in the structure theory of integrable systems. In this paper we consider the question of how much of the structure given by compatibility is bound to the situation of hamiltonian dynamic systems and how much of that can be transferred to a complete abstract situation where the algebraic structures under consideration are given by bilinear maps on some module over a commutative ring. Under suitable modification of the corresponding definitions, it turns out that notions like, *compatible, hereditary, invariance* and *Virasoro algebra* may be transferred to the general abstract setup. Thus the same methods being so successful in the area of integrable systems, may be applied to generate suitable abelian algebras and hierarchies in very general algebraic structures.

1. Introduction

In her work on integrable systems, starting with the pioneering papers [6], [7], [8] and culminating in her account on Dirac structures [1] Irene Dorfman not only paid special attention to those algebraic structures which allow the generation of abelian substructures, but also created some of the most powerful methods to generate dynamic systems having large abelian symmetry groups. In this context, also the paper [2], which in its ideas certainly is one of the crucial contributions for integrability in multidimensions, should be taken into account (compare [5] for an application of similar ideas).

With her work on compatible hamiltonian pairs, Irene Dorfman strongly influenced the perspectives of the whole field. The ideas she shaped in her early work now infiltrate under a variety of different notions and methods the whole field. For example, her ideas can be found in connection with *hereditary* or *Nijenhuis operators*, *Virasoro algebras* and *master-symmetries*.

It seems a fundamental problem to check how far these ideas eventually may reach into other areas in order to generate invariant structures in those fields which do not have access to the infinitesimal aspects underlying the study of dynamical systems. If that were the case, then one day methods similar to those known from integrability, may be slightly modified, may be applied to *time discrete systems, automata, invariant substructures* of general algebras and other areas not yet targeted for the far reaching methods coming from the established area of integrable systems.

The present paper may be a small contribution towards enlarging our notions and methods to a wider area of application. For, how the results of this paper are applied in the classical situation of Hamiltonian systems, see [4]. The main message of the results of the present paper is that *compatibility* is more a property of homomorphisms with respect to bilinear structures than a property connected to vector fields.

We start our considerations by defining hereditary structures in more or less arbitrary algebraic structures such that the crucial results about generating abelian substructures out of one or several invariants may be obtained. Then we show that a general notion of compatibility of homomorphisms in abstract algebraic structures may be characterized by this notion of hereditariness. Thus the power of compatible hamiltonian structures is made available to a wider area of applications which do not need the usual ingredients of the underlying structure of tensor bundles and Lie algebra modules.

Thereafter the notion of compatible deformations of products is introduced and the paper is concluded by introducing Virasoro algebras in general algebras and showing that these are just another aspect of the notions presented so far.

2. Hereditary Structures and Invariance in General Algebras

Fix a commutative ring $\mathbf{F}$. Let $\mathcal{L}$ be some module over $\mathbf{F}$ and consider $(\mathcal{L}, \bullet)$, where $\bullet$ is some binary bilinear operator on $\mathcal{L}$. We call $(\mathcal{L}, \bullet)$ the **reference algebra**, $(\mathcal{L}, \bullet)$ is not necessarily an associative algebra. For short, binary bilinear operators in modules over $\mathbf{F}$ are called *products*. Two elements $a, b \in \mathcal{L}$ are said to *commute* if

$$a \bullet b = b \bullet a . \qquad (2.1)$$

Let furthermore Λ be another module over $\mathbf{F}$ and consider a linear

$$\Theta : \Lambda \to \mathcal{L} . \qquad (2.2)$$

from Λ into the reference algebra $\mathcal{L}$. We call a product $[\ ,\ ]$ in Λ a Θ-**product** if Θ is a homomorphism into $(\mathcal{L}, \bullet)$, i.e. if

$$\Theta[a, b] = (\Theta a) \bullet (\Theta b) \text{ for all } a, b \in \Lambda. \qquad (2.3)$$

To emphasize that some product is a Θ-product we write $[\ ,\ ]_\Theta$ instead of $[\ ,\ ]$. Here, the symbols $[\ ,\ ]$ or $[\ ,\ ]_\Theta$ should not be confused with

Lie algebras; also the algebra under consideration is not assumed to be antisymmetric. When we consider an antisymmetric algebras we shall write $[\![\,,\,]\!]$ instead.

A linear $\Phi : \mathcal{L} \to \mathcal{L}$ is said to be **hereditary** if

$$[a,b]_\Phi := (\Phi a) \bullet b + a \bullet (\Phi b) - \Phi(a \bullet b) \tag{2.4}$$

defines a Φ-product in $\mathcal{L}$. Recalling that then $\Phi[a,b]_\Phi = (\Phi a) \bullet (\Phi b)$, we see that Φ is hereditary if and only if

$$\Phi^2(a \bullet b) + (\Phi a) \bullet (\Phi b) = \Phi\{(\Phi a) \bullet b + a \bullet (\Phi b)\} \text{ for all } a, b \in \mathcal{L}. \tag{2.5}$$

A linear map $\Phi : \mathcal{L} \to \mathcal{L}$ is said to be **invariant** with respect to $k \in \mathcal{L}$ if

$$0 = k \bullet \Phi(b) - \Phi(b) \bullet k - \Phi(k \bullet b) + \Phi(b \bullet k) \text{ for all } b \in \mathcal{L}. \tag{2.6}$$

Any linear map $\Phi : \mathcal{L} \to \mathcal{L}$ which is **left invariant**

$$\Phi(k \bullet b) = k \bullet (\Phi b) \text{ for all } b \in \mathcal{L}. \tag{2.7}$$

with respect to $k \in \mathcal{L}$ as well as **right invariant**

$$\Phi(b \bullet k) = \Phi(b) \bullet k \text{ for all } b \in \mathcal{L}. \tag{2.8}$$

is said to be **super-invariant**. Any Φ which is super-invariant is also invariant. In order to work with operators Φ, for $k \in \mathcal{L}$ we introduce a map L_k by

$$L_k(\Phi)(b) := k \bullet \Phi(b) - \Phi(b) \bullet k - \Phi(k \bullet b) + \Phi(b \bullet k) \text{ for all } b \in \mathcal{L}. \tag{2.9}$$

Hence, Φ is k-invariant if and only if

$$L_k(\Phi) = 0. \tag{2.10}$$

Formally, for $a, b \in \mathcal{L}$ instead of $a \bullet b$ we write

$$L_a(b) := a \bullet b - b \bullet a; \tag{2.11}$$

hence

$$L_a(\Phi \bullet b) = L_a(\Phi) \bullet b + \Phi(L_a(b)) \text{ for all } b \in \mathcal{L}. \tag{2.12}$$

Using this, one easily finds that L_a is a derivation with respect to operator products, i.e.

$$L_a(\Phi\Psi) = L_a(\Phi)\Psi + \Phi L_a(\Psi). \tag{2.13}$$

However, one should observe that L_a in general is not a derivation on $(\mathcal{L}, \bullet)$. Another important observation is that when Φ is hereditary then

$$\Phi(L_a\Phi) = L_{\Phi(a)}\Phi. \tag{2.14}$$

This is easily seen by direct computation.

Lemma 1. *Let Φ be hereditary and invariant with respect to k. Then Φ is invariant with respect to $\Phi(k)$. If Φ is invertible then it is invariant with respect to $\Phi^{-1}(k)$ as well. The set $\{k|\Phi$ invariant with respect to $k\}$ of all elements which leave Φ invariant is an invariant subset under the application of Φ (and of Φ^{-1} if Φ is invertible).*

Proof. By invariance of Φ with respect to k, we know (see (2.10) and (2.12)) that

$$L_k \Phi a = \Phi L_k a \quad \text{for all } a \in \mathcal{L} .   \tag{2.15}$$

From this and a twofold application of (2.14) we find, for arbitrary $a \in \mathcal{L}$,

$$L_{\Phi k}(\Phi a) = \Phi L_k(\Phi a) = \Phi(\Phi L_k a) = \Phi L_{\Phi k} a;   \tag{2.16}$$

hence

$$L_{\Phi k}(\Phi a) = \Phi L_{\Phi k} a   \tag{2.17}$$

which proves the invariance with respect to $\Phi(k)$.

Observation 1.   Let Φ be hereditary and let it be left invariant with respect to k. Then Φ is left invariant with respect to $\Phi(k)$. If Φ is invertible then it is left invariant with respect to $\Phi^{-1}(k)$ as well. The set $\{k|\Phi$ left invariant with respect to $k\}$ is invariant under the application of Φ (and of Φ^{-1} if Φ is invertible). The same results hold for right invariance.

Proof. Replace a by k in (2.5). Since Φ is left invariant with respect to k the first and fourth term cancel and the equality reads

$$\Phi(k) \bullet (\Phi b) = \Phi((\Phi k) \bullet b) \text{ for all } b.   \tag{2.18}$$

This yields the left invariance with repect to $\Phi(k)$. In case Φ is invertible, we replace a in (2.5) by $\Phi^{-1}(k)$ and apply Φ^{-1} to the remaining two terms. The proof for right invariance is similar.

Consequence 1. Let Φ be hereditary and let it be super invariant with respect to k. Then Φ is super invariant with respect to $\Phi(k)$. If Φ is invertible then it is super invariant with respect to $\Phi^{-1}(k)$ as well. The set $\{k|\Phi$ super invariant with respect to $k\}$ is invariant under the application of Φ (and of Φ^{-1} if Φ is invertible).

Theorem 1. *Let Φ be a hereditary map which is invariant with respect to k. Then $\{\Phi^n k|n \in \mathbb{N}_0\}$ is an abelian subset of $(\mathcal{L}, \bullet)$. In case Φ is invertible then $\{\Phi^n k|n \in \mathbb{Z}\}$ is abelian as well.*

Proof. From Lemma 1 we obtain by induction that Φ is invariant with respect to any $\Phi^m k$ and $\Phi^n k$. Consider

$$
\begin{aligned}
(\Phi^m k) \bullet (\Phi^n k) - (\Phi^n k) \bullet (\Phi^m k) &= L_{\Phi^m k}(\Phi^n k) = \Phi^m L_k(\Phi^n k) \\
&= \Phi^{m+n} L_k(k) \\
&= \Phi^{m+n}(k \bullet k - k \bullet k) \\
&= 0
\end{aligned}
$$

for all suitable m, n, where (2.14) and the invariance of Φ has been used. This proves that $(\Phi^m k)$ and $(\Phi^n k)$ commute. For invertible Φ, in this argument Φ^{-1} has to replace Φ.

Theorem 2. *Let Φ be a hereditary map which is super-invariant with respect to k_1 and k_2. Then for arbitrary $n, m \in \mathbb{N}$ (or $\in \mathbb{Z}$ if Φ is invertible) we have*

$$\Phi^n(k_1) \bullet \Phi^m(k_2) = \Phi^{n+m}(k_1 \bullet k_2) \tag{2.19}$$

Proof. From consequence 1 we obtain by induction that Φ is super-invariant with respect to any $\Phi^m k_1$ and $\Phi^n k_2$. Hence

$$(\Phi^m k_1) \bullet (\Phi^n k_2) = \Phi^n(\Phi^m(k_1) \bullet k_2) = \Phi^n(\Phi^m(k_1 \bullet k_2)) = \Phi^{n+m}(k_1 \bullet k_2) \tag{2.20}$$

for all suitable m, n. For invertible Φ, in this argument Φ^{-1} has to replace Φ.

Remark 1. Let Φ be hereditary and let a_1 and a_2 be eigenvectors of Φ (i.e. $\Phi a_i = \lambda_i a_i$, $\lambda_i \in \mathbf{F}$, $i = 1, 2$). Then for these a_i relation (2.5) is equivalent to

$$(\Phi - \lambda_1)(\Phi - \lambda_2)(a_1 \bullet a_2) = 0 .$$

Hence, in case an operator Φ has a spectral resolution and all the corresponding spectral projections are algebra homomorphisms, then this operator is hereditary.

Remark 2. One easily sees that Φ is left invariant with respect to k if and only if

$$[k, b]_\Phi = \Phi(k) \bullet b \quad \text{for all } b \in \mathcal{L} \tag{2.21}$$

and right invariant if and only if

$$[b, k]_\Phi = b \bullet \Phi(k) \quad \text{for all } b \in \mathcal{L} . \tag{2.22}$$

Using the definition of hereditariness, we see that a hereditary invertible Φ is super-invariant with respect to k if and only if it is k-invariant with respect to $(\mathcal{L}, [\,,\,]_\Phi$

3. Compatibility

Now, let us return to the general situation of maps from Λ into $\mathcal{L}$ where Λ is a module over $\mathbf{F}$. Assume that in Λ we have $\Theta-$ and $\Psi-$products $[\,,\,]_\Theta$ and $[\,,\,]_\Psi$, respectively. These products are said to be **compatible** if

$$[a, b] := [a, b]_\Theta + [a, b]_\psi \tag{3.23}$$

defines a $(\Theta + \Psi)$-product.

Lemma 2.   *Let* $[,]_\Theta$ *and* $[,]_\Psi$ *be* $\Theta-$ *and* $\Psi-$ *products, respectively. These products are compatible if and only if*

$$\Theta[a,b]_\Psi + \Psi[a,b]_\Theta = (\Theta a) \bullet (\Psi b) + (\Psi a) \bullet (\Theta b) \text{ for all } a, b \in \Lambda . \quad (3.24)$$

Proof.   Observe that $\Theta[a,b]_\Theta = (\Theta a) \bullet (\Theta b)$ and $\Psi[a,b]_\Psi = (\Psi a) \bullet (\Psi b)$. So, (3.24) is obviously equivalent to

$$(\Theta + \Psi)\{[a,b]_\Theta + [a,b]_\Psi\} = ((\Theta + \Psi)a) \bullet ((\Theta + \Psi)b) , \quad (3.25)$$

which proves the claim.

Observation 2.   Let $\lambda \in \mathbf{F}$. Obviously, $[,]_{\lambda\Theta}$ defined by $[a,b]_{\lambda\Theta} := \lambda[a,b]_\Theta$ is a $(\lambda\Theta)$-product whenever $[,]_\Theta$ is a Θ-product. Now, replacing in (3.24) Ψ and $[,]_\Psi$ by $\lambda\Psi$ and $[,]_{\lambda\Psi}$, respectively, we see that (3.24) remains valid. In other words, (3.24) is linear in Ψ (as well as in Θ). Hence, if $[,]_{\Theta_1}$ and $[,]_{\Theta_2}$ are compatible, and if both are compatible with $[,]_\Theta$ then $[,]_{\lambda\Theta_1} + [,]_{\sigma\Theta_2}$ is always compatible with $[,]_\Theta$.

Observation 3.   Consider the case that the reference algebra is equal to Λ, i.e. $\Lambda = \mathcal{L}$ and put $\Theta = I, \Psi = \Phi$. Furthermore, assume that $[,]_\Theta$ is the given product in $(\mathcal{L}, \bullet)$, and that $[,]_\Psi = [,]$ is a second product such that $\Phi : (\mathcal{L}, [,]) \to (\mathcal{L}, \bullet)$ is a homomorphism. Then (3.24) holds if and only if $[,]$ is the product defined in (2.3). Hence, Φ is hereditary if and only if $(\mathcal{L}, \bullet)$ and $(\mathcal{L}, [,])$ are compatible.

In order to shorten our notions we call Ψ and Θ compatible if their Ψ- and Θ- products, $[,]_\Psi$ and $[,]_\Theta$ are compatible. By application of this notion to the special case of hereditary operators Φ_1, Φ_2, we see that Φ_1 and Φ_2 are compatible if and only if $\Phi_1 + \Phi_2$ is again hereditary.

Theorem 3.   *Consider maps* $\Theta, \Psi : \Lambda \to \mathcal{L}$ *and their corresponding products* $[,]_\Theta$ *and* $[,]_\Psi$. *Assume that* Ψ *is invertible. Then* Θ *and* Ψ *are compatible if and only if* $\Phi = \Theta\Psi^{-1}$ *is hereditary.*

Proof.   Define, by use of $[,]_\Psi$ and the invertible Ψ, a second product in $\mathcal{L}$ by

$$[a,b] = \Psi[(\Psi^{-1}a), (\Psi^{-1}b)]_\Theta \text{ for } a, b \in \mathcal{L}.$$

Then $\Phi : (\mathcal{L}, [,]) \to (\mathcal{L}, \bullet)$ is a homomorphism. We obtain

$$\begin{aligned} (I + \Phi)(a \bullet b + [a,b]) &= (\Theta + \Psi)\Psi^{-1}((a \bullet b) + [a,b]) \\ &= (\Theta + \Psi)([(\Psi^{-1}a), (\Psi^{-1}b)]_\Psi \\ &\quad + [(\Psi^{-1}a), (\Psi^{-1}b)]_\Theta) . \end{aligned}$$

For general $a, b \in \mathcal{L}$ this is equal to

$$((\Theta + \Psi)\Psi^{-1}a) \bullet ((\Theta + \Psi)\Psi^{-1}b) = ((I + \Phi)a) \bullet ((I + \Phi)b)$$

if and only if Θ and Ψ are compatible. Hence I and Ψ are compatible if and only if Θ and Ψ are compatible. By use of observation 3 we obtain the required result.

Theorem 4. *Let* Φ, Ψ *be compatible hereditary operators and assume that* Φ *and* Ψ *commute. Then* $\Phi\Psi$ *is hereditary.*

Proof. For completeness we go through the proof although it is almost the same as in [3] (where the situation was more special). Since Ψ and Φ are hereditary we observe (by use of (2.5) and commutativity) that

$$
\begin{aligned}
(\Psi\Phi a) \bullet (\Psi\Phi b) &= -\Psi^2((\Phi a) \bullet (\Phi b)) + \Psi\{(\Psi\Phi a) \bullet (\Phi b) + (\Phi a) \bullet (\Psi\Phi b)\} \\
&= \Psi^2\Phi^2(a \bullet b) - \Psi^2\Phi\{a \bullet (\Phi b) + (\Phi a) \bullet b)\} \\
&\quad -\Psi\Phi^2\{(\Psi a) \bullet b + a \bullet (\Psi b)\} \\
&\quad +\Psi\Phi\{(\Phi\Psi a) \bullet b + (\Psi a) \bullet (\Phi b) \\
&\quad +(\Phi a) \bullet (\Psi b) + a \bullet (\Psi\Phi b)\} \,.
\end{aligned}
$$

Define a product $[\, , \,]_{\Psi\Phi}$ as in (2.3) and insert the last expression into

$$A_{\Phi\Psi}(a, b) = (\Phi\Psi a) \bullet (\Phi\Psi b) - \Phi\Psi[a, b]_{\Psi\Phi} \,.$$

This yields

$$
\begin{aligned}
A_{\Phi\Psi}(a, b) &= 2(\Phi\Psi)^2(a \bullet b) - (\Phi\Psi)\Psi\{[a, b]_\Phi + \Phi(a \bullet b)\} \\
&\quad -(\Phi\Psi)\Phi\{[a, b]_\Psi + \Psi(a \bullet b)\} \\
&\quad +\Psi\Phi\{(\Psi a) \bullet (\Phi b) + (\Phi a) \bullet (\Psi b)\} \\
&= \Phi\Psi\{-\Psi[a, b]_\Phi - \Phi[a, b]_\Psi + (\Psi a) \bullet (\Phi b) + (\Phi a) \bullet (\Psi b)\}
\end{aligned}
$$

which vanishes because of the compatibility of Φ and Ψ and by virtue of (3.24). Hence we have $A_{\Phi\Psi}(a, b) = 0$ which gives that $\Phi\Psi : (\mathcal{L}, [\, , \,]_{\Phi\Psi}) \rightarrow (\mathcal{L}, \bullet)$ must be a homomorphism.

Corollary 1. *Let* Φ *be hereditary. Then any polynomial in* Φ *is hereditary.*

Proof. Assume that any polynomial $P(\Phi)$ of degree $\leq N$ in Φ is hereditary (which is certainly true for $N = 1$). Obviously, Φ commutes with $P(\Phi)$ and both are compatible. Thus $\Phi P(\Phi)$ must be hereditary.

From compatibility, with I we conclude that any $\alpha I + \beta \Phi P(\Phi)$ is hereditary. Since any polynomial of degree $(N + 1)$ can be written in this form we finish the proof by induction.

Remark 3. It should be observed that the notion of compatibility and therefore the notion of hereditariness are preserved under isomorphisms with respect to their respective reference algebras. To be precise: If $T : (\mathcal{L}, \bullet) \to (\mathcal{L}_1, \bullet)$ is an isomorphism and if $\Theta, \Psi : \Lambda \to \mathcal{L}$ are compatible then $T\Theta, T\Psi : \Lambda \to (\mathcal{L}_1, \bullet)$ are compatible as well.

Let me add some remarks on nonlinear deformations. Compatibility, as we have defined it, is the tangential structure of a corresponding compatibility notion for continuous deformations of products. Assume that we have a one-parameter family of products $[\ , \ ]_\lambda$ in Λ and a family of maps $\Theta(\lambda) : \Lambda \to \mathcal{L}$. Assume further that topologies are given such that the occurring quantities are differentiable with respect to λ. We denote

$$\Theta'(\lambda) = \frac{d}{d\lambda}\Theta(\lambda) \tag{3.26}$$

$$[a, b]' = \frac{d}{d\lambda}[a, b]_\lambda \tag{3.27}$$

and we assume that $[\ , \ ]_\lambda$ is, for any λ, a $\Theta'(\lambda)$ product. We call $(\Theta(\lambda), [\ , \ ]_\lambda)$ a compatible deformation if $[\ , \ ]_\lambda$ always is a $\Theta(\lambda)$ product. One easily verifies the following

Remark 4. $(\Theta(\lambda), [\ , \ ]_\lambda)$ is a compatible deformation if and only if $\Theta(\lambda)$ and $\Theta'(\lambda)$ are always compatible.

4. Antisymmetric Algebras

The algebra $(\mathcal{L}, \bullet)$ is said to be *antisymmetric* if

$$a \bullet b = -b \bullet a \text{ for all } a, b \in \mathcal{L} . \tag{4.28}$$

For any algebra $(\mathcal{L}, \bullet)$ there is a corresponding *antisymmetrization* defined by

$$[\![a, b]\!] := \frac{1}{2}(a \bullet b - b \bullet a) . \tag{4.29}$$

Observe, that $(\mathcal{L}, [\![\ , \ ]\!])$ not necessarily is a Lie algebra since we did not assume that $(\mathcal{L}, \bullet)$ is associative.

In case $(\mathcal{L}, \bullet)$ itself is already antisymmetric then it is equal to its antisymmetrization. Any homomorphism between two algebras induces a homomorphism between their antisymmetrizations. So any automorphism Φ on $(\mathcal{L}, \bullet)$ canonically defines an automorphism on $(\mathcal{L}, [\![\ , \ ]\!])$. Obviously, for antisymmetric algebras the notions *right invariance* and *left invariance* coincide.

Remark 5. Φ is in $(\mathcal{L}, \bullet)$ super-invariant with respect to k if and only if Φ is in $(\mathcal{L}, [\![\ , \ ]\!])$ invariant with respect to k.

Proof. This is easily seen from the following identities

$$[\![k, \Phi a]\!] \;=\; L_k(\Phi a) \tag{4.30}$$

$$\Phi [\![k, a]\!] \;=\; \Phi L_k(a). \tag{4.31}$$

As a consequence, the notions invariance and super-invariance coincide for antisymmetric algebras.

5. Virasoro Algebras

We show that hereditary operators uniquely correspond to Virasoro algebras. For that we consider some antisymmetric algebra $(\mathcal{L}, [\![\ , \ ]\!])$. A set $\mathcal{L}_V := \{K_n, \tau_n \mid n \in \mathbb{N}_0\}$ in $\mathcal{L}$ is called the positive part of a Virasoro algebra (or **Virasoro algebra** for short) if there is some $\rho \in \mathbf{F}$ such that for all $n, m \in \mathbb{N}_0$

$$[\![K_n, K_m]\!] \;=\; 0 \tag{5.32}$$

$$[\![\tau_n, K_m]\!] \;=\; (m + \rho)\, K_{m+n} \tag{5.33}$$

$$[\![\tau_n, \tau_m]\!] \;=\; (m - n)\tau_{n+m}. \tag{5.34}$$

Obviously, $\mathcal{L}_V$ is a subalgebra of $(\mathcal{L}, [\![\ , \ ]\!])$. We consider in $(\mathcal{L}, [\![\ , \ ]\!])$ the operator Φ defined by

$$\Phi K_n : \;=\; K_{n+1} \tag{5.35}$$

$$\Phi \tau_n : \;=\; \tau_{n+1} \tag{5.36}$$

Theorem 5. Φ *is hereditary.*

Proof. We consider the algebra $[\ , \ ]_\Phi$ as defined in (2.4)

$$[a, b]_\Phi := (\Phi a) \bullet b + a \bullet (\Phi b) - \Phi(a \bullet b) . \tag{5.37}$$

Then we find

$$
\begin{aligned}
\Phi[K_n, K_m]_\Phi : \;&=\; \Phi\{ [\![K_{n+1}, K_m]\!] + [\![K_n, K_{m+1}]\!] - \Phi [\![K_n, K_m]\!] \} \\
&=\; 0 \\
\Phi[\tau_n, K_m]_\Phi : \;&=\; \Phi\{ [\![\tau_{n+1}, K_m]\!] + [\![\tau_n, K_{m+1}]\!] - \Phi [\![\tau_n, K_m]\!] \} \\
&=\; (m + 1 + \rho)\Phi K_{n+m+1} = [\![\tau_{n+1}, K_{m+1}]\!] \\
&=\; [\![\Phi \tau_n, \Phi K_m]\!] \\
\Phi[\tau_n, \tau_m]_\Phi : \;&=\; \Phi\{ [\![\tau_{n+1}, \tau_m]\!] + [\![\tau_n, \tau_{m+1}]\!] - \Phi [\![\tau_n, \tau_m]\!] \} \\
&=\; (m - n)\Phi \tau_{n+m+1} = [\![\tau_{n+1}, \tau_{m+1}]\!] \\
&=\; [\![\Phi \tau_n, \Phi \tau_m]\!]
\end{aligned}
$$

In order to show that hereditary operators also lead to Virasoro algebras we start with $\tau_0, K_0 \in \mathcal{L}$ and some Φ such that for some $\rho \in \mathbf{F}$

$$L_{\tau_0}\Phi = \Phi, \quad L_{\tau_0}K_0 = \rho K_0 \tag{5.38}$$

Then we define

$$K_n : = \Phi^n K_0 \tag{5.39}$$
$$\tau_n : = \Phi^n \tau_0 \tag{5.40}$$

Theorem 6. $\mathcal{L}_V := \{K_n, \tau_n \mid n \in \mathbb{N}_0\}$ *is a Virasoro algebra.* $L_{\tau_n}\Phi$ *is hereditary.*

Proof. The Virasoro algebra property is easily proved by induction with the use of (2.14). The same relation shows then

$$L_{\tau_n}\Phi = \Phi^n \tag{5.41}$$

References

[1] I. Dorfman. *Dirac Structures and Integrability of Nonlinear Equations.* Nonlinear Science: Theory and Applications. John Wiley and sons, Chichester, New York, Brisbane, Toronto, Singapore, 1993.

[2] I. Ya. Dorfman and A. S. Fokas. Hamiltonian theory over noncommutative rings and integrability in multidimensions. *J. of Math. Phys.*, 33:2504–2514, 1992.

[3] B. Fuchssteiner. The Lie algebra structure of nonlinear evolution equations admitting infinite dimensional abelian symmetry groups. *Progr. Theor. Phys.*, 65:861–876, 1981.

[4] B. Fuchssteiner. Hamiltonian structure and integrability. *In: Nonlinear Systems in the Applied Sciences, Mathematics in Science and Engineering Vol. 185 Academic Press, C. Rogers and W. F. Ames eds.*, pages 211–256, 1991.

[5] B. Fuchssteiner and A. Roy Chowdhury. A new approach to the quantum KdV. *Solitons and Chaos*, page 14, 1992.

[6] I.M.Gelfand and I.Y. Dorfman. Hamiltonian operators and algebraic structures related to them. *Funktsional'nyi Analiz i Ego Prilozheniya*, 13:13–30, 1974.

[7] I.M.Gelfand and I.Y. Dorfman. The Schouten bracket and Hamiltonian operators. *Funktsional'nyi Analiz i Ego Prilozheniya*, 14:71–74, 1980.

[8] I.M.Gelfand and I.Y. Dorfman. Hamiltonian operators and infinite-dimensional Lie-algebras. *Funktsional'nyi Analiz i Ego Prilozheniya*, 15:23–40, 1981.

Department of Mathematics
AUTOMATH University of Paderborn
D 33098 Paderborn, Germany
benno@uni-paderborn.de

Received November, 1995

A Theorem of Bochner, Revisited

F. Alberto Grünbaum[1] and Luc Haine[2]

Contents

1. Introduction

Many hierarchies of the theory of solitons possess symmetries which do not belong to the hierarchy itself. These symmetries are known under the various names of additional, master or conformal symmetries. They were discovered by Fokas, Fuchssteiner and Oevel [9], [10], [25], Chen, Lee and Lin [4] and Orlov and Schulman [26]. They are intimately related to the bihamiltonian nature of the equations of the theory of solitons which was pioneered in the work of Magri [23] and Gel'fand and Dorfman [11].

---

[1]The first author was supported in part by NSF Grant # DMS94-00097 and by AFOSR under Contract AFO F49629-92.
[2]The second author is a Research Associate for FNRS.

In the last few years, the characterization of the partition function of
2D-quantum gravity as a tau-function of the KdV hierarchy flowing out of
some fixed point of its conformal symmetries [3], [7], [12], [19], [20], [28] has
generated a tremendous amount of activity on the subject. For a nice introduction to this circle of ideas, see [27]. Yet it seems that today very few
explicit solutions to the additional symmetries of the theory of solitons are
known. The classical result of Bochner and its *extension to doubly infinite
matrices* that concern us in this paper can be seen as characterizing special
fixed points of the master symmetries fields associated with the Toda lattice
hierarchy. Successive applications of the Darboux transformation to these
fixed points are expected to yield invariant manifolds of the hierarchy or
its master symmetries. The relevance of these manifolds to an apparently
unrelated problem can be traced back to the complete solution of the "bispectral problem" by J.J. Duistermaat and F. A. Grünbaum [8]. In [8] it is
seen that all the solutions of this problem are obtained by repeated application of the Darboux process starting at special points and that "half of
the solutions" correspond to the rational solutions of the KdV hierarchy.
The link of the other half of the solutions with the KdV hierarchy remained
rather mysterious until F. Magri and J. P. Zubelli [24] showed that they
organize into integral manifolds of the conformal symmetries of the KdV
hierarchy.

In this paper we pose a bispectral problem associated with the Toda
lattice hierarchy and describe explicitly all its "basic solutions" from which
all other solutions are expected to be obtained by repeated application
of the Darboux process. We identify these basic solutions as fixed points
of (in general) some (sub)algebras of the Virasoro algebra formed by the
Toda master symmetries. Finally, we illustrate by four simple examples
our contention that repeated applications of the Darboux transformation
from some of these fixed points will provide integral manifolds of the Toda
hierarchy or its master symmetries. We have computed many more examples which illustrate our thesis. Those will be reported elsewhere [16]. In
[14] we have shown that all *known* solutions of the bispectral problem in
the context of orthogonal polynomials can be obtained by applications of
the Darboux process to carefully chosen "fixed points". The restriction to
orthogonal polynomials appears unnecessary in this context and motivates
our desire to *revisit and extend* the result of Bochner.

The precise problem we pose and solve in this paper is the following:

Determine all instances of doubly infinite tridiagonal matrices Q

$$Q = \begin{pmatrix} \cdot & \cdot & & & & \\ & \cdot & & & & \\ & a_{-2} & b_{-1} & 1 & & & \\ & & a_{-1} & b_0 & 1 & & \\ & & & a_0 & b_1 & 1 & \\ & & & & a_1 & b_2 & 1 \\ & & & & & a_2 & b_3 & 1 \\ & & & & & & \cdot & \cdot & \cdot \\ & & & & & & & \cdot & \cdot & \cdot \end{pmatrix}$$

such that at least one family of functions $f_n(k)$, $n \in \mathbb{Z}$, given by

$$\begin{aligned} &Q(\ldots, f_{-1}(k), f_0(k), f_1(k), f_2(k), \ldots)^t \\ &= k(\ldots, f_{-1}(k), f_0(k), f_1(k), f_2(k), \ldots)^t \end{aligned} \tag{1}$$

satisfies a differential equation of the form

$$\begin{aligned} B(k, d/dk)f_n(k) &= h_2(k)d^2 f_n(k)/dk^2 + h_1(k)df_n(k)/dk + h_0(k)f_n(k) \\ &= \theta_n f_n(k), \quad n \in \mathbb{Z} \end{aligned} \tag{2}$$

around some $k = k_0$.

It is, of course, possible to adopt a different form for Q. This one leads—in the case $a_0 = 0$—to *monic* orthogonal polynomials.

Three "historical" remarks are in order.

a) If one imposes the further condition that $f_n(k)$ should be a set of polynomials orthogonal with respect to a measure, our problem was raised and solved by S. Bochner [2] in 1929. If one is interested in polynomial solutions, it is natural to put $f_0(k) = 1$ and $f_{-1}(k) = 0$ and to replace the matrix Q by the semi-infinite matrix obtained by chopping all the columns to the left of b_1 and all the rows above b_1. If one then makes the assumption that all (remaining) a_i, $i = 1, 2, 3, \ldots$ are nonzero then the solution of the problem above—equivalent to the one posed by Bochner—is given (up to affine changes of coordinates) by the Hermite, Laguerre, Jacobi and Bessel polynomials.

b) An even earlier source of similar considerations comes from the work of Laguerre [21], who studied orthogonal polynomials whose weight function $w(k)$ has a logarithmic derivative given by a rational function. These polynomials p_n were found to satisfy second order differential equations with coefficients that, in general, *depend* on n. The *exceptional* cases when these coefficients are independent of n give exactly the "classical orthogonal polynomials" characterized by Bochner. A nice exposition of some of this material can be found in [22].

c) If the tridiagonal matrix is replaced by the Schroedinger operator $-d^2/dx^2 + V(x)$ then the answer to the corresponding question is completely straightforward: up to a shift of the independent variable $V(x)$ is either $c/x^2 + d$ or $cx + d$. Reference [8] is devoted to determining all the $V(x)$ that appear if one allows an arbitrary *finite* order differential operator $B(k, d/dk)$ in (2).

Here we limit ourselves to answering the "characterization" question in the second order case, and intend to return to the fuller question in a future publication [13]. One can get a flavor of the higher order situation (in the context of orthogonal polynomials) by glancing at [14], or at Section 8 of the present paper in which we have shown that all the known examples can be accounted for by means of appropriate applications of the "Darboux process". The far reaching power of this method can also be gleaned from [15], where we see examples of its use within the context of q-orthogonal polynomials and q-difference equations.

Returning to the case of a *doubly infinite matrix* Q, we will consider only the case when all a_i's are assumed to be nonzero. This allows us to determine all $f_n(k)$ in terms of $f_0(k)$ and $f_1(k)$.

If $f_0(k)$ is not (locally) identically zero we can always "normalize" the eigenfunctions of Q and assume $f_0(k) = 1$.

If $f_0(k)$ is identically zero we see that all $f_n(k)$ with $n = 1, 2, 3, \ldots$ are a polynomial of degree $n - 1$ times $f_1(k)$, while all $f_n(k)$ with $n = -1, -2, -3, \ldots$ are given by a polynomial of degree $-n - 1$ times the same $f_1(k)$. In this case by conjugating B by f_1, we are back into some of the polynomial cases considered by Bochner. In short, if we are interested in going beyond the results of Bochner we can, and do indeed, assume that

$$f_0(k) = 1.$$

We make no assumption on $f_1(k)$ or $f_{-1}(k)$. In fact the entire sequence f_n, $n \in \mathbb{Z}$, is determined by $f_1(k)$. We have, for $n \geq 2$,

$$
\begin{aligned}
f_n &= f_1(k^{n-1} + \text{lower order polynomial}) \\
&- a_1(k^{n-2} + \text{lower order polynomial})
\end{aligned}
$$

and for $n > 0$

$$
f_{-n} = -f_1(k^{n-1}/\textstyle\prod_{i=-n+1}^{0} a_i + \text{lower order polynomial}) \\
+ (k^n/\textstyle\prod_{i=-n+1}^{0} a_i + \text{lower order polynomial}).
$$

Notice that the statement of the problem is purely local: we assume that (2) holds around a point $k = k_0$, and thus we normalize $f_0(k)$ around that point. It turns out that this local assumption leads to an operator B with coefficients defined in a larger region.

Our strategy is the following:

(A) We obtain a set of necessary conditions on Q and the sequence θ_n for (1) and (2) to hold. This allows us to determine these objects up to the choice of the following free parameters

$$a_0, a_1, b_1, r = b_2 - b_1, u, v \text{ with } \theta_n = un^2 + vn.$$

(B) We show that for any such choice, the conditions (1), (2) impose a differential equation in $f_1(k)$. Furthermore, we see that *any* solution of this equation gives rise to a set (Q, θ_n, f_n, B) such that (1), (2) hold. The resulting operators B are all conjugate to each other. Therefore in the language of [8], see particularly Corollary 2.2, all these solutions are "rank two" cases. This gives a complete solution to the question posed above since we are able to explicitly solve the differential equation for $f_1(k)$ in terms of "classical functions": in the generic case this involves any solution of the *hypergeometric* equation for arbitrary choice of its parameters a, b, c.

Section 2 discusses the necessary condition. Section 3 is devoted to obtaining the differential equation for $f_1(k)$. In Section 4 we prove the "bispectral property," i.e. (1) and (2), and in Section 5 we obtain f_1 explicitly. Section 6 shows how the cases discussed by Bochner fit in the larger picture. In Section 7 we show that the solutions of our problem are fixed points of some subalgebras of the master symmetries vector fields associated with the Toda lattice hierarchy. Finally, in Section 8, we present four simple examples of explicit solutions of the Virasoro and Toda flows, obtained by application of the Darboux transformation to some of the "basic solutions" of the bispectral problem.

2. $(\text{ad } Q)^3(\Theta) = 0$ and its Solution

Rewrite (1) and (2) with $f \equiv (\ldots, f_{-1}(k), f_0(k), f_1(k), \ldots)^t$ and $\Theta = \text{diag-onal matrix } (\ldots, \theta_{-1}, \theta_0, \theta_1, \ldots)$ as

$$\begin{aligned} Qf &= kf \\ Bf &= \Theta f. \end{aligned}$$

Recall the lemma given in [14], following the basic observation in [8].

Lemma 1. *If B has order m and both (1) and (2) hold, then*

$$(\text{ad } Q)^{m+1}(\Theta) = 0.$$

Proof. One obtains immediately from (1) and (2) that

$$[Q, \Theta]f = [-k, B]f, \tag{3}$$

using that Q and B, respectively Θ and k, commute with each other. Replacing Θ, respectively B, in (3) by $(\text{ad } Q)^{r-1}(\Theta)$, respectively

$(-\text{ad } k)^{r-1}(B)$, one obtains by induction on r that

$$(\text{ad } Q)^r(\Theta)f = (-\text{ad } k)^r(B)f, \quad \text{for all } r \in \mathbb{Z}_{>0}. \qquad (4)$$

Because ad k decreases the order of a linear differential operator in k, we get from (4) with $r = m + 1$

$$(\text{ad } Q)^{m+1}(\Theta)f = 0.$$

As k varies, the $f(k)$ are linearly independent vectors so that the (finite band) operator $(\text{ad } Q)^{m+1}(\Theta)$ has infinite dimensional kernel; hence it must vanish identically.

In our case since $m = 2$ we must have $(\text{ad } Q)^3(\Theta) = 0$.

Equating the diagonals of $(\text{ad } Q)^3(\Theta) = 0$, starting with the upper one, we obtain at the $(n, n + 3)$-th entry the equations:

$$\theta_{n+2} - 3\theta_{n+1} + 3\theta_n - \theta_{n-1} = 0, \quad n \in \mathbb{Z}, \qquad (5)$$

whose general solution is

$$\theta_n = n^2 u + nv + w, \qquad (6)$$

with u, v, w free parameters. Since we can shift the θ_n's by an arbitrary constant, we may always assume that $w = 0$. Equating the $(n, n + 2)$-th and the $(n, n + 1)$-th entries to zero, we obtain

$$(\theta_{n+1} - \theta_{n+2})b_{n+2} + (\theta_{n+1} - \theta_{n-1})b_{n+1} + (\theta_{n-2} - \theta_{n-1})b_n = 0 \qquad (7)$$

and

$$(\theta_n - \theta_{n+2})a_{n+1} + 4(\theta_n - \theta_{n-1})a_n + (\theta_{n-3} - \theta_{n-1})a_{n-1} \qquad (8)$$
$$+ (\theta_n - \theta_{n-1})(b_{n+1} - b_n)^2 = 0, \quad n \in \mathbb{Z},$$

where the θ_n's, $n \in \mathbb{Z}$, are given by (6).

Using (5) and (6) we see that the general solution of (7) depends on two free parameters b_1 and $r = b_2 - b_1$

$$b_n = \frac{(n-1)(3u+v)[(n-1)u+v]r}{[(2n-3)u+v][(2n-1)u+v]} + b_1, \quad n \in \mathbb{Z}. \qquad (9)$$

Going now into the equation for a_n we see, after some labor, that the general solution for a_n is given by

$$a_n = \tilde{a}_n + \tilde{\tilde{a}}_n, \quad n \in \mathbb{Z} \qquad (9')$$

with

$$\tilde{a}_n = \frac{a1_n}{a2_n}$$

$$a1_n = n(v+(n-1)u)((n-1)(3u+v)^2(nu+v)r^2+4(2u+v)((2n-1)u+v)^2a_1)$$

$$a2_n = 4(v + (2n - 2)u)(v + (2n - 1)u)^2(v + 2nu)$$

and, finally

$$\tilde{\tilde{a}}_n = \frac{(-a_0)(n - 1)(v - 2u)(v + nu)}{(v + (2n - 2)u)(v + 2nu)}.$$

The expressions above make it clear that it is safer to assume that

$$v + nu \text{ is nonzero for all integers } n.$$

3. The Equation for $f_1(k)$

Now that we have determined the matrix Q, define the family of functions $f_n(k)$ by the recursion relation resulting from (1), namely,

$$f_n(k) = (k - b_n)f_{n-1}(k) - a_{n-1}f_{n-2}(k)$$

with $f_0(k) = 1$ and $f_1(k)$ arbitrary.

Put

$$B(k, d/dk)g = h_2\frac{d^2g}{dk^2} + h_1\frac{dg}{dk} + h_0g.$$

From now on we will assume (without loss of generality) that $b_1 = 0$. This can always be achieved by shifting the variable k.

Now we recall that (1) and (2) are both supposed to hold, and collect some consequences. From $Bf_0(k) = \theta_0 f_0(k) = 0$ we get, since $f_0(k) = 1$, that $h_0 = 0$. By computing $Bf_i(k) = \theta_i f_i(k)$, $i = 1, 2$ we get

$$-f_1(v + u) + h_2\frac{d^2f_1}{dk^2} + h_1\frac{df_1}{dk} = 0$$

and

$$-(f_1(k-r)-a_1)(2v+4u)+h_2(\frac{d^2f_1}{dk^2}(k-r)+2\frac{df_1}{dk})+h_1(\frac{df_1}{dk}(k-r)+f_1) = 0.$$

A (local) solution $f_1 \equiv 0$ is possible only if $a_1(v + 2u) = 0$, and under our assumption this means $a_1 = 0$. This case gives then $f_i = 0$ for all $i = 1, 2, 3, \ldots$ and we get (directly from (1)) for f_{-i} a family of polynomials of degree i in k and we are back into Bochner's situation (in the reversed direction).

If we assume that f_1 is not (locally) identically zero we see that the determinant of the system of equations in h_1, h_2 given above is nonzero except when

$$f_1(k) = a/(c - k)$$

for some arbitrary pair of constants a, c. The case of zero determinant is analyzed first.

The equations above force the following quantities to vanish

$$2(a + a_1)(v + 2u), \quad a_1(r - c)(v + 3u).$$

The first condition pins down, under our assumptions, the value of a. If we take $a_1 = 0$ we are, once again back in the case of Bochner (in the reversed direction). If a_1 is not zero, then the value of c is pinned down too and we get

$$f_1(k) = a_1/(k - r).$$

Now using $Bf_3 = \theta_3 f_3$ one concludes that

$$a_1 = -\frac{(r^2 - 2a_0)v + (r^2 + 4a_0)u}{4v + 4u}$$

which gives (from (1)) $f_i(k) = 0$ for all values of the index $i = 2, 3, 4, \ldots$. Furthermore $f_1, f_0, f_{-1}, \ldots, f_{-i}, \ldots$ are given by $f_1(k)$ times a polynomial of degree $1 + i$. Conjugating by f_1 we are back in one of Bochner's cases.

From now on we assume that the determinant of the system given by imposing $Bf_n(k) = \theta_n f_n(k)$ for $n = 0, 1, 2$ is nonzero.

We get

$$
\begin{aligned}
h_0 \;=\;& 0, \\
h_1 \;=\;& -((\tfrac{d^2 f_1}{dk^2}(f_1(r - k) + 2a_1) + 2f_1\tfrac{df_1}{dk})v \\
&+ (\tfrac{d^2 f_1}{dk^2}(f_1(3r - 3k) + 4a_1) + 2f_1\tfrac{df_1}{dk})u)/(f_1\tfrac{d^2 f_1}{dk^2} - 2(\tfrac{df_1}{dk})^2), \\
h_2 \;=\;& ((\tfrac{df_1}{dk}(f_1(r - k) + 2a_1) + f_1^2)v \\
&+ (\tfrac{df_1}{dk}(f_1(3r - 3k) + 4a_1) + f_1^2)u)/(f_1\tfrac{d^2 f_1}{dk^2} - 2(\tfrac{df_1}{dk})^2).
\end{aligned}
$$

Setting

$$Bf_3(k) = \theta_3 f_3(k)$$

we get

$$
\begin{aligned}
\tfrac{d^2 f_1}{dk^2} \;=\;& (((2f_1(\tfrac{df_1}{dk})^2 k - 2f_1^2\tfrac{df_1}{dk})r + (2f_1^2\tfrac{df_1}{dk} - 2a_1(\tfrac{df_1}{dk})^2)k \\
&+ (8a_1 - 4a_0)f_1(\tfrac{df_1}{dk})^2 - 6a_1 f_1\tfrac{df_1}{dk} - 2f_1^3)v \\
&+ ((6f_1(\tfrac{df_1}{dk})^2 k - 6f_1^2\tfrac{df_1}{dk})r - 4f_1(\tfrac{df_1}{dk})^2 k^2 \\
&+ (2a_1(\tfrac{df_1}{dk})^2 + 6f_1^2\tfrac{df_1}{dk})k + (8a_1 + 8a_0)f_1(\tfrac{df_1}{dk})^2 - 10a_1 f_1\tfrac{df_1}{dk} \\
&- 2f_1^3)u)/(((f_1^2 k + a_1 f_1)r - 2a_1 f_1 k + (4a_1 - 2a_0)f_1^2 + 2a_1^2)v \\
&+ ((3f_1^2 k + 3a_1 f_1)r - 2f_1^2 k^2 - 2a_1 f_1 k + (4a_1 + 4a_0)f_1^2 + 4a_1^2)u)
\end{aligned}
$$

and using this expression the coefficients in B simplify into

$$h_0 = 0,$$
$$h_1 = (((f_1 \tfrac{df_1}{dk} k - f_1^2)r + (f_1^2 - a_1 \tfrac{df_1}{dk})k$$
$$+ \quad (4a_1 - 2a_0)f_1 \tfrac{df_1}{dk} - 2a_1 f_1)v + ((3f_1 \tfrac{df_1}{dk} k - 3f_1^2)r - 2f_1 \tfrac{df_1}{dk} k^2$$
$$+ \quad (a_1 \tfrac{df_1}{dk} + 3f_1^2)k + (4a_1 + 4a_0)f_1 \tfrac{df_1}{dk} - 4a_1 f_1)u)/(a_1 \tfrac{df_1}{dk} + f_1^2),$$
$$h_2 = -(((f_1^2 k + a_1 f_1)r - 2a_1 f_1 k + (4a_1 - 2a_0)f_1^2 + 2a_1^2)v$$
$$+ \quad ((3f_1^2 k + 3a_1 f_1)r - 2f_1^2 k^2 - 2a_1 f_1 k$$
$$+ \quad (4a_1 + 4a_0)f_1^2 + 4a_1^2)u)/(2a_1 \tfrac{df_1}{dk} + 2f_1^2).$$

Again this imposes a separate treatment for the case

$$df_1(k)/dk = -f_1^2(k)/a_1.$$

The general solution of this equation is given by $f_1(k) = a_1/(k + c)$ with a constant c. This can be reduced to one of the "Bochner" cases considered above.

When we impose

$$Bf_4(k) = \theta_4 f_4(k),$$

we conclude that one of the following expressions has to hold

a) $f_1 = \dfrac{a_1}{k - r}$

b) $\dfrac{df_1}{dk} = \dfrac{(f_1 r - 2f_1 k + 2f_1^2 + 2a_1)v + (3f_1 r - 2f_1 k + 4a_1)u}{(kr + 2a_1 - 2a_0)v + (3kr - 2k^2 + 4a_1 + 4a_0)u}.$

We treat them separately.

Case a) gives $f_i(k) = 0$ for $i = 2, 3, 4, \ldots$ and the functions $f_i(k)$ with $i = 1, 0, -1, -2, \ldots$ turn out to be of the form $f_1(k)$ times a polynomial of degree $1 - i$. Once again, by conjugating B with $f_1(k)$ we can assume that we are in the polynomial case handled by Bochner.

Case b) is the only one that gives a chance to get away from polynomials. It leads to

$$h_0 = 0,$$
$$h_1 = (2f_1 - k)v + ku,$$
$$h_2 = -\dfrac{(kr + 2a_1 - 2a_0)v + (3kr - 2k^2 + 4a_1 + 4a_0)u}{2}.$$

Now something remarkable happens: as long as $f_1(k)$ satisfies the equation in b) we automatically get

$$Bf_n(k) = \theta_n f_n(k), \quad n \in \mathbb{Z}.$$

This will be proved in Section 4.

A choice of a particular solution f_1 of the Riccati equation in b) determines the operator $B = B(f_1)$ according to the formulas above. If one chooses a different solution $\tilde{f}_1$ one can see that

$$B(\tilde{f}_1) = gB(f_1)g^{-1}$$

with

$$(\ln g)' = \frac{f_1 - \tilde{f}_1}{h_2}\, v\;.$$

In particular the form of B is independent of f_1 exactly when $v = 0$.

Observe that, as in [8], when B is required to be an operator of order 2, we only have "rank two" situations.

If we had not assumed $b_1 = 0$, then the equation for f_1 in case b) would read

$$P(k)\frac{df_1}{dk} + 2vf_1^2 + [(2b_1 + 3r - 2k)u + (2b_1 + r - 2k)v]f_1 + 2a_1(v + 2u) = 0 \quad (10)$$

with $P(k)$ given in Section 4.

4. The Operator $B(k, d/dk)$

Here we prove that all solutions of $(\text{ad } Q)^3(\Theta) = 0$ determined in Section 2 give rise to a "bispectral case" if f_1 is an arbitrary solution of the equation in case b) of Section 3. The main tool is given below in the form of a differentiation formula. This formula is deduced from a string equation which is an integrated form of $(\text{ad } Q)^3(\Theta) = 0$. In this section we do not assume that $b_1 = 0$.

4.1. The String Equation

We observe that equations (7) and (8) in Section 2 can both be "integrated" once. Indeed, we can write (7) as

$$[(\theta_{n+1} - \theta_{n+2})b_{n+2} + (\theta_n - \theta_{n-1})b_{n+1}]$$
$$-[(\theta_n - \theta_{n+1})b_{n+1} + (\theta_{n-1} - \theta_{n-2})b_n] = 0,$$

which is equivalent to

$$(\theta_n - \theta_{n+1})b_{n+1} + (\theta_{n-1} - \theta_{n-2})b_n = \beta, \quad (11)$$

with

$$\beta = -(rv + (4b_1 + 3r)u). \quad (12)$$

Then, using (5) and (11), (8) becomes

$$[(\theta_n - \theta_{n+2})a_{n+1} + (\theta_n - \theta_{n-2})a_n - 2ub_{n+1}^2 - \beta b_{n+1}]$$
$$-[(\theta_{n-1} - \theta_{n+1})a_n + (\theta_{n-1} - \theta_{n-3})a_{n-1} - 2ub_n^2 - \beta b_n] = 0$$

or equivalently

$$(\theta_{n-1} - \theta_{n+1})a_n + (\theta_{n-1} - \theta_{n-3})a_{n-1} - 2ub_n^2 - \beta b_n = \alpha, \qquad (13)$$

with

$$\alpha = (rb_1 - 2a_1 + 2a_0)v + (2b_1^2 + 3rb_1 - 4a_1 - 4a_0)u. \qquad (14)$$

Equations (6), (11) and (13) can be recast in the form of a "string type" equation, which is of independent interest, since it reveals some deep connections of our problem with the Virasoro algebra (see Section 7). In the special case corresponding to the Hermite polynomials (see Section 6) $\alpha = -1$, $v = 1$, $\beta = u = 0$, this equation reduces to the famous string equation $[P, Q] = I$, which has recently played a prominent role in the study of the Hermitian matrix model in relation with $2d$-quantum gravity, see [6], [19], [20], [28] and references therein.

For $n \in \mathbb{Z}$, put

$$\sum_{j=-\infty}^{n} = \sum_{j=1}^{n} \text{ if } n > 0, \text{ or } 0 \text{ if } n = 0, \text{ or } -\sum_{j=0}^{n+1} \text{ if } n < 0. \qquad (15)$$

Let $P = (P_{ij})$ be an infinite tridiagonal matrix defined by

$$P_{nn+1} = 2(nu + v), \quad P_{nn-1} = -2una_n, \text{ and}$$
$$P_{nn} = 2(nu + v)b_{n+1} + n\beta + 2u \sum_{j=-\infty}^{n} b_j - 2vb_1 , \qquad (16)$$

$Q = (Q_{ij})$ the infinite tridiagonal matrix introduced in Section 1:

$$Q_{nn+1} = 1, \quad Q_{nn} = b_{n+1}, \quad Q_{nn-1} = a_n,$$

and let I denote the identity matrix. We have

Proposition 1. *The equation* $(\text{ad } Q)^3(\Theta) = 0$ *can be integrated in the form of a string equation*

$$\alpha I + \beta Q + 2uQ^2 = [Q, P]. \qquad (17)$$

Proof. Equating the lower and the two upper diagonals of (17), one obtains identities which are identically satisfied. The identities obtained by equating the first subdiagonal and the main diagonal reduce to (11) and (13).

Notation. Anticipating on the relation between the string equation and the Virasoro algebra to be explained in Section 7, we shall denote equations (11) and (13) respectively by $L_{-1}a_n = 0$ and $L_{-1}b_n = 0$ with

$$
\begin{aligned}
L_{-1}a_n &= 2a_n[\beta + 2u((n+1)b_{n+1} - (n-1)b_n) + (v-u)(b_{n+1} - b_n)], \\
L_{-1}b_n &= \alpha + \beta b_n + 2u(b_n^2 + (2n+1)a_n - (2n-3)a_{n-1}) \\
&\quad + 2(v-u)(a_n - a_{n-1}).
\end{aligned}
$$

$$(18)$$

4.2. A Differentiation Formula

Define

$$ A = P(k)\frac{d}{dk} + 2vf_1, $$

with

$$ P(k) = \alpha + \beta k + 2uk^2, $$

α and β as in (12) and (14), and f_1 an arbitrary solution of the equation (10) in case b) of Section 3. Let $f_n(k)$, $n \in \mathbb{Z}$, be a solution of (1) generated by $f_0 = 1$ and this choice of f_1.

Lemma 2. *The operator A is represented by the tridiagonal matrix P (16) in the basis of $f_n(k)$, $n \in \mathbb{Z}$:*

$$ Af_n = P_{nn+1}f_{n+1} + P_{nn}f_n + P_{nn-1}f_{n-1}. \tag{19} $$

Proof. The case $n = 0$ is trivially satisfied. For $n = 1$, substituting $f_2 = (k - b_2)f_1 - a_1$, gives the Riccati equation (10) satisfied by f_1. Assume that we have proved the formula for $0 \le k \le n$, $n \ge 1$, we establish it for $k = n + 1$. By the induction hypothesis, using (19) for $k = n$ and $n - 1$ and the three term recursion relation satisfied by the f_n's, we obtain by straightforward computation

$$
\begin{aligned}
Af_{n+1} &= P_{n+1\,n+2}f_{n+2} + P_{n+1\,n+1}f_{n+1} \\
&\quad + (L_{-1}b_{n+1} + P_{n+1n})f_n + (L_{-1}a_n)f_{n-1},
\end{aligned}
$$

with $L_{-1}b_{n+1} = L_{-1}a_n = 0$, because of the string equations (18). Similarly, assuming that (19) is true for $n - 1 \le k \le 1$, $n \le 1$, one shows that

$$
\begin{aligned}
a_{n-1}Af_{n-2} &= (L_{-1}b_n + a_{n-1}P_{n-2\,n-1})f_{n-1} \\
&\quad + (L_{-1}a_{n-1} + a_{n-1}P_{n-2\,n-2})f_{n-2} + a_{n-1}P_{n-2\,n-3}f_{n-3},
\end{aligned}
$$

which completes the proof.

4.3. Proof of Bispectrality

If Q, Θ is a solution of $(\text{ad } Q)^3(\Theta) = 0$, and f_n, $n \in \mathbb{Z}$, are obtained from $Qf = kf$, with f_1 a solution of the equation at the end of Section 3, one has

$$B(k, \frac{d}{dk})f_n = (n^2 u + nv)f_n,$$

with

$$B(k, \frac{d}{dk}) = \frac{P(k)}{2}\frac{d^2}{dk^2} + (k(u - v) + (b_1 + 2f_1)v - b_1 u)\frac{d}{dk}.$$

The proof follows from the differentiation formula given in (4.2). Using the three term recursion relation satisfied by the f_n's and substituting

$$\sum_{j=-\infty}^{n} b_j = \frac{n[(2b_1 + (n-1)r)v + (2(2n-1)b_1 + 3(n-1)r)u]}{2[v + (2n-1)u]},$$

(19) can be rewritten as follows:

$$A_n f_n = \gamma_n f_{n+1} + \delta_n f_n, \tag{20}$$

with

$$\begin{aligned} A_n &= A + 2nku, \quad \gamma_n = 2(v + 2nu), \\ \delta_n &= n[rv + (2b_1 + \frac{(3nu - (n-2)v)r}{v + (2n+1)u})u]. \end{aligned}$$

Apply A_n to both sides of (20) and use that $A_n = A_{n+1} - 2ku$ combined with the three term recursion relation to get

$$\begin{aligned} \gamma_n(\gamma_{n+1} - 2u)kf_{n+1} &= A_n^2 f_n - (\delta_n + \delta_{n+1} - \gamma_{n+1}b_{n+2})A_n f_n \\ -(\gamma_{n+1}\delta_n b_{n+2} - \delta_n \delta_{n+1} - a_{n+1}\gamma_n\gamma_{n+1})f_n. \end{aligned}$$

By this relation one eliminates f_{n+1} from (20) and one obtains a second order differential equation satisfied by f_n which, after some labor, can be written as

$$P(k)[B(k, \frac{d}{dk})f_n - (n^2 u + nv)f_n] = 0.$$

5. Solving the Equation for $f_1(k)$

In this section we solve the equation given in case b) of Section 3.

5.1. The Case $v = 0$

In the case $v = 0$ the equation satisfied by f_1 simplifies to

$$\frac{df_1}{dk} = \frac{f_1(3r - 2k) + 4a_1}{3kr - 2k^2 + 4a_1 + 4a_0}.$$

5.1.1. The Case $a_0 + a_1 = 0$. The general solution is given by

$$\frac{4a_0}{3r} + \frac{8a_0 k \log(\frac{2k-3r}{k})}{9r^2} + c\,k$$

with an arbitrary choice of the constant c.

When $r = 0$, the solution is $-a_0/k + ck$.

5.1.2. The Case $a_0 + a_1$ Non-Zero. The general solution is

$$c\sqrt{-3kr + 2k^2 - 4(a_1 + a_0)} \left(\frac{\sqrt{9r^2 + 32(a_1 + a_0)} - 3r + 4k}{-\sqrt{9r^2 + 32(a_1 + a_0)} - 3r + 4k} \right)^{\frac{3r}{2\sqrt{9r^2 + 32(a_1 + a_0)}}} + \frac{a_1 k}{a_1 + a_0}$$

for an arbitrary choice of the constant c.

5.2. The Case of Non-Zero v

Defining

$$\mu = \frac{2v}{(kr + 2a_1 - 2a_0)v + (3kr - 2k^2 + 4a_1 + 4a_0)u}$$

and introducing the function $w(k)$ via the formula

$$f_1 = -(dw/dk)/(\mu w),$$

the initial nonlinear Riccati equation is converted into the second order linear differential equation for w

$$-(krv + 2a_1 v - 2a_0 v + 3kru - 2k^2 u + 4a_1 u + 4a_0 u)^2 d^2w/dk^2$$
$$-2k(v - u)(krv + 2a_1 v - 2a_0 v + 3kru - 2k^2 u + 4a_1 u + 4a_0 u)dw/dk +$$
$$-4a_1 v(v + 2u)w = 0.$$

$$(21)$$

5.2.1. u Non-Zero. If we assume that u is not zero, the leading coefficient has two double roots, denoted by k_1 and k_2. In terms of k_1, k_2 the equation is

$$-(k - k_1)^2(k - k_2)^2 u^2 \frac{d^2w}{dk^2} + k(k - k_1)(k - k_2)u(v - u)\frac{dw}{dk} - a_1 v(v + 2u)w = 0.$$

There are two cases to be discussed separately.

Case a)

If $k_1 \neq k_2$, our equation has regular singular points at k_1 and k_2 and the roots of the indicial equation at $k = k_1$ include s_1 (as given below). Likewise the indicial equation at $k = k_2$ includes s_2 (as given below) as one of its roots.

If we replace w by

$$y(k - k_1)^{s_1}(k - k_2)^{s_2}$$

with s_1, s_2 given by

$$s_1 = \frac{\sqrt{k_1^2 v^2 + a_1(-4v^2 - 8uv) - 2k_1 k_2 uv + k_2^2 u^2} + k_1 v - k_2 u}{2k_1 u - 2k_2 u}$$

$$s_2 = \frac{\sqrt{k_2^2 v^2 + a_1(-4v^2 - 8uv) - 2k_1 k_2 uv + k_1^2 u^2} - k_2 v + k_1 u}{2k_1 u - 2k_2 u}$$

the function y satisfies the following hypergeometric equation

$$\frac{dy}{dk}(k(-\tfrac{v}{u} + 2s_2 + 2s_1 + 1) - 2k_1 s_2 - 2k_2 s_1)$$

$$-\frac{y(s_2 + s_1)(v - (s_2 + s_1)u)}{u} + \frac{d^2 y}{dk^2}(k - k_1)(k - k_2) = 0.$$

In fact the process can be reversed, and any solution of this equation gives one w, thus one y and eventually one f_1. Notice that μ has for poles $k = k_1$ and $k = k_2$.

We write down the expression for f_1 in terms of any solution of the hypergeometric equation written in its standard form.

Take

$$a = s_2 + s_1 \, ,$$

$$b = -\frac{v - s_2 u - s_1 u}{u} \, ,$$

$$c = \frac{k_1 v + 2k_2 s_1 u - 2k_1 s_1 u - k_1 u}{(k_2 - k_1)u}$$

We claim that in terms of Gauss' hypergeometric function, the function

$$_2F_1(a, b, c, (k - k_1)/(k_2 - k_1))$$

solves the equation for y given earlier. The same is true for any other solution of the standard hypergeometric equation.

Put $k = (k_2 - k_1)x + k_1$ and $F(x) = y(k)$. From the expression for f_1

$$f_1 = -\frac{1}{\mu}\frac{d}{dk}\log w$$

and $w = y(k - k_1)^{s_1}(k - k_2)^{s_2}$, using

$$\frac{v}{u} = a - b, \quad s_1 = \frac{a(b - c)}{b - a - 1}, \quad s_2 = \frac{a(c - a - 1)}{b - a - 1},$$

it follows that

$$f_1 = \frac{k_2 - k_1}{a - b}\left(x(x - 1)\frac{d}{dx}\log F(x) + \frac{a(b - c)}{b - a - 1}(x - 1) + \frac{a(c - a - 1)}{b - a - 1}x\right),$$

with $F(x)$ an arbitrary non-zero solution of the hypergeometric equation

$$x(1 - x)F''(x) + (c - (a + b + 1)x)F'(x) - abF(x) = 0.$$

From (9) and (9') we find that the functions $\tilde{f}_n = f_n/(k_2 - k_1)^n$ satisfy a three term recursion relation $Q\tilde{f} = x\tilde{f}$ with

$$a_n = \frac{(n + a - 1)(n - b)(n - c + a)(n + c - b - 1)}{(2n - b + a - 2)(2n - b + a - 1)^2(2n - b + a)}$$

$$b_n = \frac{1}{2}\left(\frac{(b + a - 1)(2c - b - a - 1)}{(2n - b + a - 3)(2n - b + a - 1)} + 1\right),$$

and, from the expressions for h_0, h_1 and h_2 in case b) of Section 3, we get

$$x(x - 1)\frac{d^2\tilde{f}_n}{dx^2} + \left((b - a + 1)x + 2(a - b)\tilde{f}_1 - \frac{bc + ac - c - 2ab}{b - a - 1}\right)\frac{d\tilde{f}_n}{dx}$$
$$= n(n + a - b)\tilde{f}_n.$$

Case b)

We now consider the case $k_1 = k_2$. From equation (21) we see that this happens when

$$a_0 = \frac{r^2v^2 + (6r^2 + 16a_1)uv + (9r^2 + 32a_1)u^2}{16u(v - 2u)},$$

from which it follows that

$$k_1 = k_2 = \frac{r(v + 3u)}{4u}.$$

If we make the change of variable

$$x = \frac{1}{k - k_1}$$

and put $\tilde{w}(x) = w(k)$ we get

$$u^2x\frac{d^2\tilde{w}}{dx^2} + u(k_1(v - u)x + v + u)\frac{d\tilde{w}}{dx} + a_1v(v + 2u)x\tilde{w} = 0,$$

and the further replacement

$$\tilde{w} = x^{-\frac{1}{2}\left(1+\frac{v}{u}\right)} e^{\frac{k_1 x}{2}\left(1-\frac{v}{u}\right)} y$$

gives

$$\frac{d^2 y}{dx^2} = \left(\frac{c^2}{4} + \frac{k_1(v^2 - u^2)}{2u^2 x} + \frac{v^2 - u^2}{4u^2 x^2}\right) y$$

with

$$c^2 = \frac{(k_1^2 - 4a_1)v^2 - (2k_1^2 + 8a_1)uv + k_1^2 u^2}{u^2}.$$

If one puts $z = cx$ and $W(z) = y(x)$ this becomes Whittaker's equation

$$\frac{d^2 W}{dz^2} + \left(-\frac{1}{4} + \frac{\lambda}{z} + \frac{u^2 - v^2}{4u^2 z^2}\right) W = 0,$$

with

$$\lambda = \frac{k_1(u^2 - v^2)}{2u^2 c}.$$

Putting $u = 1$, it follows that the function f_1 is given by

$$f_1 = \left(\frac{v+1}{2v}\right)(k - k_1) - \frac{\lambda c}{v(v+1)} - \frac{c}{v}\left(\frac{d}{dz}\log W(z)\right)\left(\frac{c}{k - k_1}\right),$$

with $W(z)$ an arbitrary non-zero solution of Whittaker's equation. If we make the translation $\tilde{k} = k - k_1$, the f_n's satisfy $Qf = \tilde{k}f$ with

$$a_n = -\frac{c^2}{4}\frac{(v + 2n - 2\lambda - 1)(v + 2n + 2\lambda - 1)}{(v + 2n - 2)(v + 2n - 1)^2(v + 2n)}$$

$$b_n = \frac{2\lambda c}{(v + 2n - 3)(v + 2n - 1)}$$

and the second order equation

$$\tilde{k}^2 \frac{d^2 f_n}{d\tilde{k}^2} + \left((1 - v)\tilde{k} + \frac{2(f_1 v^2 + f_1 v + \lambda c)}{v + 1}\right)\frac{df_n}{d\tilde{k}} = n(n + v)f_n.$$

5.2.2. $u = 0$. In this case the equation (21) above becomes

$$-(k^2 r^2 + (4a_1 - 4a_0)kr + 4a_1^2 - 8a_0 a_1 + 4a_0^2)\frac{d^2 w}{dk^2}$$
$$-(2k^2 r + (4a_1 - 4a_0)k)\frac{dw}{dk} - 4a_1 w = 0.$$

Again, there are two cases.

Case a)

Assume that $r \neq 0$. If we make the change of variable

$$k = \frac{x - 2a_1 + 2a_0}{r}$$

and put $\tilde{w}(x) = w(k)$ we get

$$-\frac{4a_1 \frac{d\tilde{w}}{dx}}{r^2 x} + \frac{4a_0 \frac{d\tilde{w}}{dx}}{r^2 x} + \frac{4a_1 \tilde{w}}{r^2 x^2} + \frac{d^2 \tilde{w}}{dx^2} + \frac{2 \frac{d\tilde{w}}{dx}}{r^2} = 0$$

and the further replacement

$$\tilde{w} = \frac{e^{-\frac{x}{r^2}} y}{x^{\frac{2(a_0 - a_1)}{r^2}}}$$

gives

$$\frac{d^2 y}{dx^2} = \frac{(x^2 - 4a_1 x + 4a_0 x - 2a_1 r^2 - 2a_0 r^2 + 4a_1^2 - 8a_0 a_1 + 4a_0^2) y}{r^4 x^2}.$$

If one puts $z = \frac{2x}{r^2}$ and $W(z) = y(x)$ this becomes the Whittaker equation

$$\frac{d^2 W}{dz^2} + \left(-\frac{1}{4} + \frac{\lambda}{z} + \frac{1/4 - \nu^2}{z^2} \right) W = 0,$$

with

$$\lambda = \frac{2(a_1 - a_0)}{r^2}, \quad \nu^2 = \frac{1}{4} - \frac{2(a_1 + a_0)}{r^2} + \frac{4(a_1 - a_0)^2}{r^4}.$$

The function f_1 is given by

$$f_1 = \frac{r}{2} \left(-\lambda + \frac{z}{2} - z \frac{d}{dz} \log W(z) \right).$$

The functions $\tilde{f}_n = 2f_n/r$ satisfy $Q\tilde{f} = z\tilde{f}$ with

$$a_n = \frac{(2n + 2\lambda - 2\nu - 1)(2n + 2\lambda + 2\nu - 1)}{4}$$

$$b_n = 2(n + \lambda - 1)$$

and the second order equation

$$-z \frac{d^2 \tilde{f}_n}{dz^2} - (z - 2\lambda - 2\tilde{f}_1) \frac{d\tilde{f}_n}{dz} = n\tilde{f}_n.$$

Case b)

When $r = 0$ and $a_0 \neq a_1$, equation (21) becomes

$$\frac{d^2w}{dk^2} + \frac{k}{a_1 - a_0}\frac{dw}{dk} + \frac{a_1}{(a_1 - a_0)^2}w = 0.$$

The replacement

$$w = e^{-\frac{k^2}{4(a_1 - a_0)}}\, y,$$

gives

$$\frac{d^2y}{dk^2} - \frac{k^2 - 2(a_1 + a_0)}{4(a_1 - a_0)^2}\, y = 0.$$

If one puts $k = cx$, $c = \sqrt{a_1 - a_0}$ and $W(x) = y(k)$, this becomes the Weber–Hermite equation

$$\frac{d^2W}{dx^2} + \left(\nu + \frac{1}{2} - \frac{x^2}{4}\right)W = 0,$$

with

$$\nu = \frac{a_0}{a_1 - a_0}.$$

The function f_1 is given by

$$f_1 = \frac{k}{2} - c\left(\frac{d}{dx}\log W(x)\right)\left(\frac{k}{c}\right).$$

The f_n's satisfy $Qf = kf$ with

$$a_n = c^2(n + \nu), \quad b_n = 0,$$

and the second order equation

$$-c^2\frac{d^2 f_n}{dk^2} + (2f_1 - k)\frac{df_n}{dk} = nf_n.$$

Finally, if $a_0 = a_1$, all b_n's are zero and all a_n's are equal, say to 1. This forces $f_1 = 1/k$ or $f_1 = (1/2)(k \pm \sqrt{k^2 - 4})$. The first case is easily excluded and the second one gives $h_1 = \pm\sqrt{k^2 - 4}$ and $h_2 = 0$.

6. The Cases of Bochner

Here we see how the cases determined by Bochner [2] fit as special cases of the results given above.

Both the Hermite and the Laguerre cases are special cases of 5.2.2 ($v \neq 0$, $u = 0$). Generically the Jacobi and Bessel cases are special cases of

5.2.1 ($v \neq 0$, $u \neq 0$) and particular instances of these are special cases of
5.1 ($v = 0$). The details are given below.

We start from the expressions in case b) of Section 3

$$\frac{df_1}{dk} = \frac{(f_1 r - 2f_1 k + 2f_1^2 + 2a_1)v + (3f_1 r - 2f_1 k + 4a_1)u}{(kr + 2a_1 - 2a_0)v + (3kr - 2k^2 + 4a_1 + 4a_0)u},$$

$$h_0 = 0,$$
$$h_1 = (2f_1 - k)v + ku,$$

$$h_2 = -\frac{(kr + 2a_1 - 2a_0)v + (3kr - 2k^2 + 4a_1 + 4a_0)u}{2}.$$

The case $u = 0$ (and nonzero v) gives

$$\frac{df_1}{dk} = \frac{f_1 r - 2f_1 k + 2f_1^2 + 2a_1}{kr + 2a_1 - 2a_0},$$

$$h_0 = 0,$$
$$h_1 = (2f_1 - k)v,$$

$$h_2 = -\frac{(kr + 2a_1 - 2a_0)v}{2}.$$

The Case of Hermite Polynomials. We have

$$r = 0, \quad f_1 = k, \quad a_0 = 0, \quad a_1 = \tfrac{1}{2}$$

and we get

$$-1/2 \, d^2 H_n(k)/dk^2 + kdH_n(k)/dk = nH_n(k), \text{ for } n \geq 0.$$

This case amounts to the choice

$$\nu = 0, \quad c^2 = \tfrac{1}{2} \quad \text{and} \quad W(x) = e^{-\frac{x^2}{4}}$$

in case b) of 5.2.2.

The Case of Laguerre Polynomials. We have

$$r = 2, \quad f_1 = k, \quad a_0 = 0, \quad a_1 = a + 1$$
$$h_0 = 0, \quad h_1 = kv, \quad h_2 = (-k - a - 1)v$$

and if we put

$$k + a + 1 = z$$

we get

$$-zd^2 L_n(z)/dz^2 + (z - a - 1) \, dL_n(z)/dz = nL_n(z), \text{ for } n \geq 0.$$

This case amounts to the choice

$$v = \lambda - \frac{1}{2} = \frac{a}{2} \quad \text{and} \quad W(z) = z^{\frac{a+1}{2}} e^{-\frac{z}{2}}$$

in case a) of 5.2.2.

We now consider the case of $u \neq 0$, normalized to $u = 1$.

The Case of Jacobi Polynomials. Since we have

$$
\begin{aligned}
h_0 &= 0, \\
h_1 &= (2f_1 - k)v + ku, \\
h_2 &= -\frac{(kr + 2a_1 - 2a_0)v + (3kr - 2k^2 + 4a_1 + 4a_0)u}{2}
\end{aligned}
$$

we see that plugging in

$$a_0 = 0, \ u = 1, \ f_1 = k$$

and setting

$$z = c_1 k + c_2$$

we get for B the operator

$$h_2 d^2/dk^2 + h_1 d/dk = (z^2 - 1)d^2/dz^2 + (z - c_2)(1 + v)d/dz$$

provided we choose c_1 and c_2 as follows:

$$c_1 = -\frac{4c_2}{r(v+3)}, \quad c_2^2 = \frac{r^2(v+3)^2}{r^2v^2 + 6r^2v + 16a_1v + 9r^2 + 32a_1}.$$

Recall that in the Jacobi case one has

$$a_1 = \frac{4(a+1)(b+1)}{(b+a+2)^2(b+a+3)}$$

$$r = \frac{4(a-b)}{(b+a+2)(b+a+4)}.$$

Now the operator above is the standard Jacobi operator provided we put

$$c_2 = \frac{b-a}{b+a+2}, \quad v = b+a+1.$$

If $v \neq 0$, denoting by $\tilde{a}$, $\tilde{b}$, $\tilde{c}$ the parameters in the standard form of the hypergeometric equation, this case amounts to the choice

$$\tilde{a} = \tilde{c} = b+1, \qquad \tilde{b} = -a,$$

$$F(x) = (x-1)^a \quad \text{and} \quad x = \frac{z+1}{2}$$

in case a) of 5.2.1.

The Case of Bessel Polynomials. One has

$$a_0 = 0, \quad u = 1, \quad f_1 = k$$

and

$$r = \frac{4b}{a(a+2)}, \quad a_1 = -\frac{b^2}{a^2(a+1)}, \quad v = a-1$$

and we get

$$h_2 d^2/dk^2 + h_1 d/dk = (k - b/a)^2 d^2/dk^2 + k(1+v)d/dk$$

and by putting $\tilde{k} = k - b/a$ we get for B the expression

$$\tilde{k}^2 d^2/d\tilde{k}^2 + (a\tilde{k} + b)d/d\tilde{k}.$$

When $v \neq 0$, this case amounts to the choice

$$v = 2\lambda + 1 = a - 1, \quad c = -b, \quad W(z) = z^{1-\frac{a}{2}}e^{\frac{c}{z}}$$

in case b) of 5.2.1.

7. The Virasoro Constraints

In the last few years the partition function of the Hermitian matrix model and its so-called double scaling limit, the Kontsevich integral, have been characterized as highest weight vectors of some representations of the Virasoro algebra, see [18] and references therein. This result can be derived from the fact that these partition functions are tau-functions of the Toda lattice and the KdV hierarchies, flowing from some fixed point of the so-called master (conformal) symmetries associated with these hierarchies. In this section we want to show that the solutions of our problem can be characterized as fixed points of (in general) some (sub)algebras of the Virasoro algebra formed by the Toda master symmetries fields. Similar results have been obtained in [1] and [17], in the special cases of the Hermite, Laguerre and Jacobi polynomials. Here we like to stress that the Virasoro constraints are nothing but an integrated form of the equation $(\mathrm{ad}\, Q)^3(\Theta) = 0$. We intend to explore the representation theoretic implications of our result in some further publication, see [17] for an account in the case of Laguerre polynomials.

The Toda lattice hierarchy is defined by

$$T_j : \frac{\partial Q}{\partial t_j} = [(Q^{j+1})_+, Q], \quad j = -1, 0, 1, 2, \ldots, \tag{22}$$

where $(Q^{j+1})_+$ denotes the upper part of Q^{j+1}, including the diagonal; $j = -1$ is the identically zero flow and is included for notational convenience,

$j = 0$ is the usual Toda lattice flow. The Toda lattice hierarchy is known to admit two hamiltonian formulations. As a consequence ([1], [5], [16], see also [6], [24] for an exposition of these ideas in the context of the KdV hierarchy), one can define so-called master symmetries vector fields

$$V_j : \frac{\partial Q}{\partial t_j} = [B^{(j)}, Q] + Q^{j+1}, \ j = -1, 0, 1, 2, \ldots, \tag{23}$$

satisfying the commutation relations of the Virasoro algebra:

$$[V_i, V_j] = (j - i)V_{i+j}. \tag{24}$$

The Toda flows commute between themselves

$$[T_i, T_j] = 0, \tag{25}$$

and can be generated, via commutators, by the master symmetries fields:

$$[V_i, T_j] = (j + 1)T_{i+j}. \tag{26}$$

Below we list the first few of the master symmetries fields:

- $V_{-1}a_n = 0$, $V_{-1}b_n = 1$

- $V_0 a_n = 2a_n$, $V_0 b_n = b_n$

- $V_1 a_n = 2a_n[(n+1)b_{n+1} - (n-1)b_n]$
  $V_1 b_n = b_n^2 + (2n+1)a_n - (2n-3)a_{n-1}$

- $V_2 a_n = a_n[(3-2n)(a_{n-1} + b_n^2) + 2a_n + (2n+3)(a_{n+1} + b_{n+1}^2) + 2(b_{n+1} - b_n)\sum_{j=-\infty}^{n} b_j]$
  $V_2 b_n = b_n(b_n^2 + (2n+1)a_n - (2n-5)a_{n-1}) + 2(n+1)a_n b_{n+1} - 2(n-2)a_{n-1}b_{n-1} + 2(a_n - a_{n-1})\sum_{j=-\infty}^{n} b_j,$

where the symbol $\sum_{j=-\infty}^{n}$ is defined as in (15). The corresponding matrices $B^{(j)}$ are given by:

- $B^{(-1)} = 0$

- $B^{(0)}_{nn} = n + x$, $B^{(0)}_{mn} = 0$, $m \neq n$

- $B^{(1)}_{nn+1} = n + y$, $B^{(1)}_{nn-1} = -a_n(n+1-y)$

  $B^{(1)}_{nn} = nb_{n+1} + \sum_{j=-\infty}^{n} b_j + y(b_{n+1} - 1) + z$

  all other entries zero

- $B_{nn+2}^{(2)} = n + p,\ B_{nn-2}^{(2)} = -a_{n-1}a_n(n+1-p)$

  $B_{nn+1}^{(2)} = n(b_{n+1} + b_{n+2}) + \sum_{j=-\infty}^{n} b_j + p(b_{n+1} + b_{n+2} - b_1 - b_2) + q$

  $B_{nn-1}^{(2)} = -a_n[(n+1)(b_n + b_{n+1}) + \sum_{j=-\infty}^{n+1} b_j - p(b_n + b_{n+1} - b_1 - b_2) - q]$

  $B_{nn}^{(2)} = (n-2)(a_n + a_{n+1} + b_{n+1}^2) + 2\sum_{j=-\infty}^{n+1} a_j + \sum_{j=-\infty}^{n+1} b_j^2 + b_{n+1}(\sum_{j=-\infty}^{n+1} b_j) +$

  $p(b_{n+1}^2 - b_{n+1}(b_1 + b_2) + b_1 b_2 + a_{n+1} + a_n - a_1) + q(b_{n+1} - b_1) + h$

  all other entries zero,

where x, y, z, p, q, h denote arbitrary parameters.

With these preliminaries, we can state the result of this section. Define

$$L_j = \alpha V_j + \beta V_{j+1} + 2uV_{j+2} \\ + [(r - 2b_1)v + (2b_1 + 3r)u]T_j + 2(v - u)T_{j+1},$$

$j = -1, 0, 1, \ldots$. From the commutation relations (24), (25) and (26), one immediately obtains that the L_j's form a subalgebra of the algebra of master symmetries:

$$[L_i, L_j] = (j - i)(\alpha L_{i+j} + \beta L_{i+j+1} + 2uL_{i+j+2}). \tag{27}$$

Proposition 2. *The solutions of* $(\mathrm{ad}\ Q)^3(\Theta) = 0$ *can be characterized as the solutions of the Virasoro type constraints*

$$L_j a_n = L_j b_n = 0, \quad n \in \mathbb{Z}, \quad j = -1, 0, 1, \ldots \tag{28}$$

Proof. From the string equation (17) we deduce

$$\alpha Q^{j+1} + \beta Q^{j+2} + 2uQ^{j+3} = [Q, PQ^{j+1}], \quad j = -1, 0, 1, \ldots \tag{29}$$

We observe that by picking

$$x = 0, \quad y = z = p = 1, \quad q = 3b_1 + r, \quad h = 2a_0 + a_1 + 2b_1^2$$

in the definition of $B^{(j)}$, we have

$$P = -2vb_1 I + \beta B^{(0)} + 2uB^{(1)} + 2(v - u)Q_+ \tag{30}$$

and

$$PQ = -2va_0 I + \alpha B^{(0)} + \beta B^{(1)} + 2uB^{(2)} \\ + [(r - 2b_1)v + (2b_1 + 3r)u]Q_+ + 2(v - u)(Q^2)_+. \tag{31}$$

The identity (30) is obtained by a straightfoward computation, remembering the definition of P in (16), while (31) amounts to checking the formulas:

$$4u \sum_{j=-\infty}^{n} b_j + 2((2n+1)u + v)b_{n+1} + (2n+1)\beta + ((r-2b_1)v + (2b_1 + 3r)u) = 0$$

and

$$4u \sum_{j=-\infty}^{n-1} a_j + 2u \sum_{j=-\infty}^{n} b_j^2 + \beta \sum_{j=-\infty}^{n} b_j + 4u(na_n + a_0) + 2v(a_n - a_0) + n\alpha = 0,$$

which are derived by adding up respectively (11) and (13).

From the definitions (22) and (23) of the Toda and the master symmetries flows, it is now clear that the identity (29) for $j = -1$ and $j = 0$ is equivalent to $L_{-1}a_n = L_{-1}b_n = 0$ and $L_0 a_n = L_0 b_n = 0$. If $u \neq 0$, from the commutation relations (27), we deduce automatically that $L_j a_n = L_j b_n = 0$, for all $j \geq 1$. By continuity, it follows that these identities continue to hold when $u = 0$, completing the proof.

8. Examples of Virasoro and Toda Flows

When the tridiagonal matrix Q is replaced by a Schroedinger operator, F. Magri and J. P. Zubelli [24] have shown that *all* the solutions of the bispectral problem described in [8], organize into manifolds which are integral manifolds of either the KdV hierarchy (rank 1 case) or its master symmetries fields (rank 2 case). The case of the KdV hierarchy had been done in [8]. Our forthcoming joint work [16] will give an account of similar results in the context of the "discrete-continuous" bispectral problem.

Below we present the simplest examples of an integral curve and an integral surface of the vector fields L_j, $j = -1, 0, 1, \ldots$, obtained respectively by one and two applications of the Darboux transformation to the Laguerre polynomials (with $a = 1$) and the Jacobi polynomials (with $a = b = 1$). We refer the reader to [14] for a convenient formulation of the Darboux process in the context of semiinfinite tridiagonal matrices. There we showed that all the instances of orthogonal polynomials described back in 1938 by H. Krall as the only cases which are eigenfunctions of a fourth order differential operator, can be obtained by application of the Darboux process to some of the classical orthogonal polynomials. Finally we exhibit some rational solutions to the Toda lattice hierarchy itself, obtained by application of the Darboux transformation to a specific biinfinite tridiagonal matrix.

The Krall–Laguerre Curve. The (monic) Laguerre polynomials correspond to

$$a_n = n(n + a), \quad b_n = 2n + a - 1.$$

In this case $u = a_0 = 0$, $a_1 = b_1 = 1 + a$ and $r = 2$. Hence, from (12), (14) and (28), $\alpha = 0$, $\beta = -2v$ and the corresponding semiinfinite matrix Q satisfies (remember that $T_{-1} = 0$):

$$(V_{j+1} - T_{j+1} + aT_j)(Q) = 0, \quad j = -1, 0, 1, \ldots \tag{32}$$

In [14] we have shown that the curve $\tilde{Q}(R)$ obtained by one application of the Darboux transformation to the Laguerre polynomials with $a = 1$, is given by:

$$\tilde{a}_n = \frac{n^2(R + n - 1)(R + n + 1)}{(R + n)^2}$$

$$\tilde{b}_n = \frac{(2n - 1)R^2 + 4n(n - 1)R + (n - 1)n(2n - 1)}{(R + n - 1)(R + n)}, \tag{33}$$

where R denotes the free parameter of the Darboux transformation. The corresponding (monic) polynomials are orthogonal on $[0, \infty[$ for the weight $e^{-k} + (1/R)\delta(k)$, with $\delta(k)$ the Dirac delta function. They are eigenfunctions of a fourth order differential operator $B(k, d/dk)$. From (33) one computes that the vector field $L_{-1} = V_0 - T_0$ is tangent to the curve $\tilde{Q}(R)$ and that the other vector fields $L_j = V_{j+1} - T_{j+1}$, $j = 0, 1, 2, \ldots$, vanish along that curve. Precisely one has

$$(R\frac{d}{dR} + V_0 - T_0)(\tilde{Q}(R)) = 0 \tag{34}$$

and

$$(V_{j+1} - T_{j+1})(\tilde{Q}(R)) = 0, \quad j = 0, 1, 2, \ldots \tag{35}$$

In fact, by the commutation relations (27), it is enough to check (35) for $j = 0, 1$. Observe that when R tends to infinity (34) and (35) reduce to (32) with $a = 0$, in agreement with (33) which then becomes $\tilde{a}_n = n^2$, $\tilde{b}_n = 2n - 1$.

The Krall–Legendre Surface. The (monic) Jacobi polynomials correspond to:

$$a_n = \frac{4n(n + a + b)(n + a)(n + b)}{(2n + a + b - 1)(2n + a + b)^2(2n + a + b + 1)}$$

$$b_n = \frac{b^2 - a^2}{(2n + a + b - 2)(2n + a + b)}.$$

Specializing the constraints (28) to this case, one finds (see Section 6 for the identification of the parameters) that the corresponding semiinfinite matrix Q satisfies:

$$(V_{j+2} - V_j + (a + b)T_{j+1} + (a - b)T_j)(Q) = 0, \quad j = -1, 0, 1, \ldots \tag{36}$$

In [14] we have shown that two appropriate successive applications of the Darboux transformation to the Jacobi polynomials with $a = b = 1$, lead to a surface of semiinfinite matrices $\tilde{Q}(R, S)$ (R, S, the free parameters of the Darboux process), with entries given by

$$\tilde{a}_n = \frac{n^2 Q(n-1)Q(n+1)}{(4n^2 - 1)Q(n)^2}$$

$$\tilde{b}_n = \frac{(S-R)[3n^4 - 6n^3 + (R+S+3)n^2 - (R+S)n - RS]}{Q(n-1)Q(n)} \tag{37}$$

with

$$Q(n) = n^4 + (R + S - 1)n^2 + RS.$$

The corresponding (monic) polynomials are orthogonal on $[-1, +1]$ for the weight $1/2 + (1/R)\delta(k+1) + (1/S)\delta(k-1)$, obtained by adding discrete masses to the weight function of the Legendre polynomials. They are eigenfunctions of a sixth order differential operator $B(k, d/dk)$ which reduces to a fourth order one only if the choice $R = S$ is made.

One computes that the vector fields $L_j = V_{j+2} - V_j$, $j = -1, 0, 1, \ldots$ are tangent to the surface $\tilde{Q}(R, S)$. Precisely:

$$[2(S\frac{\partial}{\partial S} + (-1)^j R\frac{\partial}{\partial R}) + V_{j+2} - V_j]\tilde{Q}(R, S) = 0, \quad j = -1, 0, 1, \ldots \tag{38}$$

In particular $L_{-1} = L_1 = L_3 = \ldots$ and $L_0 = L_2 = L_4 = \ldots$ along $\tilde{Q}(R, S)$. Again, by the commutation relations (27), it suffices to check (38) for $j = -1, 0$. Observe also that when R and S tend to infinity, (38) reduces to (36) with $a = b = 0$, in agreement with (37) which then becomes $\tilde{a}_n = n^2/(4n^2 - 1)$, $\tilde{b}_n = 0$.

Some Rational Solutions to the Toda Flows. Consider the *biinfinite* tridiagonal matrix given by

$$a_n = 1, \quad b_n = -2, \quad n \in \mathbb{Z}$$

corresponding to $a_0 = a_1 = 1$, $b_1 = -2$, $r = 0$. One application of the Darboux transformation, with free parameter t_0, leads to

$$\tilde{a}_n(t_0) = \frac{\tau_{n-1}(t_0)\tau_{n+1}(t_0)}{\tau_n^2(t_0)}$$

$$\tilde{b}_n(t_0) = -2 + \frac{\partial}{\partial t_0} \log \frac{\tau_n(t_0)}{\tau_{n-1}(t_0)},$$

with $\tau_n(t_0) = t_0 + n - 1$. The matrix $\tilde{Q}(t_0)$ with entries $\tilde{a}_n(t_0)$, $\tilde{b}_n(t_0)$ provides a solution to the Toda flow T_0 in (22). One checks easily that $(\text{ad } \tilde{Q}(t_0))^3(\Theta) = 0$ with $\theta_n = n^2 + (2t_0 - 1)n$ and that we have a "rank two" bispectral situation.

A second application of the Darboux transformation, for appropriate choices of the parameters t_0 and t_1, leads to a matrix $\tilde{\tilde{Q}}(t_0, t_1)$ with entries

$$\tilde{\tilde{a}}_n(t_0, t_1) = \frac{\tau_{n-1}(t_0,t_1)\tau_{n+1}(t_0,t_1)}{\tau_n^2(t_0,t_1)}$$
$$\tilde{\tilde{b}}_n(t_0, t_1) = -2 + \frac{\partial}{\partial t_0}\log\frac{\tau_n(t_0,t_1)}{\tau_{n-1}(t_0,t_1)} ,$$

with

$$\tau_n(t_0, t_1) = 2(2t_1 - t_0)^3 - 3(2n - 3)(2t_1 - t_0)^2$$
$$-6(n-2)(n-1)t_0 + 6(2n^2 - 6n + 5)t_1 - (n-2)(n-1)(2n-3) .$$

As a function of t_0 and t_1, $\tilde{\tilde{Q}}(t_0, t_1)$ solves the first and second Toda flows T_0 and T_1 in (22). One can check that $\tilde{\tilde{Q}}(t_0, t_1)$ possesses two different one-dimensional families of eigenfunctions which are also eigenfunctions of *nonconjugate* fourth order differential operators. It seems natural to conjecture that by continuing the process, one obtains the analogue of the rational solutions of the KdV hierarchy in [8] providing the "rank one" solutions to the bispectral problem. It may be appropriate to remark that the first application of the Darboux process above corresponds, just as in the continuous case, to *translation* as a look at $\tau_n(t_0)$ will confirm. In this regard the usual KdV flow corresponds to the second Toda flow.

References

[1] M. Adler and P. van Moerbeke, Matrix integrals, Toda symmetries, Virasoro constraints and orthogonal polynomials, *Duke Mathematical Journal* **80**, 3 (1995), 863–911.

[2] S. Bochner, Über Sturm–Liouvillesche Polynomsysteme, *Math. Z.* **29** (1929), 730–736.

[3] E. Brézin and V. A. Kazakov, Exactly solvable field theories of closed strings, *Phys. Lett.* B **236** (1990), 144–150.

[4] H. H. Chen, Y. C. Lee and J. E. Lin, On a new hierarchy of symmetries for the Kadomtsev–Petviashvili equation, *Physica* **9D** (1983), 439–445.

[5] P. Damianou, Master symmetries and R-matrices for the Toda Lattice, *Lett. Math. Phys.* **20** (1990), 101–112.

[6] L. Dickey, Additional symmetries of KP, Grassmannian and the string equation, *Modern Physics Letters A* **8** no. 13 (1993), 1259–1272.

[7] M. R. Douglas and S. H. Shenker, Strings in less than one dimension, *Nuclear Phys.* B **335** (1990), 635–654.

[8] J.J.. Duistermaat and F. A. Grünbaum, Differential equations in the spectral parameter, *Commun. Math. Phys.* **103** (1986), 177–240.

[9] A. S. Fokas and B. Fuchssteiner, The hierarchy of the Benjamin-Ono equation, *Phys. Lett.* **86A** (1981), 341–345.

[10] B. Fuchssteiner, Mastersymmetries, higher order time-dependent symmetries and conserved densities of nonlinear evolution equations, *Progress of Theoretical Physics* **70** no. 6 (1983), 1508–1522.

[11] I. M. Gel'fand and I. Ya. Dorfman, Hamiltonian operators and algebraic structures related to them, *Funkts. Anal. Prilozhen.* **13** no. 4 (1979), 13–30, English translation: *Functional Anal. Appl.* **13** no. 4 (1980), 248–262.

[12] D. J. Gross and A. A. Migdal, Nonperturbative two-dimensional quantum gravity, *Phys. Rev. Lett.* **64** (1990), 127–130.

[13] F. A. Grünbaum and L. Haine, in preparation.

[14] F. A. Grünbaum and L. Haine, Orthogonal polynomials satisfying differential equations: The role of the Darboux transformation, in: D.Levi and P.Winternitz (eds.), *Symmetries and Integrability of Difference Equations*, CRM Proceedings & Lecture Notes **9**, (American Mathematical Society, Providence, 1996), to appear.

[15] F. A. Grünbaum and L. Haine, The q-version of a theorem of Bochner, *Jour. of Computational & Appl. Math.* **68** (1996), 103–114.

[16] F. A. Grünbaum, L. Haine, F. Magri and J. P. Zubelli, in preparation.

[17] L. Haine and E. Horozov, Toda orbits of Laguerre polynomials and representations of the Virasoro algebra, *Bull. Sc. Math.*, 2e série, t. 117, no. 4 (1993), 485–518.

[18] C. Itzykson and J. B. Zuber, Combinatorics of the Modular Group II, the Kontsevich Integrals, *Int. Journ. Mod. Phys. A*, vol. **7**, no. 23 (1992), 5661–5705.

[19] V. G. Kac and A. Schwarz, Geometric interpretation of the partition function of 2D gravity, *Phys. Lett. B* **257** (1991), 329–334.

[20] M. Kontsevich, Intersection theory on the moduli space of curves and the matrix Airy function, *Commun. Math. Phys.* **147** (1992), 1–23.

[21] E. Laguerre, Sur la réduction en fractions continues d'une fraction qui satisfait à une équation différentielle linéaire du premier ordre dont les coefficients sont rationnels, *J. de Math. Pures et Appliquées* **1** (1885), 135–165, Oeuvres II, 685–711, Chelsea (1972).

[22] A. P. Magnus and A. Ronveaux, Laguerre and orthogonal polynomials in 1984, in *Polynômes Orthogonaux et Applications, Proceedings, Bar-le-Duc 1984*, Lecture Notes Math. **1171**, xxvii–xxxii.

[23] F. Magri, A simple model of the integrable Hamiltonian equation, *Journal of Mathematical Physics* **19** No. 5 (1978), 1156–1162.

[24] F. Magri and J. P. Zubelli, Differential equations in the spectral parameter, Darboux transformations and a hierarchy of master symmetries for KdV, *Commun. Math. Phys.* **141** (1991), 329–351.

[25] W. Oevel and B. Fuchssteiner, Explicit formulas for symmetries and conservation laws of the Kadomtsev–Petviashvili equation, *Phys. Lett.* **88A** (1982), 323–327.

[26] A. Orlov and E. Schulman, Additional symmetries for integrable equations and conformal algebra representations, *Lett. Math. Phys.* **12** (1986), 171–179.

[27] P. van Moerbeke, Integrable foundations of string theory, in *Lectures on Integrable Systems*, Editors O. Babelon, P. Cartier, Y. Kosmann-Schwarzbach, World Scientific (1994), 163–267.

[28] E. Witten, Two-dimensional gravity and intersection theory on moduli space, *Surveys in Differential Geometry* **1** (1991), 243–310.

F. Alberto Grünbaum
Department of Mathematics
University of California
Berkeley, CA, 94720

Luc Haine
Department of Mathematics
Université Catholique de Louvain
1348 Louvain-la-Neuve, Belgium

Received November, 1995

Obstacles to Asymptotic Integrability

Y. Kodama and A. V. Mikhailov

Abstract

We study nonintegrable effects appearing in the higher order corrections of an asymptotic perturbation expansion for a given nonlinear wave equation, and show that the analysis of the higher order terms provides a sufficient condition for asymptotic integrability of the original equation. The nonintegrable effects, which we call "obstacles" to the integrability, are shown to result in an inelasticity in soliton interaction. The main technique used in this paper is an extension of the normal form theory developed by Kodama and the approximate symmetry approach proposed by Mikhailov. We also discuss the case of the KP equation with the higher order corrections, a quasi-two dimensional extension of weakly dispersive nonlinear waves.

1. Introduction

For a given PDE, asymptotic expansions such as the envelope (or quasi-monochromatic) approximation and the long wave limit can be used for testing the integrability of the PDE. The simplest test of this type was formulated by Calogero [3]. He suggested expanding the PDE until the first nonlinear correction appears and checking whether this correction is integrable. It is also well-known [3] [19] that for a wide class of nonlinear dispersive wave equations, the leading order nonlinear equation in an aymptotic expansion turns out to be given by an integrable equation, such as the Korteweg–de Vries (KdV) equation in the long wave limit (the weakly dispersive limit), and the nonlinear Schrödinger (NLS) equation in the envelope approximation (the strongly dispersive limit). This implies that most of the nonlinear dispersive wave equations are integrable at the nontrivial leading orders in the asymptotic sense. However, as we know, the set of integrable systems may be so small that we expect to find a nonintegrable effect in the higher order corrections in the expansion. This has been pointed out in [11] for the weakly dispersive limit (the KdV case) and in

[8] for the strongly dispersive limit (the NLS case) by means of the normal form technique and also by the approximate symmetry approach in [15].

In this paper, we give a comprehensive study of the nonintegrable effects, which we call the "obstacles" to integrability, appearing in higher order corrections, and show that the analysis of the higher order corrections provides a sufficient condition for the asymptotic integrability of a given nonlinear wave equation. What we mean by the term "asymptotic integrability" is that the equation with higher order corrections can be shown to be integrable up to a certain order in an asymptotic sense. Existence of higher local symmetries and conservation laws may be taken as a definition of integrability [16]. In this sense the obstacles discovered provide us with the necessary conditions of integrability (see examples in Section 2).

The content of this paper is as follows: In Section 2, we consider the envelope approximation of a nonlinear dispersive wave equation, and derive a general form of the NLS equation with higher order corrections. We then look for the symmetries of this perturbed NLS equation in the class of differential polynomials, and find the obstacles to the existence of the higher order corrections up to certain orders (Proposition 2.1). There is no obstacle in the first order correction, however in the second and higher order corrections several obstacles appear. These obstacles also give the conditions for the nonexistence of higher order corrections to the NLS integrals (Proposition 2.2).

To illustrate the above procedure, we start here with the case of a nonlinear Klein–Gordon equation as a simple model. In Section 3, we introduce a near identity transformation given by the Lie transformation [11] for the NLS equation with higher order corrections. The purpose of this transformation is to simplify the equation for an analysis of the higher order corrections. The simplified equation is called the "normal form" of this perturbed equation. The main result in this section is: if there is no obstacles in the higher order corrections of symmetries (or integrals), then there exists a near identity transformation such that the perturbed equation is mapped to the NLS equation with its higher symmetries up to this order (Theorem 3.1). Namely, the original wave equation is asymptotically integrable up to this order and is approximated by a NLS hierarchy.

In Section 4, we define the specific form of the normal form based on a requirement that the normal form also keeps some of the NLS integrals. This leads to an existence of special solutions to the normal form in the same form as those of the NLS equation, such as the one-soliton solution. We then study the interaction problem of a two soliton solution, and show that the obstacles result in an inelasticity in the interaction (Theorem 4.1). Thus the obstacles are the secular (or resonant) terms in an asymptotic expansion. In Section 5, we extend our discussion to the case of the Kadomtsev–Petviashvili (KP) equation as a model of a two-dimensional near integrable equation. Here the nonlocality in the higher order corrections plays an essential role in asymptotic expansion and the normal form technique. Similar results on the normal form for this case are also obtained (Proposition 5.2).

Section 6 is devoted to a concluding remark and discussion. We also provides basic properties of the space of differential polynomials in Appendix A, and a detail computation in the envelope approximation in Appendix B.

2. Asymptotic Expansions and Symmetries

In this section we focus our attention on the envelope approximation. Let us consider a nonlinear PDE of the following form for a function $f = f(x,t)$,

$$L \cdot f + N(f) = 0. \tag{2.1}$$

Here L is a linear differential operator of time variable t and a one space variable x with constant coefficients, and $N(f)$ represents the nonlinear part, $N(f) = N_2(f) + N_3(f) + \cdots$ where $N_2(f), N_3(f), \ldots$ are quadratic, cubic,... functions in f and its derivatives. We assume that f is small (of order $\epsilon \ll 1$) and therefore that it satisfies the corresponding linear equation with small nonlinear corrections. The basic idea of the envelope approximation is to represent the solution of (2.1) in the form of a plane wave with a slowly varying amplitude. Because of the nonlinearity higher (and zero) harmonics contribute in the expansion of f as the higher orders, i.e.

$$
\begin{aligned}
f = {}& \epsilon A e^{i[\omega(k)t - kx]} + \epsilon A^* e^{-i[\omega(k)t - kx]} + \epsilon^2 A_0 + \\
& + \sum_{m=2}^{\infty} \epsilon^m \left(A_m e^{im[\omega(k)t - kx]} + A_m^* e^{-im[\omega(k)t - kx]} \right),
\end{aligned} \tag{2.2}
$$

where the linear dispersion relation $\omega = \omega(k)$ is defined from the linear equation $L \cdot e^{i[\omega(k)t - kx]} = 0$. Here we have assumed that only the master mode with amplitude A is resonant, this is, the dispersion relation is non-degenerate ($L \cdot e^{in[\omega(k)t - kx]} \neq 0$ if $n \neq 0, \pm 2, \pm 3, \ldots$), so all other modes are not in resonance and we shall treat them as slave modes. The requirement that all amplitudes have a slow dependence on x and t, so that they are functions of $\epsilon^p x, \epsilon^q t$ for some $p, q > 0$, leads uniquely to determine all slave amplitudes in terms of the master amplitude. This results in an equation for the master amplitude, and there are no obstacles to this procedure in any order of ϵ.

Let us illustrate the envelope expansion by taking the nonlinear Klein–Gordon equation as an example

$$(\partial_t^2 - \partial_x^2 + 1) \cdot f + af^2 + bf^3 + cf^4 + df^5 = 0, \tag{2.3}$$

where a, b, c, d are real constants. The dispersion relation corresponding to (2.3) is nondegenerate, since $\omega(k) = \sqrt{1 + k^2}$. For simplicity we here take $k = 0$. It then follows from (2.2) and (2.3) that all amplitudes in the expansion are functions of slow variables $\tau = \epsilon^2 t$ and $\xi = \epsilon x$, and it is a self-consistent assumption [19]. Substituting f of (2.2) into (2.3), and taking into account that $A = A(\xi, \tau)$ and $A_k = A_k(\xi, \tau)$, we collect coefficients at each harmonic and set them to be zero separately. In order to get the second nonlinear correction for the master amplitude equation, one has to keep terms up to order ϵ^5 inclusively. The resulting equations for the slave and master amplitudes can be easily solved by iterations (remember ϵ is a small parameter; see Appendix B for details). Thus all slave amplitudes are expressed in terms of A, A^* and its derivatives. Finally the equation for the master amplitude reads[1]

$$
\begin{aligned}
2iA_\tau \;=\; & A_{\xi\xi} + \left(\frac{10}{3}a^2 - 3b\right)|A|^2 A + \\
& + \epsilon^2 \left\{ \frac{1}{4}A_{\xi\xi\xi\xi} + \left(\frac{62}{9}a^2 - 3b\right)|A|^2 A_{\xi\xi} + \right. \\
& + \left(\frac{19}{9}a^2 - \frac{3}{2}b\right)A^* A_\xi^2 + \left(\frac{34}{3}a^2 - 3b\right)|A_\xi|^2 A + 4a^2 A_{\xi\xi}^* A^2 + \\
& + \left. \left(\frac{485}{54}a^4 - \frac{173}{6}a^2 b - \frac{15}{8}b^2 + 28ac - 10d\right)|A|^4 A \right\} + O(\epsilon^4).
\end{aligned}
$$

$$(2.4)$$

We would like to point out that a correction of order ϵ^k is a homogeneous differential polynomial of the scaling weight $k + 3$.

The leading order of (2.4) is the NLS equation which is integrable and possesses infinitely many local conservation laws and symmetries [17]. The ϵ^2-order correction may violate integrability. If we take a nonzero wave vector k in (2.2), then corrections of order ϵ appear as well, but those corrections do not carry any information about the nonintegrability at this order (Proposition 3.2, see also the next section). We consider the expansion (2.2) of (f, f_t) in terms of (A, A^*) as a change of variables which is formally invertible. Thus, if the original equation in (f, f_t) possesses symmetries or conservation laws, then the corresponding expansion in (A, A^*) will retain such properties. For instance, the equation (2.3) has two obvious integrals of motion, momentum and energy,

$$
P \;=\; \int f_x f_t \, dx ,
$$

$$(2.5)$$

---

[1] We here assume that $10a^2 \neq 9b$, othewise another scaling has to be used (see for example [3]).

$$H = \int \left(\frac{1}{2} f_t^2 + \frac{1}{2} f_x^2 + \frac{1}{2} f^2 + \frac{a}{3} f^3 + \frac{b}{4} f^4 + \frac{c}{5} f^5 + \frac{d}{6} f^6 \right) dx . \quad (2.6)$$

Substituting f of (2.2) into (2.5) and (2.6) we obtain integrals of (2.4)

$$P = 2i\epsilon^2 \int A A_\xi^* \, d\xi +$$

$$+ i\epsilon^4 \int \left\{ A_\xi A_{\xi\xi}^* - \left(\frac{22}{9} a^2 - 3b \right) |A|^2 A A_\xi^* \right\} d\xi + O(\epsilon^6) , \quad (2.7)$$

$$H = 2\epsilon \int |A|^2 \, d\xi +$$

$$+ \epsilon^3 \int \left\{ 2|A_\xi|^2 - \left(\frac{37}{9} a^2 - \frac{9}{2} b \right) |A|^4 \right\} d\xi + O(\epsilon^5) . \quad (2.8)$$

We here want to mention that there are no oscillating harmonics in (2.7) and (2.8).

The symmetries corresponding to (2.5) and (2.6) are nothing but the translations in x and t. In terms of the variables A, A^* they correspond to $A_{\tau_1} = A_\xi$ and (2.4) respectively. Equation (2.4) has one more obvious symmetry $A \to e^{i\theta} A$. The origin of this symmetry is in our assumption that solutions of the nonlinear Klein–Gordon equation (2.3) can be characterized by two different time scales (fast scale t and slow scale $\tau = \epsilon^2 t$) which we treat as independent variables. This symmetry can be expressed in terms of f and its derivatives as an asymptotic series in ϵ. Each independent variable (or scale) provides us with a symmetry. Let us address the question as to whether there exist other symmetries of equation (2.4) in the class of differential polynomials. This is in fact a key idea used in [16] for the test of integrability. One more motivation for this problem is the following: If we expect that the original equation is integrable, i.e. possesses higher symmetries, then the corresponding envelope equation has the same property.

Assuming that $10a^2 > 9b$ and after obvious rescalings, we can rewrite (2.4) in the form:

$$q_\tau = iq_{\xi\xi} + 2i|q|^2 q + i\epsilon^2 (a_1^2 q_{\xi\xi\xi\xi} + a_2^2 |q|^2 q_{\xi\xi} + a_3^2 q^* q_\xi^2 +$$

$$+ a_4^2 q^2 q_{\xi\xi}^* + a_5^2 |q_\xi|^2 q + a_6^2 |q|^4 q) + O(\epsilon^4) . \quad (2.9)$$

The leading term of (2.9) is the NLS equation, and its first few symmetries are given by [17]

$$K_0^0 = iq , \quad K_1^0 = q_\xi , \quad K_2^0 = iq_{\xi\xi} + 2i|q|^2 q , \quad K_3^0 = q_{\xi\xi\xi} + 6|q|^2 q_\xi ,$$

$$K_4^0 = iq_{\xi\xi\xi\xi} + i8|q|^2 q_{\xi\xi} + i6q^* q_\xi^2 + i2q^2 q_{\xi\xi}^* + i4|q_\xi|^2 q + i6|q|^4 q. \quad (2.10)$$

The flows K_0^0 and K_1^0 commute with (2.9), as we have mentioned. Other flows do not commute with (2.9) in general. Of course, the flow K_2^0 can be deformed at order ϵ^2, and we obtain just (2.9) to provide the commutativity. Let us then consider a problem: Is it possible to find ϵ^2-order deformations of the flows $K_3^0, K_4^0, \cdots$ in the class of the differential polynomials in q and q^*, which commute with the equation (2.9)? The answer is nontrivial: *The deformations of the flows $K_n^0, n = 3, 4, \ldots$, commuting with (2.9) exist if and only if the coefficients a_k^2 satisfy the following constraint*

$$12a_1^2 - a_2^2 - 2a_3^2 - 4a_4^2 + a_5^2 + 2a_6^2 = 0. \tag{2.11}$$

For K_3^0 and K_4^0 this statement can be verified by direct computation. In the next section we shall prove that the condition (2.11) is sufficient for the existence of approximate symmetries in order ϵ^2 for any K_n^0 of the hierarchy.

In terms of the coefficients $a, \ldots, d$ in (2.4) or (2.3), the above condition means [15]

$$22a^2b - 3b^2 - 28ac + 10d = 0. \tag{2.12}$$

Two known cases of the integrable nonlinear Klein–Gordon equation [21] given by

$$f_{tt} - f_{xx} + \sin(f) = 0 \quad \text{and} \quad f_{tt} - f_{xx} + \frac{1}{3}e^f - \frac{1}{3}e^{-2f} = 0$$

of course satisfy this constraint. We also note that the equations

$$f_{tt} - f_{xx} + e^f - 1 = 0 \quad \text{and} \quad f_{tt} - f_{xx} + \sin(f) + \lambda \sin(2f) = 0, \ (\lambda \neq 0)$$

are known to be nonintegrable and this can be also seen from the failure of the test of integrability (2.12).

We now consider the general case where the right hand side of the equation (2.9) is given by

$$q_\tau = K = K^0 + \epsilon K^1 + \epsilon^2 K^2 + \epsilon^3 K^3 + \cdots \tag{2.13}$$

which emerges from the envelope expansion of (2.1), and assumes the form

$$K^0 = K_2^0 = iq_{\xi\xi} + 2i|q|^2q, \quad K^1 = a_1^1 q_{\xi\xi\xi} + a_2^1|q|^2q_\xi + a_3^1 q^2 q_\xi^*,$$

$$K^2 = ia_1^2 q_{\xi\xi\xi\xi} + a_2^2|q|^2 q_{\xi\xi} + a_3^2 q^* q_\xi^2 + a_4^2 q^2 q_{\xi\xi}^* + a_5^2|q_\xi|^2 q + a_6^2|q|^4 q,$$

$$K^3 = a_1^3 q_{\xi\xi\xi\xi\xi} + a_2^3 q^* q_\xi q_{\xi\xi} + a_3^3 q_\xi^* q_\xi^2 + a_4^3|q|^2 q_{\xi\xi\xi} + a_5^3 qq_\xi^* q_{\xi\xi} +$$

$$+ a_6^3 qq_\xi q_{\xi\xi}^* + a_7^3|q|^4 q_\xi + a_8^3|q|^2 q^2 q_\xi^* + a_9^3 q^2 q_{\xi\xi\xi}^*, \cdots. \tag{2.14}$$

Each correction K^n is a homogeneous differential polynomial $K^n \in \mathcal{E}_{n+3}$ (see Appendix A for the definition of the space $\mathcal{E}_{n+3}$). We say that a differential polynomial $K_k = K_k^0 + \epsilon K_k^1 + \epsilon^2 K_k^2 + \cdots$ is an *approximate symmetry of order n* of equation (2.13) if the commutator (i.e. the Lie bracket defined in Appendix A) of K and K_k gives

$$[K, K_k] = O(\epsilon^{n+1}). \tag{2.15}$$

We then find the obstacles for the existence of approximate symmetry:

Proposition 2.1. *Equation* (2.13) *has approximate symmetry* K_3

1. *of order* 1 *for any coefficients* $a_1^1, \cdot, a_3^1$ *in* (2.13);

2. *of order* 2 *if and only if the coefficients* $a_1^1, \cdot, a_3^1$ *and* $a_1^2, \ldots, a_6^2$ *in* (2.13) *satisfy*

$$
\begin{aligned}
\mu_1^2 \equiv{} & 24a_1^2 - 2a_2^2 - 4a_3^2 - 8a_4^2 + 2a_5^2 + 4a_6^2 - \\
& -18(a_1^1)^2 + 3a_1^1 a_2^1 - a_2^1 a_3^1 + 2(a_3^1)^2 = 0;
\end{aligned} \tag{2.16}
$$

3. *of order* 3 *if and only if the condition* (2.16) *and the following two conditions are satisfied,*

$$
\begin{aligned}
\mu_1^3 \equiv{} & -4a_2^3 + 12a_4^3 + 4a_5^3 - 8a_6^3 - 4a_8^3 + 24a_9^3 + 72a_1^2 a_1^1 - \\
& -18a_2^2 a_1^1 + 12a_6^2 a_1^1 - 2a_2^2 a_3^1 - 12a_4^2 a_3^1 + + a_5^2 a_3^1 + 4a_6^2 a_3^1 - \\
& -54(a_1^1)^3 + 9(a_1^1)^2 a_2^1 - a_2^1 (a_3^1)^2 + 2(a_3^1)^3 = 0,
\end{aligned} \tag{2.17}
$$

and

$$
\begin{aligned}
\mu_2^3 \equiv{} & -240a_1^3 + 8a_2^3 - 16a_3^3 + 48a_4^3 - 8a_7^3 - 4a_8^3 + 24a_9^3 + \\
& +576a_1^2 a_1^1 - 60a_2^2 a_1^1 + 24a_4^2 a_1^1 - 48a_1^2 a_2^1 - 4a_4^2 a_2^1 + 2a_5^2 a_2^1 + \\
& +4a_6^2 a_2^1 + 48a_1^2 a_3^1 - 4a_3^2 a_3^1 - 28a_4^2 a_3^1 + 4a_5^2 a_3^1 + 12a_6^2 a_3^1 + \\
& +54(a_1^1)^2 a_2^1 - 36(a_1^1)^2 a_3^1 + 6a_1^1 a_2^1 a_3^1 - (a_2^1)^2 a_3^1 - \\
& -6a_1^1 (a_3^1)^2 - a_2^1 (a_3^1)^2 - 324(a_1^1)^3 + 6(a_3^1)^3 = 0.
\end{aligned} \tag{2.18}
$$

The requirement of the existence of approximate symmetry K_3 of order 4 gives five more conditions $\mu_1^4 = \cdots = \mu_5^4 = 0$ in addition to (2.16),(2.17) and (2.18). All these conditions are obstacles to asymptotic integrability.

The NLS equation has an infinite sequence of local conservation laws (conserved densities) $L_{K_2^0}(\rho_k^0) \in \mathrm{Im}(D)$ [17] such as

$$
\begin{aligned}
\rho_0^0 = |q|^2, \quad \rho_1^0 = iqq_\xi^*, \quad \rho_2^0 = |q_\xi|^2 - |q|^4, \\
\rho_3^0 = iqq_{\xi\xi\xi}^* + 3iq^2 q^* q_\xi^*, \cdots.
\end{aligned} \tag{2.19}
$$

A similar question can be raised: Whether it is possible to find a differential polynomial $\rho_n = \rho_n^0 + \epsilon\rho_n^1 + \epsilon^2\rho_n^2 + \cdots$ such that

$$
\frac{\partial}{\partial\tau}\rho_n = L_K\rho_n \in \mathrm{Im}(D) + O(\epsilon^{k+1}), \tag{2.20}
$$

where L_K is the vector field associated with the flow K of (2.13) (see Appendix A for the definition). We call ρ_n satisfying (2.20) an *approximate integral of order* k. Then we have:

Proposition 2.2. *Equation* (2.13) *has*

1. *approximate integrals ρ_0 of order 4, ρ_1 of order 3, ρ_2 of order 2, and ρ_n of order 1 for all n;*

2. *approximate integrals ρ_3, ρ_4, ρ_5 of order 2 if and only if the condition* (2.16) *is satisfied;*

3. *approximate integral ρ_2 of order 3 if and only if the condition* (2.17) *is satisfied;*

4. *approximate integral ρ_3 of order 3 if and only if the conditions* (2.16) *and* (2.18) *are satisfied;*

5. *approximate integrals ρ_4, ρ_5 of order 3 if and only if the conditions* (2.16), (2.17) *and* (2.18) *are satisfied.*

The proof of both propositions is straightforward: we look for corrections in the form of homogeneous differential polynomials in the most general form and find conditions of solvability of the linear equations for the coefficients of these polynomials that emerge from (2.15) and (2.20) respectively. We solve quite different equations for these propositions, but the obstacles for the solvability turn out to be the same!

An immediate consequence of part 1 in Proposition 2.2 is that the perturbed NLS equation (2.13) is asymptotically integrable up to order ϵ, but not to the higher orders. Also, we note here that in the general form of the perturbed NLS equation (2.13) these approximate integrals ρ_n exist only up to finite orders for all n. However, for physically interesting equations like the nonlinear Klein–Gordon equation, we expect to have at least the first three approximate integrals ρ_0, ρ_1 and ρ_2 for all orders as the physical conservation laws such as mass, momentum and energy. Then, because of the existence of ρ_2 in this case, the envelope expansion satisfies the condition (2.17) (part 3 of Proposition 2.2), so that there is only one obstacle at order ϵ^3.

3. Near Identity Transformation and Normal Forms

Here we consider a near identity transformation of the NLS equation with higher order corrections (2.13) aiming to simplify it, so that it may be analyzed by a known method (e.g. the spectral transformation method). Since the near identity transform is invertible in the asymptotic sense, the original and the transformed equations can be considered equivalent. Namely, we here study a classification of near integrable systems in the asymptotic sense.

The basic idea of the normal form theory is to remove all the nonresonant (nonsecular) terms in the higher order corrections for the simplifica-
tion. Obvious resonant terms are given by symmetries of the NLS equation.
The main objective here is to show that a perturbed equation can be transformed into a linear combination of symmetries and well defined obstacles.
We call this transformed equation the "normal form" of the perturbed NLS
equation (2.13).

The higher order corrections (2.14) of the NLS equation are differential
polynomials of q, q^* and their derivatives. A near identity transformation
$q = u + O(\epsilon)$ leaves the leading order of (2.13) unaltered, but affects the corrections. We require the transformed equation to be also given in terms of
the differential polynomials of u, u^* and their derivatives. For this purpose
we find the Lie transformation to be useful [11].

The Lie transform may be given in the form

$$q = \exp(L_\phi)u = u + L_\phi u + \frac{1}{2!}(L_\phi)^2 u + \frac{1}{3!}(L_\phi)^3 u + \cdots, \tag{3.1}$$

where L_ϕ is a vector field associated with the generating function

$$\phi = \epsilon \phi^1 + \epsilon^2 \phi^2 + \epsilon^3 \phi^3 + \cdots, \tag{3.2}$$

and ϕ^k are functions of u, u^* which may contain nonlocal terms as described
below. With (3.1) we also have

$$F(q, q^*, q_\xi, q_\xi^*, \cdots) = \exp(L_\phi)F(u, u^*, u_\xi, u_\xi^*, \cdots)$$

for any differential polynomial F.

The Lie transformation (3.1) maps the equation (2.13),

$$q_\tau = L_K q \equiv K^0 + \epsilon K^1 + \epsilon^2 K^2 + \cdots, \tag{3.3}$$

into a similar equation for the transformed variable u,

$$u_\tau = L_G u \equiv G^0 + \epsilon G^1 + \epsilon^2 G^2 + \cdots, \tag{3.4}$$

where the functions $K = K(q, q^*)$ and $G = G(u, u^*)$ are assumed to be the
differential polynomials of their arguments. Substituting (3.1) into (3.3)
leads to

$$L_G \exp(L_\phi)u = \exp(L_\phi)L_K u,$$

from which we obtain the transformation formula for the vector fields as
the adjoint action,

$$L_G = Ad_{\exp(L_\phi)}L_K \equiv \exp(L_\phi)L_K \exp(-L_\phi) \tag{3.5}$$

or

$$L_G = L_K + [L_\phi, L_K] + \frac{1}{2!}[L_\phi, [L_\phi, L_K]] + \cdots.$$

This can be also written on the functions (see (A.7) in Appendix A),

$$G = K + [\phi, K] + \frac{1}{2!}[\phi, [\phi, K]] + \cdots . \tag{3.6}$$

Note here that the functions K, G and ϕ are all expressed in terms of the transformed variables u and u^*. Expanding (3.6) in the powers of ϵ yields

$$
\begin{align}
G^0 &= K^0 , \tag{3.7}\\
G^1 &= K^1 - [K^0, \phi^1] , \tag{3.8}\\
G^2 &= K^2 - [K^0, \phi^2] + [\phi^1, K^1] + \frac{1}{2!}[\phi^1, [\phi^1, K^0]] , \tag{3.9}\\
G^3 &= K^3 - [K^0, \phi^3] + [\phi^1, K^2] + [\phi^2, K^1] + \frac{1}{2!}[\phi^1, [\phi^1, K^1]] + \\
&\quad + \frac{1}{2!}[\phi^2, [\phi^1, K^0]] + \frac{1}{2!}[\phi^1, [\phi^2, K^0]] + \frac{1}{3!}[\phi^1, [\phi^1, [\phi^1, K^0]]] . \tag{3.10}
\end{align}
$$

Giving an appropriate expression for ϕ^k, these expressions determine the functions $G^0, G^1, \cdots$ in each order. If the functions $\phi^1, \phi^2, \cdots$ and $K^0, K^1, \cdots$ are differential polynomials, then so are the $G^0, G^1, \cdots$. It is clear that the scaling weight of ϕ^k is equal to $k+1$. Since the only requirement is that the result of the transformation be a differential polynomial, the function ϕ^k may contain nonlocal terms. In order to determine the form of the generating function ϕ^k, we first note:

Proposition 3.1. *Let $A, F \in \mathcal{E}$ and $B \in \mathcal{H}$ such that $[A, F] = 0$ and $L_F B \in Im(D)$. Then $[AD^{-1}B, F] \in \mathcal{E}$.*

(A proof of the proposition is given in Appendix A.) In other words, if A is a symmetry and B is a conserved density of the NLS equation i.e. $F = K_2^0 \in \mathcal{H}$, then ϕ^k may include the nonlocal term $AD^{-1}B$, which results in a local transformation.

Now let us consider the case of the NLS equation with higher corrections (2.13) in each order. We first discuss the case at $O(\epsilon)$. The function ϕ^1 has the scaling weight 2 and can be chosen in the form

$$\phi^1 = i\alpha_1^1 u_\xi + i\alpha_2^1 u D^{-1}|u|^2 , \tag{3.11}$$

where α_1^1, α_2^1 are real constants. It follows from (3.8) and (2.14) that

$$G^1 = a_1^1 u_{\xi\xi\xi} + (a_3^1 + 4\alpha_1^1 - 2\alpha_2^1)u^2 u_\xi^* + (a_2^1 - 2\alpha_2^1)|u|^2 u_\xi .$$

We then obtain:

Proposition 3.2. *The ϵ-order correction K^1 of the general form (2.14) can be transformed into a symmetry of the NLS equation,*

$$G^1 = a_1^1 K_3^0 = a_1^1 (iq_{\xi\xi\xi} + 6i|q|^2 q_\xi) \tag{3.12}$$

by the near identity transformation (3.1) *generated by* (3.11) *with*

$$\alpha_1^1 = (a_2^1 - 6a_1^1 - a_3^1)/4, \quad \alpha_2^1 = (a_2^1 - 6a_1^1)/2. \tag{3.13}$$

Namely, there is no obstacle to the integrability at this order. Note in particular that if there is no linear dispersion correction in K^1, then K^1 can be totally removed at this order, i.e. $G^1 = 0$.

Now we move to the case at $O(\epsilon^2)$. The ϵ^2-order correction K^2 in (2.14) has in general six independent differential monomials of the scaling weight 5. It follows from (3.9) that the scaling weight of ϕ^2 must be equal to 3. There are only two differential monomials of this weight, namely $u_{\xi\xi}$ and $|u|^2u$. As we have learned in the previous case, nonlocal terms also may give rise to extra freedom. According to Proposition 3.1, the following nonlocal terms $uD^{-1}(uu_\xi^*)$ and $u_\xi D^{-1}|u|^2$ may contribute in ϕ^2. Thus we take

$$\phi^2 = \alpha_1^2 u_{\xi\xi} + \alpha_2^2|u|^2u + \alpha_3^2 uD^{-1}(uu_\xi^*) + \alpha_4^2 u_\xi D^{-1}|u|^2 , \tag{3.14}$$

and (3.9) yields

$$G^2 = i(\tilde{a}_1^2 u_{\xi\xi\xi\xi} + \tilde{a}_2^2|u|^2 u_{\xi\xi} + \tilde{a}_3^2 u_\xi^2 u^* + \tilde{a}_4^2 u^2 u_{\xi\xi}^* + \tilde{a}_5^2|u_\xi|^2 u + \tilde{a}_6^2|u|^4 u), \tag{3.15}$$

where the coefficients are

$$
\begin{aligned}
\tilde{a}_1^2 &= a_1^2, \ \tilde{a}_2^2 = a_2^2 + 2a_4^2 + 3\alpha_2^1 a_1, \ \tilde{a}_3^2 = a_3^2 - 4\alpha_1^2 + 2\alpha_2^2 + 3\alpha_2^1 a_1, \\
\tilde{a}_4^2 &= a_4^2 + 2\alpha_2^2 + 2\alpha_3^2 - 2\alpha_1^1 a_3 - 4(\alpha_1^1)^2 + 2\alpha_1^1 \alpha_2^1, \\
\tilde{a}_5^2 &= a_5^2 - 8\alpha_1^2 + 4\alpha_2^2 + 2\alpha_3^2 + 2\alpha_4^2 - 2\alpha_1^1 a_2 + 6\alpha_2^1 a_1 + 2\alpha_1^1 \alpha_2^1, \\
\tilde{a}_6^2 &= a_6^2 + 4\alpha_2^2 + 3\alpha_3^2 + \frac{1}{2}\alpha_2^1 a_2 - \frac{3}{2}\alpha_2^1 a_3 + (\alpha_2^1)^2 - 3\alpha_1^1 \alpha_2^1. \tag{3.16}
\end{aligned}
$$

Suppose we have chosen coefficients α_1^1 and α_2^1 as in (3.13), so that the transformed equation (3.4) is integrable up to order ϵ, i.e. $G^1 = a_1^1 K_3^0$. We then try to find the coefficients $\alpha_1^2, \ldots, \alpha_4^2$ in (3.14) by solving the equations (3.16) such that G^2 in the transformed equation (3.4) becomes $G^2 = a_1^2 K_4^0$, that is, the asymptotic integrability up to $O(\epsilon^2)$. However, this is not possible in general, since the system of equations (3.16) is overdetermined with 6 equations for 4 unknown variables. The Fredholm alternative then gives the two conditions among the coefficients $\alpha_1^2, \ldots, \alpha_4^2$ for the solvability of (3.16). They are also the invariants under the transformation generated by (3.14), and are given by

$$
\begin{aligned}
\nu_1^2 &= \tilde{a}_1^2 = a_1^2, \tag{3.17} \\
\nu_2^2 &= 24\tilde{a}_1^2 - 2\tilde{a}_2^2 - 4\tilde{a}_3^2 - 8\tilde{a}_4^2 + 2\tilde{a}_5^2 + 4\tilde{a}_6^2 \\
&= 24a_1^2 - 2a_2^2 - 4a_3^2 - 8a_4^2 + 2a_5^2 + 4a_6^2 - \\
&\quad -18(a_1^1)^2 + 3a_1^1 a_2^1 - a_2^1 a_3^1 + 2(a_3^1)^2. \tag{3.18}
\end{aligned}
$$

These conditions imply that the function G^2 can be expressed by only two independent functions. In particular we write G^2 in the form

$$G^2 = a_1^2 K_4^0 + i(\hat{a}_2^2 |u|^2 u_{\xi\xi} + \hat{a}_3^2 u_\xi^2 u^* + \hat{a}_4^2 u^2 u_{\xi\xi}^* + \hat{a}_5^2 |u_\xi|^2 u + \hat{a}_6^2 |u|^4 u) \,, \quad (3.19)$$

where $\hat{a}_2^2, \ldots, \hat{a}_6^2$ are any real constants satisfying the condition (3.18) with $\tilde{a}_1^2 = 0$ and $\hat{a}_k^2 = \tilde{a}_k^2$ for $k = 2, \ldots, 6$, e.g. $\hat{a}_2^2 = -\nu_2^2/2$, and $\hat{a}_3^2 = \cdots = \hat{a}_6^2 = 0$. It then follows that the second order correction K^2 can be transformed into the symmetry K_4^0 if and only if $\nu_2^2 = 0$. The invariant ν_2^2 indeed coincides with the obstacle μ_1^2 in (2.16), and it leads to the fact that the transformed equation possesses approximate symmetries $K_n = K_n^0$ of order 2. Since the Lie transform maps commuting flows into commuting ones (it follows from (3.5) i.e. Lie algebra isomorphism), the inverse transform of the flow K_n^0 provides us with the approximate symmetry of equation (2.13). Namely the equation (2.13) has approximate symmetries K_n of order 1 (no conditions), and of order 2 if the condition (2.16) is satisfied. We call the transformed equation (3.4) with (3.12) and (3.15) the *normal form* of the equation (2.13). Thus we have:

Theorem 3.1. *(Normal form theorem) There exists a Lie transformation (3.1) generated by (3.2) with (3.11) and (3.14) such that the perturbed NLS equation (2.13) is transformed into the normal form given by*

$$u_\tau = K_2^0 + \epsilon a_1^1 K_3^0 + \epsilon^2 (a_1^2 K_4^0 + \mu_1^2 R^2) + O(\epsilon^3) \,. \quad (3.20)$$

Here the function R^2 is given by

$$R^2 = i(\hat{b}_2^2 |u|^2 u_{\xi\xi} + \hat{b}_3^2 u_\xi^2 u^* + \hat{b}_4^2 u^2 u_{\xi\xi}^* + \hat{b}_5^2 |u_\xi|^2 u + \hat{b}_6^2 |u|^4 u) \,,$$

and the coefficients $\hat{b}_1^2, \ldots, \hat{b}_6^2$ are arbitrary real constants satisfying the condition,

$$-2\hat{b}_2^2 - 4\hat{b}_3^2 - 8\hat{b}_4^2 + 2\hat{b}_5^2 + 4\hat{b}_6^2 = 1 \,. \quad (3.21)$$

This theorem can be formally extended for arbitrary order (we will report this in a future communication).

With the Lie transform (3.1), we now discuss the approximate symmetries and integrals in terms of the so-called master symmetry. Let us first recall [7] that the master symmetry M^0 of the NLS equation is given by

$$M^0 = i\xi u_{\xi\xi} + 2i\xi |u|^2 u + 2i u_\xi + 2i u D^{-1}(|u|^2) \,,$$

and the whole hierarchy of the NLS symmetries can be generated as $(n0)$

$$K_{n+1}^0 = \frac{(-1)^{n+1}}{n} [M^0, K_n^0] \,.$$

Similarly, the hierarchy of local conservation laws can be obtained as

$$\rho^0_{n+1} + \frac{(-1)^n}{n+2} L_{M^0}(\rho^0_n) \in \text{Im}(D).$$

Then an *approximate master symmetry* M of order k is defined as a (non-local) flow satisfying one of the following conditions:

$$K_{n+1} - \frac{(-1)^{n+1}}{n}[M, K_n] = O(\epsilon^{k+1}),\tag{3.22}$$

$$\rho_{n+1} + \frac{(-1)^n}{n+2}L_M(\rho_n) + O(\epsilon^{k+1}) \in \text{Im}(D).\tag{3.23}$$

If there is no obstacle at the order of ϵ^2, i.e. $\mu^2_1 = 0$, then we can transform the equation (2.13) into an asymptotically integrable equation

$$u_\tau = K^0_2 + \epsilon a^1_1 K^0_3 + \epsilon^2 a^2_1 K^0_4 + O(\epsilon^3).\tag{3.24}$$

The master symmetry M^0 of the NLS equation is the approximate master symmetry of order 2 of the above equation (3.24). Then the corresponding inverse Lie transform

$$L_M = Ad^{-1}_{\exp(L_\phi)}L_{M^0} = \exp(-L_\phi)L_{M^0}\exp(L_\phi)$$

provides us with the approximate master symmetry M of order 2 for (2.13). Now the coincidence of conditions for the corrections of symmetries and conservation laws looks very natural. Moreover, if the obstacle (2.16) vanishes, then there exists ϵ^2-order deformation of the whole hierarchy of higher symmetries and conservation laws.

At order ϵ^3 we find the invariants of the Lie transformation as

$$\begin{aligned}
\nu^3_1 &\equiv \tilde{a}^3_1 = a^3_1,\\
\nu^3_2 &\equiv -4\tilde{a}^3_2 + 12\tilde{a}^3_4 + 4\tilde{a}^3_5 - 8\tilde{a}^3_6 - 4\tilde{a}^3_8 + 24\tilde{a}^3_9 = \mu^3_1,\\
\nu^3_3 &\equiv -240\tilde{a}^3_1 + 8\tilde{a}^3_2 - 16\tilde{a}^3_3 + 48\tilde{a}^3_4 - 8\tilde{a}^3_7 - 4\tilde{a}^3_8 + 24\tilde{a}^3_9 = \mu^3_2.
\end{aligned}$$

The problem of listing the invariants of the Lie transformation generated by ϕ_n can be reduced to the solvability conditions of the equation[2]

$$ad_{K^0}\Phi_n \equiv [K^0, \Phi_n] = F^n,$$

---

[2]In general, equation for G^n has the following form

$$G^n = K^n + \Delta_n - ad_{K^0}\phi_n,$$

where Δ_n is a sum of extra terms due to the transformations $\phi_{n-1}, \ldots, \phi_1$ (for instance $\Delta_2 = [\phi^1, K^1] + \frac{1}{2!}[\phi^1, [\phi^1, K^0]]$). The terms in Δ_n may contain nonlocal expressions in general. Let us define a projector P_L to a space of differential polynomials ($P_L A = A$ iff A is a differential polynomial and $P_L B = 0$ if B is a nonlocal term, i.e. B contains D^{-1}). We split the generator ϕ_n in two part $\phi_n = \phi^L_n + \phi^N_n$ where ϕ^L_n is a differential polynomial of scaling weight $n+1$ extended by terms of the form $S_k D^{-1}(\rho_{n-k+3})$, where S_k and where $F^n \in \mathcal{E}_{n+3}$ and $\Phi_n \in \hat{\mathcal{F}}_{n+1}$ is a generator of the form

$$\Phi_n = \sum_{k=1}^{\dim(\mathcal{F}_{n+1})} \alpha_k(n) e_k(n+1) + \sum_{k=1}^{n} \beta_k(n) K_k^0 D^{-1}(\rho_{n-k}^0) \,,$$

and $e_k(n+1)$ is a basis in $\mathcal{F}_{n+1}$. The linear operator ad_{K^0} maps $\mathcal{F}_{n+1}$ into $\mathcal{E}_{n+3}$. Let $ad_{K^0}^A$ be a linear operator adjoint to ad_{K^0}, which is defined on a dual space $\mathcal{E}_{n+3}^A$, and let $Y_k^A \in \mathcal{E}_{n+3}^A$ be the solutions of a homogeneous equation $ad_{K^0}^A Y_k^A = 0$. Then $\nu_k^n = \langle Y_k^A, G^n \rangle$ are the invariants of the Lie transformation. The number of the obstacles at order ϵ^n is then given by $C_n = \dim(\mathcal{E}_{n+3}) - \dim(\mathcal{F}_{n+1}) - n - 1$:

n	1	2	3	4	5	6	7	8	9	10	11	12	13
C_n	0	1	2	5	9	16	26	42	63	96	140	203	289

4. Normal Forms and Inelasticity

As we have seen in the previous section, the NLS equation with the higher order corrections (2.13) can be Lie-transformed to the normal form (3.20) which contains 5 arbitrary constants $\hat{b}_2^2, \ldots, \hat{b}_6^2$ with 1 constraint (3.21) at order ϵ^2. Here we define a specific form of the normal form by choosing explicit constants $\hat{b}_2^2, \ldots, \hat{b}_6^2$ based on the conservation laws, and then we show how the behavior of some solutiuons to (2.13) is analyzed using this form. In particular, we show that the obstacles appearing in the corrections of symmetries (or conservation laws) lead to an inelasticity in the interaction of two soliton solutions [8]. This implies that the obstacles give additional secularities in an asymptotic perturbation method. We here consider only the ϵ^2-order obstacle μ_1^2, but the higher order obstacles can be studied in a similar fashion.

4.1. Normal Forms and Soliton Solutions

One of the key points in the normal form theory is that the normal form admits simple invariants which provide a coordinate system to describe

---

ρ_{n-k+2} are a symmetry and a conserved density of NLS of scaling weights k and $n-k+2$ respectively ($\phi_n^L \in \hat{\mathcal{F}}_{n+1}$), and therefore $ad_{K^0}\phi_n^L$ is a differential polynomial. The part ϕ_n^N solves the equation $(I - P_L)(ad_{K^0}\phi_n^N - \Delta_n) = 0$, and is fixed. Then only the freedom of the transformation is in the part ϕ_n^L. To find invariants of the transformation, we have to solve a system of linear homogeneous equations $ad_{K^0}^A Y^A = 0$ where $ad_{K^0}^A$ is operator adjoint to ad_{K^0} and Y^A is a vector from a space which is dual to $\mathcal{E}_{n+3}$ with an inner product $\langle \cdot, \cdot \rangle$. Each solution of the above problem provides us with an invariant of the transformation. Indeed, let Y^A be a solution; then

$$\langle Y^A, G^n \rangle = \langle Y^A, K^n \rangle + \langle Y^A, P^L \Delta_n \rangle + \langle Y^A, P^L ad_{K^0}\phi_n^N \rangle \,.$$

some of the solutions, e.g. quasi-periodic solutions [2]. In our case, we require the normal form to possess several (but not all in general) conserved quantities of the NLS equation. This implies that the normal form keeps a finite dimensional invariant surface characterized by these quantities, so that it admits some of the solutions of the NLS equation, e.g. one-soliton solution.

We first recall that some of the explicit solutions of the NLS equation can be obtained by the variation of the integral given by

$$\mathfrak{S}[\lambda_1, \ldots, \lambda_n] \equiv \int \rho_n^0 + \lambda_1 \int \rho_{n-1}^0 + \cdots + \lambda_n \int \rho_0^0 = \text{constant}, \quad (4.1)$$

where $\lambda_1, \ldots, \lambda_n$ are arbitrary real constants (Lagrange multipliers). For example, the one-soliton solution in the general form is obtained by taking $n = 2$ as follows: The variation $\delta\mathfrak{S}[\lambda_1, \lambda_2]/\delta u^* = 0$ gives a second order ODE for $u(\xi, \cdot)$,

$$u_{\xi\xi} + 2|u|^2 u + i\lambda_1 u_\xi + \lambda_2 u = 0, \quad (4.2)$$

from which we obtain the soliton solution in the form,

$$u(\xi, \cdot) = \eta \, \text{sech}[\eta(\xi + \theta(\cdot))]e^{i\kappa\xi + i\sigma(\cdot)}. \quad (4.3)$$

Here we have used $\lambda_1 = 2\kappa$ and $\lambda_2 = \kappa^2 + \eta^2$, and θ and σ are the functions of τ. This defines a four dimensional invariant set in the solutions of the NLS equation characterized by $(\kappa, \eta, \theta, \sigma)$.

Then the requirement that the normal form possess some of the solutions of the NLS equation can be formulated as the equation $L_G\mathfrak{S}[\lambda_1, \ldots, \lambda_n] = 0$, that is, the invariance of $\mathfrak{S}[\lambda_1, \ldots, \lambda_n]$ under the flow of normal form. This can be written as

$$L_G\rho_{j_1}^0, \ldots, L_G\rho_{j_k}^0, \quad L_G\rho_n^0 \in \text{Im}(D), \quad (4.4)$$

for some $0 \leq j_1 < j_2 < \cdots < j_k < n$, where we have taken $\lambda_i \neq 0$ for $i = j_1, \cdots j_k$.

Let us first try to determine G_O^2, the obstacle part of G^2, i.e. $G_O^2 = G^2 - a_1^2 K_4^0 = \mu_1^2 R^2$ in (3.20), which admits the one-soliton solution in the general form (4.3), that is, we require $\mathfrak{S}[\lambda_1, \lambda_2]$ to be independent of the flow $L_{G_O^2}$. However it can be shown by a direct computation of (4.4) that the requirement of *all* three conditions $L_{G_O^2}\rho_k^0 \in \text{Im}(D)$ for $k = 1, 2, 3$ leads to only the trivial case $G_O^2 = 0$, i.e. $\hat{a}_2^2 = \cdots = \hat{a}_6^2 = 0$. This implies that the normal form does not admit the one-soliton solution in the general form (4.3) up to $O(\epsilon^2)$. However if we set some of the constants λ_i's to be zero, then we obtain by direct computation:

Proposition 4.1. *There are only two nontrivial cases of the coefficients $(\hat{b}_2^2, \ldots, \hat{b}_6^2)$ in $G_O^2 = \mu_1^2 R^2$ which satisfy the requirement $\mathfrak{S}[\lambda_1, \ldots, \lambda_n]$ being invariant and contain no arbitrary constants except for the normalization:*

1. $-\frac{1}{6}(1,1,0,-2,1)$ *for* $\mathfrak{S}[0,\lambda_2]$ =*constant, and*

2. $\frac{1}{18}(1,-1,0,2,3)$ *for* $\mathfrak{S}[\lambda_1,0]$ =*constant.*

Calculating variations of the invariants $\mathfrak{S}[\lambda_1,\ldots,\lambda_n]$ for these cases, we find that only case 1 in Proposition 4.1 admits a one-soliton solution in the form

$$u(\xi,\cdot) = \eta \ \mathrm{sech}[\eta(\xi + \theta(\cdot))]e^{i\sigma(\cdot)} . \qquad (4.5)$$

Here we have taken $\lambda_2 = \eta^2$ (i.e. $\lambda_1 = 2\kappa = 0$ in (4.3)). Then one can verify that the solution (4.5) satisfies the equation $G_O^2(u,u^*) = L_{G_O^2}u = 0$ as a consequence of the vector field $L_{G_O^2}$ being tangent to the invariant surface $\mathfrak{S}[0,\lambda_2]$. Namely the soliton solution (4.5) exists for the normal form up to this order, and the obstacle does not give any contribution to this solution. This explains the result in [12] where the higher order corrections to the one-soliton solution of (4.5) are obtained using *only* the symmetries of the NLS equation as the renormalizations of velocity and frequency of this soliton. In other words, all the secularities for the *one*-soliton solution in the asymptotic theory are given only by the symmetries. However, as we will show in the next subsection, the two-soliton interaction leads to an additional secularity appearing as an inelasticity of the interaction. This is a direct and important consequence of the obstacles.

It is also important to note that these G_O^2 in Proposition 4.1 are equivalent in the sense of the Lie transformation (Lie-equivalence). This means that the solution to the one case of the normal form can be described by a Lie transform of that one to the other case.

We would also like to comment on the solution to the original higher order equation (2.13). As a consequence of the existence of exact solution (4.5) of the normal form, the solution q can be written in the following form from the Lie transformation (3.1)

$$
\begin{aligned}
q &= e^{L_\phi}u = u + L_\phi u + \cdots \\
&= \sum_{n=1}^{N}\sum_{j=1}^{n} \epsilon^{n-1}\beta_j^n\eta^n \mathrm{sech}^{2j-1}[\eta(\xi + \theta(\tau))]e^{i\sigma(\tau)} + O(\epsilon^{N+1}), \quad (4.6)
\end{aligned}
$$

where β_j^n are some constants, $\theta(\tau)$ and $\sigma(\tau)$ are given by the renormalized velocity and frequency [12]

$$
\begin{aligned}
\theta(\tau) &= \sum_{n=0}\epsilon^{2n+1}(\hat{a}^{2n+1}\eta^{2n+2}\tau + \theta_0^{2n+1}), \\
\sigma(\tau) &= \sum_{n=0}\epsilon^{2n}(\hat{a}^{2n}\eta^{2n+2}\tau + \sigma_0^{2n}) ,
\end{aligned}
$$

with appropriate constants $\theta_0^{2n+1}, \sigma_0^{2n}$ and $\hat{a}^n \propto a_1^n$ in the coefficients of the linear terms in the higher order corrections.

4.2. Inelasticity in Soliton Interaction

We now study the interaction problem of solitons using the normal forms obtained in the previous subsection. As we know in general, even one-soliton initial data to the perturbed NLS equation such as (2.13) produces radiation. Then for the study of inelasticity in soliton interaction, one has to identify the radiations generated only by the interaction. This may not be possible for the general two soliton initial data. However, for a special case of two soliton initial data, one can explicitly study the problem. This is precisely the purpose of using the normal form where we have the explicit soliton solution (4.5). In [8], Kano used the G_O^2 in case 1 of Proposition 4.1 to study the interaction problem of a bound 2-soliton solution, and showed the existence of an inelasticity due to the obstacle at the order ϵ^2 (see also [11] for the case of KdV corrections where the general two soliton problem has been studied). We here give a summary of this result for a comprehensive exposition of the effects due to the obstacles.

For the purpose of demonstrating an existence of inelasticity (or secularity), it may be enough to consider the simplest case of the normal form, i.e. including only the obstacle part of G^2, so we take

$$
\begin{aligned}
u_\tau \;=\; & K_2^0 + \epsilon^2 G_O^2 = i(u_{\xi\xi} + 2|u|^2 u) + \\
& -\epsilon^2 i \frac{\mu_1^2}{6}(|u|^2 u_{\xi\xi} + u_\xi^2 u^* - 2|u_\xi|^2 u + |u|^4 u) \,.
\end{aligned} \tag{4.7}
$$

As we have shown, this equation has two *exact* integrals,

$$
I_0^0 \equiv \int \rho_0^0, \text{ and } I_2^0 \equiv \int \rho_2^0 \,, \tag{4.8}
$$

and it admits the one-soliton solution in the form (4.5), i.e.

$$
u(\xi, \tau) = \eta \, \text{sech}[\eta(\xi + \theta_0)]e^{i\eta^2\tau + i\sigma_0}. \tag{4.9}
$$

In the case of the general normal form, we know from Proposition 2.2 that those integrals I_0^0 and I_2^0 are only approximations up to orders of 4 and 2, respectively, and the following discussions on the interaction problem for two solitons should be modified. However, as we discussed, Proposition 2.2, physically interesting equations are expected to have these integrals for all orders, and for these cases the equation (4.7) may be considered to be generic (detail discussions will be given elsewhere).

As the initial data, we take a bound state of two solitons of the NLS equation with the parameters $\eta_1\eta_2$, given by

$$
u(\xi, \tau) = \frac{2}{D}\left(\frac{\eta_1 + \eta_2}{\eta_1 - \eta_2}\right)[\eta_1 \cosh(\theta_2) \, e^{i\sigma_1} + \eta_2\cosh(\theta_1) \, e^{i\sigma_2}] \tag{4.10}
$$

with

$$D = \left(\frac{\eta_1 + \eta_2}{\eta_1 - \eta_2}\right)^2 \cosh(\theta_1 - \theta_2) + \cosh(\theta_1 + \theta_2) + \frac{4\eta_1\eta_2}{(\eta_1 - \eta_2)^2} \cos(\sigma_1 - \sigma_2),$$

where $\theta_j = \eta_j\xi + \theta_{0j}$ and $\sigma_j = \eta_j^2\tau + \sigma_{0j}$. In particular, if we choose the constants $\theta_{0j} = \sigma_{0j} = 0$, then the initial profile $u(\xi, 0)$ is real and symmetric in ξ. It is also interesting to note that this function $u(\xi, 0)$ is well approximated by a *linear* superposition of two identical solitons in the form (4.5) with $\eta \equiv \eta_0$ and the separation distance $2\xi_0$,

$$u(\xi, 0) \approx \eta_0 \operatorname{sech}[\eta_0(\xi - \xi_0)] + \eta_0 \operatorname{sech}[\eta_0(\xi + \xi_0)] . \qquad (4.11)$$

Here the parameters η_1 and η_2 are expressed by η_0 and ξ_0 as [5]

$$\eta_{1,2} = \eta_0 \left(1 + \frac{2\xi_0}{\sinh(2\eta_0\xi_0)} \pm \operatorname{sech}(\eta_0\xi_0)\right) .$$

Then the bound soliton solution (4.10) behaves as if these solitons bounce around $\xi = 0$ and collide with the period τ_p given by

$$\tau_p = \frac{2\pi}{\eta_1^2 - \eta_2^2}.$$

One should however note that the bound solution (4.10) consists of *nonlinearly* interacting two solitons with amplitudes η_1 and η_2. Our purpose here is to observe an effect of the obstacle G_O^2 during the process of collision of these two approximate solitons in (4.11), and to show inelasticity in the collision.

We now show that there will be a creation of radiations and shifts of the parameters η_1 and η_2 due to the obstacle G_O^2. In order to do so, we employ the perturbed inverse scattering technique proposed in [9]. A key point of this method is to expand the solution u with respect to the (squared) eigenfunctions of the Lax operator L (the expansion theorem). Then in terms of the scattering data of the operator given by [1]

$$\Sigma = \{(\zeta_j = \kappa_j + i\eta_j, \beta_j)_{j=1}^N, \hat{r}(k) = \bar{b}(k)/a(k), \text{ for } k \in \mathbb{R}\} \qquad (4.12)$$

the integrals I_0^0 and I_2^0 in (4.8) are explicitly represented as

$$I_0^0 = 2(\eta_1 + \eta_2) - 2\Delta_0, \qquad (4.13)$$

$$I_2^0 = -\frac{2}{3}(\eta_1^3 + \eta_2^3) - 2\Delta_3, \qquad (4.14)$$

where the radiation parts Δ_{2l} are given by

$$\Delta_{2l}(\tau) = \frac{2^{2l-1}}{\pi} \int_{-\infty}^{\infty} k^{2l} \log|a(k)|^2 \, dk \leq 0. \qquad (4.15)$$

Note here that the real parts κ_j of the eigenvalues ζ_j in the scattering data can be shown to remain zero for all τ under the evolution of (4.7) due to the symmetry in the initial data and the equation (4.7) with respect to the reflection in ξ.

Then assuming the shift of the parameters, $\eta_j \to \eta_j + \Delta\eta_j(\tau)$ and the production of the radiations $\Delta_{2l}(\tau)$ due to the collision, we obtain at the leading order

$$\Delta\eta_1 = -\frac{\eta_2^2 \Delta_0 + \Delta_2}{\eta_1^2 - \eta_2^2} \geq 0, \qquad (4.16)$$

$$\Delta\eta_2 = \frac{\eta_1^2 \Delta_0 + \Delta_2}{\eta_1^2 - \eta_2^2} \leq 0. \qquad (4.17)$$

We see here that the larger parameter η_1 is increased, while the smaller one is decreased by the creation of radiations Δ_0 and Δ_2. This indicates the existence of energy exchange between solitons through the radiations as a result of inelasticity. This is consistent with the numerical observation in [8]. The radiations $\Delta_{2l}(\tau)$ may be computed from the equation for the scattering data $\hat{r}(\tau; k)$ [9],

$$\frac{d\hat{r}}{d\tau} = -4ik^2\hat{r} - i\frac{\epsilon^2\mu_1^2}{6a^2(k)} \int_{-\infty}^{\infty} \left[(G_O^2)^* \psi_1^2 - G_O^2 \psi_2^2 \right] d\xi \qquad (4.18)$$

where $\psi = (\psi_1, \psi_2)$ are the eigenfunctions of the Lax operator, $L\psi = k\psi$, and the function $a(k)$ is given by

$$a(k) = \left(\prod_{j=1}^{2} \frac{k - i\eta_j}{k + i\eta_j} \right) \exp\left[\frac{1}{2\pi i} \int_{-\infty}^{\infty} \frac{\log |a(k')|^2}{k - k'} dk' \right]. \qquad (4.19)$$

We solve the evolution of $\hat{r}$ in (4.18) by means of perturbation, i.e. we replace u in G_O^2 for the exact bound 2-soliton solution of the NLS equation, ψ_1 and ψ_2 for the corresponding eigenfunctions, and $a(k)$ for $a_0(k) = \prod_{j=1}^{2}(k - i\eta_j)/(k + i\eta_j)$. Then the function of $\Delta_{2l}(\tau)$ is given by

$$\Delta_{2l}(\tau) = \frac{2^{2l-1}}{\pi} \int_{-\infty}^{\infty} D_{2l}(\tau; k) \, dk + o(\epsilon^4), \qquad (4.20)$$

where D_{2l} can be estimated by using the identity $|a(k)|^2 + |b(k)|^2 = 1$,

$$D_{2l}(\tau; k) = k^{2l} \log |a(k)|^2 = -k^{2l} \log(1 + |\hat{r}(\tau; k)|^2) \approx -k^{2l} |\hat{r}(\tau; k)|^2$$

$$= -\epsilon^4 \left(\frac{\mu_1^2}{6}\right)^2 k^{2l} \left| \int_0^{\tau} d\tau' e^{-4ik^2\tau'} \int_{-\infty}^{\infty} \left[(G_O^2)^* \psi_1^2 - G_O^2 \psi_2^2 \right] d\xi \right|^2.$$

An explicit integration of (4.20) has been carried out in [8], and it showed a nonvanishing contribution in the radiations. In particular, note from (4.18) that the production of the radiation is due to the obstacle G_O^2 and appears only during the collision of the two solitons. The results in this section can be summarized as follows:

Theorem 4.1. *Assume that the perturbed NLS equation (2.13) has nonzero obstacle μ_1^2 and keeps two approximate integrals ρ_0 and ρ_2 of order 4. Then there exist a solitary wave solution with a parameter η in (4.5) up to order $o(\epsilon^4)$, and these solitary waves interact inelastically as producing radiations and shifting their parameters.*

5. Normal Forms of KP Deformations

The long wave approximation provides us with another class of useful asymptotic expansion. The KdV and mKdV equations rise up as universal nonlinear wave equations for one dimensional weakly dispersive problems with quadratic and cubic nonlinearities respectively. As in the case of envelope approximation, here the higher order corrections may contain obstacles for asymptotic integrability. Indeed these obstacles have been discussed earlier in the context of normal forms in [11] and symmetry deformations [15]. In the case of two spatial dimensions the KP equation,

$$u_t = KP^0 = u_{xxx} + 6uu_x - 3D^{-1}(u_{yy}) , \qquad (5.1)$$

replaces the KdV equation (D^{-1} denotes ∂_x^{-1}). It is a universal equation for weakly dispersive waves with quadratic nonlinearities propagating in a narrow cone along the x–direction (a quasi-two dimensional approximation). Equation (5.1) itself is nonlocal and if we would carry on the expansion in order to capture the higher order corrections we get even more nonlocal contributions:

$$u_t = KP^0 + \epsilon KP^1 + \epsilon^2 KP^2 + \cdots , \qquad (5.2)$$

where the higher order corrections are given by

$$
\begin{aligned}
KP^1 &= a_1 u_{xxy} + a_2 D^{-2}(u_{yyy}) + a_3 u_x D^{-1}(u_y) + a_4 uu_y , && (5.3) \\
KP^2 &= b_1 u_{xxxxx} + b_2 u_{xyy} + b_3 D^{-3}(u_{yyyy}) + b_4 u_x u_{xx} + \\
&+ b_5 uu_{xxx} + b_6 u^2 u_x + b_7 u_x D^{-2}(u_{yy}) + b_8 u D^{-1}(u_{yy}) + \\
&+ b_9 u_y D^{-1}(u_y) + b_{10} \partial_y D^{-1}(uu_y) . && (5.4)
\end{aligned}
$$

From these expressions, we see that the expansion (5.2) can be naturally characterized in terms of pseudo-differential polynomials. Let us first assign a scaling weight to the monomials in the following way:

- scaling weight of $\partial_x^n \partial_y^m u$ is equal to $n + 2m + 2$,

- scaling weight of a product is a sum of the multipliers scaling weights,

- each appearence of D^{-q} reduces the scaling weight by q.

For instance, the scaling weight of $\partial_y D^{-1}(uu_y)$ is 7, and the terms in KP^n in (5.2) have the scaling weights $5+n$ respectively. We say that a monomial is odd if it changes its sign under the involution,

$$\partial_x \to -\partial_x, \quad \partial_y \to -\partial_y, \quad D^{-1} \to -D^{-1}, \quad u \to u.$$

Thus the only odd terms appear in KP^0 and its corrections. Also one can mention that the maximum power of D^{-1} in a monomial is restricted to $k-4$ where k is the scaling weight of the monomial.

Equation (5.1) has infinite sequence of symmetries:

$$
\begin{aligned}
KP_3^0 &= u_x, \quad KP_4^0 = u_y, \quad KP_5^0 = KP^0, \\
KP_6^0 &= u_{xxy} - D^{-2}(u_{yyy}) + 2u_x D^{-1}(u_y) + 4uu_y, \\
KP_7^0 &= u_{xxxxx} - 10u_{xyy} + 5D^{-3}(u_{yyyy}) + 20u_x u_{xx} + \\
&+ 10uu_{xxx} + 30u^2 u_x - 10u_x D^{-2}(u_{yy}) - 20uD^{-1}(u_{yy}) - \\
&- 20u_y D^{-1}(u_y) - 10\partial_y D^{-1}(uu_y), \\
KP_8^0 &= u_{xxxxy} - \frac{10}{3}u_{yyy} + D^{-4}(u_{yyyyy}) + 16u^2 u_y + 12u_x u_{xy} \\
&+ 8u_y u_{xx} + 8uu_{xxy} + 12uu_x D^{-1}(u_y) + 2u_{xxx}D^{-1}(u_y) - \\
&- 6D^{-1}(u_y)D^{-1}(u_{yy}) - 4u_y D^{-2}(u_{yy}) - 4uD^{-2}(u_{yyy}) - \\
&- 2u_x D^{-3}(u_y) + 4u_x D^{-1}(uu_y) - 2\partial_y D^{-2}(uu_y).
\end{aligned}
$$

The existence of approximate symmetries of a first few orders for (5.2) is summarized as follows:

Proposition 5.1. *Equation (5.2) has approximate symmetry KP_6 of order 1 if and only if*

$$\kappa_1 \equiv 3a_1 - a_2 - a_4 = 0 \tag{5.5}$$

Unlike the case of the NLS deformation discussed in the previous sections, we thus have the obstacle (5.5) at order ϵ.

By a direct computation we also have:

Proposition 5.2. *The equation $u_t = KP^0 + \epsilon^2 KP^2$ has approximate symmetry KP_6 of order 2 if and only if the coefficients $b_1, \ldots, b_{10}$ in (5.4) satisfy the following conditions*

$$
\begin{aligned}
\kappa_1^2 &\equiv 5b_1 + b_2 + b_3 = 0, &\tag{5.6} \\
\kappa_2^2 &\equiv 20b_1 - 4b_5 + b_6 + b_{10} = 0, &\tag{5.7} \\
\kappa_3^2 &\equiv 20b_1 + 4b_3 - b_5 + b_8 + b_{10} = 0, &\tag{5.8} \\
\kappa_3^2 &\equiv 10b_1 - 2b_3 - 2b_5 - b_9 = 0, &\tag{5.9} \\
\kappa_1^2 &\equiv 2b_3 + b_7 = 0. &\tag{5.10}
\end{aligned}
$$

Note that the condition (5.6) restricts the linear part of the perturbation (5.3). This restriction coincides with the degeneracy condition for
dispersion laws [20].

Let us consider a near identity transformation defined by

$$u \to u + \epsilon \delta D^{-1}(u_y) \tag{5.11}$$

with the coordinate change

$$x \to x, \quad y \to y + \epsilon \alpha x, \quad t \to t . \tag{5.12}$$

Then choosing the constants α and δ as

$$\alpha = -\frac{a_1 + a_2}{6}, \quad \delta = -\frac{a_1 - a_2 - a_3}{6},$$

the perturbed KP equation (5.2) up to order ϵ,

$$u_t = KP^0 + \epsilon KP^1 , \tag{5.13}$$

is transformed into a *normal form*,

$$u_t = KP^0 + \epsilon \frac{a_1 - a_2}{2} KP_6^0 + O(\epsilon^2) . \tag{5.14}$$

Note in particular that together with the Lie transformation (with
$\phi_1 = \delta D^{-1}(u_y)$) we have used the near identity transformation (5.12) of the
independent variables. In principle, we could extend the group of transformations (5.12) in the following way:

$$x \to x, \quad y \to y + \epsilon \alpha x, \quad t \to t + \epsilon \beta y + \epsilon^2 \gamma x , \tag{5.15}$$

which leads to

$$\partial_t \to L_{KP^0} , \quad \partial_y \to L_{u_y} + \beta \epsilon L_{KP^0} , \quad \partial_x \to D + \alpha \epsilon L_{u_y} + \gamma \epsilon^2 L_{KP^0} , \tag{5.16}$$

and therefore

$$D^{-1} \to D^{-1} + \sum_{n=1}^{\infty} (-1)^n D^{-n-1} (\alpha \epsilon L_{u_y} + \gamma \epsilon^2 L_{KP^0})^n .$$

The transformation (5.12) at order ϵ has two invariants given by κ_1 in
(5.5) and $\kappa_2 = a_1 - a_2$, while (5.12) has only one invariant which coincides
with (5.5). Thus, if the condition (5.5) is satisfied, then the KP equation
and any equation of the form (5.13) are equivalent modulo transformations

(5.11) and (5.16). There is an obvious generalization of the above transformations of the independent variables. Namely, we can consider

$$\partial_x \quad \rightarrow \quad D + \sum_{n=1} \epsilon^n \alpha_n L_{K P_{n+3}^0}, \tag{5.17}$$

$$\partial_y \quad \rightarrow \quad L_{K P_4^0} + \sum_{n=1} \epsilon^n \beta_n L_{K P_{n+4}^0}, \tag{5.18}$$

$$\partial_t \quad \rightarrow \quad L_{K P^0} + \sum_{n=1} \epsilon^n \gamma_n L_{K P_{n+5}^0}. \tag{5.19}$$

If there are no obstacles for a particular equation so that all the conditions of integrability are satisfied, then modulo Lie transformations extended by the above described transformations of the dependent variables, this equation is equivalent to the KP equation.

6. Concluding Remarks and Discussion

Recently, there have been several publications on the same issue as discussed in the present study by means of the method of multiple scaling expansion. We here discuss a connection of this method to our result. The procedure of the method may be summalized as follows: Consider the perturbed equation (2.13), and introduce the multiple time scales $\tau_n \equiv \epsilon^n \tau$. Then the perturbed equation (2.13) becomes

$$\frac{\partial q}{\partial \tau_0} - K_2^0 = \sum_{n=1}^{\infty} \epsilon^n \left(K^n - \frac{\partial q}{\partial \tau_n} \right). \tag{6.1}$$

Now expanding q in a power series of ϵ^n,

$$q = q^0 + \epsilon q^1 + \epsilon^2 q^2 + \cdots, \tag{6.2}$$

and substituting this into (6.1), we have at the leading order $O(1)$,

$$\frac{\partial q^0}{\partial \tau_0} - K_2^0 = 0, \tag{6.3}$$

which is the NLS equation for q^0. At order ϵ, the function q^1 in (6.2) satisfies

$$\frac{\partial q^1}{\partial \tau_0} - L_{q^1} K_2^0 = K^1 - \frac{\partial q^0}{\partial \tau_1}, \tag{6.4}$$

where $L_{q^1} K_2^0$ is the linearized equation in q^1 of K_2^0 about q^0. Then removing all the secular terms in the right hand side leads to

$$\frac{\partial q^0}{\partial \tau_1} = S^1, \tag{6.5}$$

where the secular term S^1 in K^1 is given by the symmetry of the NLS equation (6.3), that is, $[K_2^0, S^1] = 0$. It turns out that the symmetry S^1 can be expressed as $S^1 = a_1^1 K_3^0$ which gives the same result as ours. At the order of ϵ^2, a similar manner leads to

$$\frac{\partial q^0}{\partial \tau_2} = S^2, \tag{6.6}$$

with S^2 being the secular term in $K^2 + L_{q^1} K^1 + (1/2)(L_{q^1})^2 K_2^0 - \partial q^1/\partial \tau_1$, i.e. $[K_2^0, S^2] = 0$. Here one should also require the *commutatibity* condition among the flows (6.5) and (6.6) as the consistency of the multiple scaling method (i.e. τ_n are all independent),

$$\frac{\partial^2 q^0}{\partial \tau_1 \partial \tau_2} = \frac{\partial^2 q^0}{\partial \tau_2 \partial \tau_1}, \tag{6.7}$$

which is $[S^1, S^2] = 0$. Then if S^2 is asssumed to be a differential polynomial in q^0 and $(q^0)^*$, it is just the symmetry of the NLS hierarchy, i.e. $S^2 = a_1^2 K_4^0$. The secularity conditions are also checked by using either one-soliton solution [14] or the asymptotic solution of the continuous wave [4]. With these procedures, one then conclude that the q^0 satisfies the NLS equation with its hierarchy as the higher order corrections,

$$\frac{\partial q^0}{\partial \tau} = \sum_{n=0}^{\infty} \epsilon^n \frac{\partial q^0}{\partial \tau_n} = K_2^0 + \sum_{n=1}^{\infty} \epsilon^n a_1^n K_{n+2}^0. \tag{6.8}$$

Namely there is *no* obstacle in the q^0 equation. However, as we have showed in Section 4 that the secular terms in the orders higher than or equal to ϵ^2 consist of not only the symmetries K_n^0 of the NLS hierarchy, but also of the obstacles. In particular, the obstacles play the important rule in soliton interaction as leading to the inelasticity. The above result (6.8) overlooked this type of secularity, and it implies that the equations for q^n with $n \geq 2$ still contain the secular terms. One should also note that the general solution F of the equation $[K_2^0, F] = 0$, which defines the secularities, can not be expressed only in terms of differential polynomials, so it might not be possible to decompose an expression into secular and nonsecular parts within the class of differential polynomials. We would also like to mention that the symmetries producing the secular terms in the expansion may *not* commute each other in general (i.e. symmetry algebra is not abelian). This has been noted even in a case of ODE asymptotic problem [13], and it implies that the method of multiple scalings with more than two scaling variables fails in general.

As a final remark, we also mention that the Lie transform (3.1) up to order ϵ has been formally extended to a form including the independent variables through the master symmetry for the KdV problem [6]. Then all the asymptotically integrable equations up to order ϵ are shown to be equivalent to the KdV equation without any corrections under this extended Lie transformation. Namely, the KdV equation gives the representative of this equivalent class over the higher order equations including even the case of higher order linear dispersion up to order ϵ. However because of the explicit dependency on the independent variables, the Lie transform is not uniformly valid and the equivalence is only a weak sense. This is similar to the well-known theorem of Grobman-Hartman in the theory of ODE [2], where the equivalence is just *topological* and not *diffeomorphic* in general because of resonance (remember the dispersion change gives a resonance). One can however obtain a uniform transformation of this kind by using a near identity transformation including the change of independent variables discussed in the section of the KP deformation.

Appendix A: Calculus on Differential Polynomials

A.1. Basic Definitions

We start with definitions of important subspaces of the linear vector space $\mathcal{D}$ of differential polynomials of $\{u, u^*, u_\xi, u_\xi^*, u_{\xi\xi}, u_{\xi\xi}^*, \ldots\}$ over $\mathbb{C}$, where $*$ denotes the complex conjugation. Each differential polynomial is assumed to depend on finite number of variables, i.e. it has a finite number of monomials. We also assume that $\mathbb{C}$ does not belong to $\mathcal{D}$. In this section we use notations $u_0, u_0^*, u_1, u_1^*, u_2, u_2^*$, etc. instead of $u, u^*, u_\xi, u_\xi^*, u_{\xi\xi}, u_{\xi\xi}^*$, etc.

To each monomial we assign a positive integer, which we call the scaling weight W of the monomial, defined as

$$W(u_k) = W(u_k^*) = k+1; \quad W(A*B) = W(A)+W(B); \quad W(\alpha) = 0 \text{ if } \alpha \in \mathbb{C}.$$

For example $W(i|u|^2 u_\xi) = 4$. We say that a differential polynomial has the scaling weight k if each of its monomials has the scaling weight k. According this grading the space $\mathcal{D}$ can be decomposed in a direct sum of subspaces $\mathcal{D}_k$ with definite scaling weights k,

$$\mathcal{D} = \bigoplus_{k=1}^{\infty} \mathcal{D}_k.$$

Each subspace $\mathcal{D}_k$ has a finite dimension. For example,

$$
\begin{aligned}
\mathcal{D}_1 &= \text{Span}_{\mathbb{C}}\{u_0, u_0^*\}, \\
\mathcal{D}_2 &= \text{Span}_{\mathbb{C}}\{u_0^2, u_1, |u_0|^2, u_1^*, u_0^{*2}\}, \\
\mathcal{D}_3 &= \text{Span}_{\mathbb{C}}\{u_0^3, u_0 u_1, |u_0|^2 u_0, u_2, u_0^* u_1, u_0 u_1^*, u_2^*, |u_0|^2 u_0^*, u_0^* u_1^*, u_0^{*3}\}.
\end{aligned}
$$

Here $\mathrm{Span}_{\mathbb{C}}\{e_1,\ldots,e_n\}$ denotes all linear combinations of basis vectors e_k with complex coefficients.

Another characteristic of monomials is a guage class. Let us transform

$$u_k \rightarrow \exp(i\theta)u_k \quad \text{and} \quad u_k^* \rightarrow \exp(-i\theta)u_k^* ,$$

then we say that a monomial has gauge n if after the transformation it gains a multiplier $\exp(in\theta)$. We say that a differential polynomial has the gauge n if each of its monomials has gauge n. The space of the differential polynomials $\mathcal{D}$ and each subspace $\mathcal{D}_k$ can be decomposed as a direct sum of subspaces with definite gauges:

$$\mathcal{D} = \bigoplus_{n=-\infty}^{\infty} \mathcal{D}^n, \quad \mathcal{D}_k = \bigoplus_{n=-k}^{k} \mathcal{D}_k^n . \tag{A.1}$$

Also we define a real form of the space (and later of the algebra) of the differential polynomials due to the involution $\ddagger$ defined as $u_k^{\ddagger} = (-1)^k u_k$, $\alpha^{\ddagger} = \alpha^*$ for $\alpha \in \mathbb{C}$. Thus $\mathcal{D}$, and each subspace $\mathcal{D}_k^n$ can be decomposed into a direct sum of two subspaces over the field of real numbers $\mathbb{R}$

$$\mathcal{D}_k^n = \mathcal{H}_k^n \bigoplus_{\mathbb{R}} \mathcal{E}_k^n ,$$

where $\mathcal{H}_k^n$ is a subspace of polynomials which are invariant, and polynomials in $\mathcal{E}_k^n$ change the sign under involution. For instance

$$u \in \mathcal{H}_1^1, \ iu \in \mathcal{E}_1^1, \ iu_{\xi\xi} + 2i|u|^2 u \in \mathcal{E}_3^1, \ |u_{\xi}|^2 + |u|^4 \in \mathcal{H}_4^0 .$$

We often use the spaces $\mathcal{E}_k^1$, $\mathcal{H}_k^1$ and $\mathcal{H}_k^0$ and denote them

$$\mathcal{E}_k = \mathcal{E}_k^1 , \ \mathcal{F}_k = \mathcal{H}_k^1 , \ \mathcal{G}_k = \mathcal{E}_k^0 , \ \mathcal{H}_k = \mathcal{H}_k^0 . \tag{A.2}$$

The spaces $\mathcal{E}_k$ rise up in the asymptotic envelope expansion (2.13) ($K^m \in \mathcal{E}_{m+3}$) and in approximate symmetries ($K_k^m \in \mathcal{E}_{m+k}$). The polynomials ρ_n^k in the definition of approximate conservation laws belong to $\mathcal{H}_{k+n+2}$. The spaces $\mathcal{F}_k$ are involved in the Lie transformation (see Section 3). Each of these linear spaces has a finite dimension, and $\dim(\mathcal{E}_k) = \dim(\mathcal{F}_k)$, $\dim(\mathcal{G}_k) = \dim(\mathcal{H}_k)$. These dimensions are found to be:

k	1	2	3	4	5	6	7	8	9
$\dim(\mathcal{E}_k)$	1	1	2	3	6	9	16	24	39
$\dim(\mathcal{H}_k)$	0	1	2	4	6	11	16	27	40

k	10	11	12	13	14	15	16
$\dim(\mathcal{E}_k)$	58	90	131	197	283	413	586
$\dim(\mathcal{H}_k)$	63	92	141	202	299	426	614

For convenience we give the bases vectors of a first few spaces $\mathcal{E}_k$

$$
\begin{aligned}
\text{Basis}(\mathcal{E}_1) &= \{iu\}, \\
\text{Basis}(\mathcal{E}_2) &= \{u_1\}, \\
\text{Basis}(\mathcal{E}_3) &= \{iu_2, i|u|^2 u\}, \\
\text{Basis}(\mathcal{E}_4) &= \{u_3, |u|^2 u_1, u^2 u_1^*\}, \\
\text{Basis}(\mathcal{E}_5) &= \{iu_4, iu_1^2 u^*, i|u|^2 u_2, i|u|^4 u, i|u_1|^2 u, iu^2 u_2^*\}, \\
\text{Basis}(\mathcal{E}_6) &= \{u_5, u_1 u_2 u^*, |u|^2 u_3, |u|^4 u_1, |u_1|^2 u_1, uu_2 u_1^*, \\
&\qquad |u|^2 u^2 u_1^*, uu_1 u_2^*, u^2 u_3^*\}
\end{aligned}
$$

and $\text{Basis}(\mathcal{F}_k) = i\text{Basis}(\mathcal{E}_k)$.

A.2. Operators on Differential Polynomials

Let us define a linear differential operator D as

$$
D = \sum_{n=0} \left(u_{n+1} \frac{\partial}{\partial u_n} + u_{n+1}^* \frac{\partial}{\partial u_n^*} \right). \tag{A.3}
$$

This is nothing but the total differentiation in ξ. It is clear that operator D defines a monomorphism (i.e. a homomorphism with empty kernal space, $\text{Ker}(D) = \emptyset$)

$$
D: \quad \mathcal{D}_k \to \mathcal{D}_{k+1}; \quad D: \quad \mathcal{H}_k^n \to \mathcal{E}_{k+1}^n; \quad D: \quad \mathcal{E}_k^n \to \mathcal{H}_{k+1}^n.
$$

We shall say that a differential polynomial A is a total derivative if $A \in \text{Im}(D)$.

A linear operator of the form $L = \sum_{n=0} \left(\psi_n \frac{\partial}{\partial u_n} + \psi_n^* \frac{\partial}{\partial u_n^*} \right)$, where ψ_n are differential polynomials is called a vector field on $\mathcal{D}$ (so D is an example of a vector field). It is easy to check that a commutator of vector fields (as differential operators) is a vector field.

Proposition A.1. *Let ψ_n be differential polynomials. Then a vector field $L = \sum_{n=0} \left(\psi_n \frac{\partial}{\partial u_n} + \psi_n^* \frac{\partial}{\partial u_n^*} \right)$ commutes with D if and only if $\psi_n = D^n(\psi)$ for some function $\psi(= \psi_0)$.*

Proof. It follows from $[D, L](u_k) = 0$ for each k that

$$
\psi_{k+1} = \sum_{n=0} \left(u_{n+1} \frac{\partial \psi_k}{\partial u_n} + u_{n+1}^* \frac{\partial \psi_k}{\partial u_n^*} \right) = D(\psi_k).
$$

Thus a vector field commuting with D can be characterized by one differential polynomial $\psi = \psi_0$ and we shall denote it L_ψ,

$$
L_\psi = \sum_{n=0} \left(\psi_n \frac{\partial}{\partial u_n} + \psi_n^* \frac{\partial}{\partial u_n^*} \right), \qquad \psi_n = D^n(\psi). \tag{A.4}
$$

We call the function ψ the generating function of the vector field L_ψ, and we have $\psi = L_\psi(u)$. Only this kind of vector fields will appear in further considerations. Moreover we always assume that the generating functions belong to $\mathcal{D}_k^1$, $k = 1, 2, 3, \cdots$, i.e. $\mathcal{E}_k$ or $\mathcal{F}_k$ in (A.2).

Let $\psi \in \mathcal{E}_n$ and $\phi \in \mathcal{F}_n$. Then the corresponding vector fields define monomorphisms

$$L_\psi : \mathcal{H}_k^m \to \mathcal{E}_{k+n-1}^m \quad , \quad L_\psi : \mathcal{E}_k^m \to \mathcal{H}_{k+n-1}^m,$$
$$L_\phi : \mathcal{H}_k^m \to \mathcal{H}_{k+n-1}^m \quad , \quad L_\phi : \mathcal{E}_k^m \to \mathcal{E}_{k+n-1}^m.$$

The Lie bracket between two functions $\phi, \psi \in \mathcal{D}^1$ is defined as

$$[\phi, \psi] = L_\phi(\psi) - L_\psi(\phi) \tag{A.5}$$

The Lie bracket has the following properties:

- Skew-symmetry: $[\phi, \psi] = -[\psi, \phi]$;

- Bi-linearity (over $\mathbb{R}$): $[\alpha\phi_1 + \beta\phi_2, \psi] = \alpha[\phi_1, \psi] + \beta[\phi_2, \psi], \alpha, \beta \in \mathbb{R}$;

- Jacoby identity: $[\omega, [\phi, \psi]] + [\phi, [\psi, \omega]] + [\psi, [\omega, \phi]] = 0$.

In other words, the Lie bracket defines the structure of a Lie algebra on $\mathcal{D}^1$. The listed properties immediately follow from the following proposition:

Proposition A.2. *Let $\psi, \phi \in \mathcal{D}^1$ and $\alpha, \beta \in \mathbb{R}$. Then*

$$L_{\alpha\phi+\beta\psi} = \alpha L_\phi + \beta L_\psi \tag{A.6}$$

and

$$[L_\psi, L_\phi] = L_{[\psi,\phi]} . \tag{A.7}$$

Proof. The proof is straightforward (see for instance [18]).

With differential equations,

$$u_\tau = F, \quad F \in \mathcal{D}^1 , \tag{A.8}$$

we associate a vector field L_F. This vector field represents action of ∂_τ on $\mathcal{D}$. Indeed, for any $f \in \mathcal{D}$ we have $f_\tau = L_F(f)$.

We say that a function $K \in \mathcal{D}^\infty$ is a *symmetry* of (A.8), if $[F, K] = 0$, or in other words, if the corresponding vector fields commute. In this case two equations (A.8) and $u_\eta = K$ are compatible, and a simultaneous solution depending on two variables τ, η exists. Symmetries form a linear subspace in $\mathcal{D}^\infty$, and moreover, they form a Lie algebra with the product given by the Lie bracket (A.5).

We say that a function $\rho \in \mathcal{D}^0$ is a *density of a local conservation law* (or a conserved density) of (A.8), if ρ_τ is a total derivative, that is, $L_F(\rho) \in \mathrm{Im}(\mathrm{D})$. If $\rho \in \mathrm{Im}(\mathrm{D})$ then the conservation law is called trivial. Only nontrivial conservation laws make sense. Therefore the natural object for the definition of the local conserved densities is a factor space $\mathcal{D}^0/\mathrm{Im}(\mathrm{D})$.

Let us discuss nonlocal extensions of above defined spaces of differential polynomials. Namely we would like to extend the space $\mathcal{D}$ by adding terms of the form $\mathrm{D}^{-1}(X)$ with $X \in \mathcal{D}^0$ to the main set of variables $u_0, u_0^*, u_1, u_1^*, u_2, u_2^*, \ldots$, where D^{-1} denotes an inverse operator to D. The operator D^{-1} is well defined on the vector X that belongs to the space $\in \mathrm{Im}(\mathrm{D})$, i.e. $X = \mathrm{D}Y$, and $\mathrm{D}^{-1}(X) = Y$. To define the action of D^{-1} on the whole space (say $\mathcal{D}_n^0$) we have to extend the source space $\mathcal{D}_{n-1}^0$. Each appearence of D^{-1} reduces the scaling weight by 1. So, for instance, the scaling weight of $\mathrm{D}^{-1}(u_0 v_2)$ is equal to 3. Let us consider a simple example. The space $\mathcal{D}_2^0$ is one-dimensional (any polynomials has a form $\alpha|u|^2$, $\alpha \in \mathbb{C}$). We could extend this space by adding $\mathrm{D}^{-1}(u_1 u^*)$ and $\mathrm{D}^{-1}(uu_1^*)$ to the basis $\{|u|^2\}$. But these three vectors are linearly dependent (indeed, $\mathrm{D}^{-1}(u_1 u^*) + \mathrm{D}^{-1}(uu_1^*) - |u|^2 = 0$). To create a basis in the extended space we can join to $\{|u|^2\}$ any linear combination $\beta \mathrm{D}^{-1}(u_1 u^*) + \gamma \mathrm{D}^{-1}(uu_1^*)$ with $\beta \neq \gamma$, i.e. $\mathrm{D}^{-1}(X)$ where X is any representative of the one dimensional factor space $\mathcal{D}_3^0/\mathrm{Im}(\mathrm{D})|_{\mathcal{D}_2^0}$. For instance the set $\{|u|^2, \mathrm{D}^{-1}(uu_1^*)\}$ forms a basis in the extended space (let us denote this extended space $\hat{\mathcal{D}}_2^0$). The spaces $\hat{\mathcal{D}}_2^0$ and $\mathcal{D}_3^0$ are isomorphic and moreover this isomorphism is given by D. Extensions $\hat{\mathcal{D}}_k^0$ of other spaces $\mathcal{D}_k^0$ are built up in a similar way, namely, we take any basis $X_1, X_2, \ldots, X_s$, $s = \dim(\mathcal{D}_{k+1}^0) - \dim(\mathcal{D}_k^0)$ of representatives in the factor space $\mathcal{D}_{k+1}^0/\mathrm{Im}(\mathrm{D})|_{\mathcal{D}_k^0}$ and join $\mathrm{D}^{-1}(X_1), \mathrm{D}^{-1}(X_2), \ldots, \mathrm{D}^{-1}(X_s)$ to a basis of $\mathcal{D}_k^0$. Such extension $\hat{\mathcal{D}}_k^0$ we call the first extension of $\mathcal{D}_k^0$. One can build up the second extension by extending $\hat{\mathcal{D}}_k^0$ in a similar manner. For convenience we write out a first few sets of the basis vectors:

$\hat{\mathcal{D}}_1^0$: $\{\mathrm{D}^{-1}(|u|^2)\}$

$\hat{\mathcal{D}}_2^0$: $\{|u|^2, \mathrm{D}^{-1}(uu_1^*)\}$

$\hat{\mathcal{D}}_3^0$: $\{u_1 u^*, uu_1^*, \mathrm{D}^{-1}(uu_2^*), \mathrm{D}^{-1}(|u|^4)\}$

$\hat{\mathcal{D}}_4^0$: $\{u_2 u^*, |u_1|^2, uu_2^*, |u|^4, \mathrm{D}^{-1}(uu_3^*), \mathrm{D}^{-1}(|u|^2 uu_1^*)\}$

$\hat{\mathcal{D}}_5^0$: $\{u_3 u^*, |u|^2 u_1 u^*, u_2 u_1^*, |u|^2 uu_1^*, u_1 u_2^*, uu_3^*, \mathrm{D}^{-1}(uu_4^*), \mathrm{D}^{-1}(|u|^2 uu_2^*),$
$\quad \mathrm{D}^{-1}(u_1^2 u^{*2}), \mathrm{D}^{-1}(u^2 u_1^{*2}), \mathrm{D}^{-1}(|u|^6)\}$.

Action of vector fields L_ϕ on nonlocal objects is obvious since $L_\phi \mathrm{D}^{-1} = \mathrm{D}^{-1} L_\phi$. Also, the generating function of vector fields may depend on nonlocal objects.

Proof of Proposition 3.1. We have to prove that $F = L_K(AD^{-1}B) - L_{AD^{-1}B}(K)$ is a differential polynomial if A is a symmetry, i.e. $[K, A] = 0$ and B is a conserved density $L_K(B) \in \text{Im}(D)$. Let us expand F:

$$F = AD^{-1}(L_K(B)) + L_K(A)D^{-1}(B) -$$
$$- \sum_{n=0} \left(D^n(AD^{-1}(B))\frac{\partial K}{\partial u_n} + D^n(A^*D^{-1}(B^*))\frac{\partial K}{\partial u_n^*} \right).$$

The first term is a differential polynomial, since $L_K(B) \in \text{Im}(D)$; the rest can be rewritten in the form $(L_K(A) - L_A(K))D^{-1}(B) + Q$ where Q is a differential polynomial and the first term vanishes since A is a symmetry.

Appendix B: Envelope Expansion of (2.3)

Substituting (2.2) into (2.3), and setting terms oscillating with different frequencies to be zero separately, we obtain the following equations:

$$A_0 = -2a|A|^2 - \epsilon^2(aA_0^2 + 2a|A_2|^2 + 3bA^2A_2^* +$$
$$+6b|A|^2A_0 + 3bA_2A^{*2} + 6c|A|^4 - A_{0yy}) + O(\epsilon^4),$$
$$3A_2 = aA^2 + \epsilon^2(2aA_0A_2 + 4iA_{2\tau} - A_{2yy} +$$
$$+2aA^*A_3 + 3bA^2A_0 + 6b|A|^2A_2 + 4c|A|^2A^2) + O(\epsilon^4),$$
$$8A_3 = 2aAA_2 + bA^3 + \epsilon^2(2aA_0A_3 + 6bAA_0A_2 + 3bA^*A_2^2 + 6b|A|^2A_3 +$$
$$+4cA^3A_0 + 12c|A|^2AA_2 + 5dA^4A^* + 6iA_{3\tau} - A_{3yy}) + O(\epsilon^4),$$
$$\cdots$$
$$2iA_\tau = A_{yy} - 2aAA_0 - 2aA^*A_2 - 3b|A|^2A - \epsilon^2(2aA_2^*A_3 + 3bAA_0^2 +$$
$$+6bA|A_2|^2 + 6bA^*A_0A_2 + 3bA^{*2}A_3 + 4cA^3A_2^* + 12c|A|^2AA_0 +$$
$$+12c|A|^2A^*A_2 + 10d|A|^4A + A_{\tau\tau}) + O(\epsilon^4).$$

These can be easily solevd and all slave harmonics can be expressed in terms of the master harmonic:

$$A_0 = -2a|A|^2 - \epsilon^2 \left\{ 2a|A|_{yy}^2 + \left(\frac{38a^3}{9} - 10ab + 6c \right)|A|^4 \right\} + O(\epsilon^4)$$

$$A_2 = \frac{a}{3}A^2 + \epsilon^2 \left\{ \frac{2a}{9}(AA_{yy} - A_y^2) + \right.$$
$$\left. + \left(\frac{59a^3}{54} - \frac{31ab}{12} + \frac{4c}{3} \right)|A|^2A^2 \right\} + O(\epsilon^4)$$

$$A_3 = \left(\frac{a^2}{12} + \frac{b}{8} \right)A^3 + \epsilon^2 \left\{ \left(\frac{17a^2}{144} + \frac{3b}{32} \right)(A^2A_{yy} - AA_y^2) + \right.$$
$$\left. + \left(\frac{79a^4}{144} - \frac{43a^2b}{48} - \frac{21b^2}{64} - \frac{3ac}{20} + \frac{5d}{8} \right)|A|^2A^3 \right\} + O(\epsilon^4)$$

$$A_4 = \left(\frac{a^3}{54} + \frac{ab}{12} + \frac{c}{15} \right) A^4 + O(\epsilon^2)$$

$$A_5 = \left(\frac{2a^4}{1296} + \frac{5a^2b}{144} + \frac{b^2}{64} + \frac{11ac}{180} + \frac{d}{24} \right) A^5 + O(\epsilon^2)$$

$$\cdots$$

With these amplitudes we obtain f in terms of derivatives up to order $O(\epsilon^6)$.

Acknowledgments. We would like to express our gratitude to A. Fokas and I. M. Gel'fand for inviting us to contribute this article to a memorial book for our friend Irene Dorfman. We would also like to thank A. Fokas, A. Degasperis and P. Santini for useful discussions and sending the preprints [6] and [4] to us. The work of Y.K. was partially supported by NSF grant DMS-9403597. The work of A.V.M. was partially supported by the Soros foundation grant MLY300 and by INTAS grant INTAS-93-166.

References

[1] M. J. Ablowitz and H. Segur, Solitons and the Inverse Scattering Transform, *SIAM Stud. in Applied Math.* **4**, SIAM Publication (1981).

[2] V. I. Arnold and Yu. S. Il'yashenko, in: *Dynamical Systems I, Encyclopaedia of Mathematical Sciences* **1** D. V. Anosov et al., eds., Springer-Verlag, Berlin (1988).

[3] F. Calogero, in: *What is integrability?*, V. E. Zakharov, ed., Springer-Verlag (1991).

[4] A. Degasperis, S. V. Manakov and P. M. Santini, Multiple-scale Perturbation beyond the Nonlinear Schroedinger equation. I, preprint, August 1995.

[5] C. Desem and P. L. Chu, *Optical Soliton – Theory and Experiment*, J. R. Taylor, ed., Cambridge University Press, Cambridge (1992), Chap. 5.

[6] A. S. Fokas and Q. M. Liu, Asymptotic Integrability of Water Waves, preprint, August 1995.

[7] B. Fuchssteiner, *Prog. Theor. Phys.* **70** (1983), 1508.

[8] T. Kano, *J. Phys. Soc. Japan* **58** (1989), 4322.

[9] D. J. Kaup and A. C. Newell, *Proc. R. Soc. London, Ser A* **361** (1978), 413.

[10] Y. Kodama, *Phys. Lett.* **107A** (1985), 245; *112A* (1985), 193; *Physica* **16D** (1985), 14.

[11] Y. Kodama, *Phys. Lett.* **123A** (1987), 276; Normal Form and Solitons, in: *Topics in Soliton Theory and Exactly Solvable Nonlinear Equations*, M. J. Ablowitz et al., eds., World Scientific (1987), 319–340.

[12] Y. Kodama, *J. Phys. Soc. Japan* **45** (1978), 311.

[13] Y. Kodama, *Phys. Lett.* **191A** (1994), 223.

[14] R. A. Kraenkel, M. A. Manna and J. G. Pereira, *J. Math. Phys.* **36** (1995), 307.

[15] A. V. Mikhailov, talks given at NEEDS-91 conference, Gallipoly, Italy (1991) and at NATO Advanced Research Workshop: Singular Limits of Dispersive waves, Lyon, France (1991), (unpublished).

[16] A. V. Mikhailov, A.B.Shabat and V.V. Sokolov, in: *What is integrability?*, V. E. Zakharov, ed., Springer-Verlag (1991), 115–184.

[17] S. P. Novikov, S. V. Manakov, L. P. Pitaevskii and V. E. Zakharov, Theory of Solitons, Consultant Bureau, New York (1984).

[18] P. J. Olver, *Applications of Lie Groups to Differential Equations*, Springer-Verlag, New York (1986).

[19] T. Taniuti, *Suppl. Prog. Theor. Phys.* **55** (1974), 1.

[20] V. E. Zakharov and Shuluman, in: *What is integrability?*, V. E. Zakharov, ed., Springer-Verlag (1991), 185–249.

[21] A. V. Zhiber and A. B. Shabat, *Soviet Phys. Docl.* **24** (5) (1979), 1104.

Y. Kodama
Department of Mathematics
Ohio State University
Columbus, OH 43210, USA

A. V. Mikhailov
L.D.Landau Institute for Theoretical Physics
Kosygina 2, GSP1, Moscow 117940, Russia

Received October, 1995

Infinitely-Precise Space-Time Discretizations of the Equation $u_t + uu_x = 0$

B. A. Kupershmidt

In memory of Irene Dorfman

Abstract

The classical Volterra system $u_{n,t} = \text{const} u_n(u_{n+1} - u_{n-1})$ is time-discretized in four different ways such that each one of the infinity of conservation laws of the Volterra system is preserved exactly. Since in the space-continuous limit the Volterra system turns into the basic nonlinear infinite-dimensional dynamical system $u_t + uu_x = 0$, the Volterra conservation laws are discretizations of the conservation laws $(u^m/m)_t + [(u^{m+1}/(m+1)]_x = 0$, $m \in \mathbf{N}$.

1. Introduction

The subject of discretizations of dynamical systems, whether for numerical purposes or theoretical ones, is not a science and it is not an art. At best, it's a crude craft. True, the problems are difficult if not intractable. The crux of the difficulty lies in approximating continuous (in time) flows of local evolution equations by discrete time-step maps also of a *local* character. Clearly, this is in general possible only in approximate ways. But suppose we relax the locality condition in constructing a discretization; what other conditions should we impose? Ideally, we would like the discrete scheme to possess the most important properties of the continuous system; but what properties are "the most important"? From the practical point of view, the behavior of solutions is of primary interest, but since we generally do not know much about them (if we did we might have decided not to bother with laboring over numerical schemes) we should pick out the next most important property: the integrals; for theoretical purposes, they are the most important. The trouble is, typical dynamical systems of Classical Physics, such as dynamics of fluids and plasmas, have only a finite (small) number of integrals, and preservation of those integrals is a conditions not strong enough to reduce significantly the freedom in constructing a discretization. This leaves us with atypical systems where there is an infinite (or large enough) number of integrals present — these are so-called *integrable systems*. If one were to find a good device for solving the discretization problem, the integrable systems is the place to start. In fact, some general results in this program have already been found. Suris [1, 2] discretized all classical Toda lattices connected with simple and affine Lie algebras, and Gibbons and myself [3] discretized all scalar lattice Lax equations. (There exists another group of results, on integrable maps [4–6], but in these maps there is no dependence on the time-step parameter Δt since the time-step is taken to be 1, so there is no relation with the problem at hand.)

In this paper I take up the inviscid Burgers equation

$$u_t + uu_x = 0. \tag{1.1}$$

It has the virtue of being the basic nonlinear equation of practical physics as well as an integrable system, since it has an infinity of conservation laws

$$(u^m/m)_t + [(u^{m+1}/(m+1)]_x = 0, \qquad m \in \mathbf{N}. \tag{1.2}$$

The space-time discretization of this equation is performed in two steps. The first step, the space discretization, is well known:

$$u_{n,t} = \text{const } u_n(u_{n+1} - u_{n-1}), \qquad n \in \mathbf{Z}\Delta x, \tag{1.3}$$

which is the classical Volterra (also Manakov and Kac–V. Moerbeke) system. I now recall why it is integrable, in the language [7] employed throughout the paper.

Consider the Lax equation

$$L_t = \text{const}[(L^2)_{\geq 0}, L] = \text{const}[L, (L^2)_{<0}], \tag{1.4}$$

where

$$L = \zeta + u\zeta^{-1}; \tag{1.5}$$

all variables such as u are functions of t and $\mathbf{Z}\Delta x$; ζ is an operator version of the shift $e^{(\Delta x)d/dx}$:

$$\zeta^s f = \Delta^s(f)\zeta^s = f^{(s)}\zeta^s, \quad f^{(s)}(n\Delta x) = f((n+s)\Delta x), \quad s, n \in \mathbf{Z}; \tag{1.6}$$

for an operator $M = \sum a_i \zeta^i$, we set

$$M_{\geq 0} = \sum_{i \geq 0} a_i \zeta^i, \quad M_{<0} = \sum_{i<0} a_i \zeta^i, \quad \text{Res}(M) = a_0. \tag{1.7}$$

From the second equality in (1.4) we obtain

$$
\begin{aligned}
L_t &= u_t \zeta^{-1} = \text{const}[L, (L^2)_{<0}] = \text{const}[\zeta + u\zeta^{-1}, u\zeta^{-1}u\zeta^{-1}] = \\
&= \text{const}[\zeta, uu^{(-1)}\zeta^{-2}] = \text{const}[u^{(1)}u - uu^{(-1)}]\zeta^{-1},
\end{aligned}
$$

so that

$$u_t = \text{const } u[u^{(1)} - u^{(-1)}], \tag{1.8}$$

which is the Volterra system (1.3). The conserved densities of the Volterra system (1.8) are [7],

$$H_m = \text{Res}(L^{2m}) / \binom{2m}{m} m, \qquad m \in \mathbf{N}, \tag{1.9}$$

which go to those of (1.2), u^m/m, when $\Delta x \to 0$.

The second step consists of time-discretization of the Volterra system (1.8). To do this, we are going to discretize the Lax flow (1.4). There are no general rules for doing such a discretization, and I'll try the method of "the bizarre ansatz" of [3]; the justification is that the method works. In fact, there will be two different discretization schemes (4 if one counts the "time-reversed" schemes), based on the first and the second equality in (1.4):

$$[1 + h(\zeta^2 + \gamma)]L = \tilde{L}[1 + h(\zeta^2 + \gamma)], \tag{1.10}$$

$$(1 + h\alpha\zeta^{-2})L = \tilde{L}(1 + h\alpha\zeta^{-2}), \tag{1.11}$$

where: $h = h(\Delta t / \Delta x)$ is a formal parameter and an unspecified function of the ratio {time-step Δt divided by the space-step Δx}, subject to the only condition

$$\left. \frac{dh}{d(\cdot)} \right|_{(\cdot)=0} \neq 0. \tag{1.12}$$

E.g., $h = \text{const} \Delta t / \Delta x$ will do; $\tilde{L} = L(t + \Delta t)$ is the value of L at the shifted time; and γ, α are unknown functions of $\{u^{(s)}\}$ and h which make {the conjugation as the discrete-time evolution} definitions (1.10) and (1.11) well-defined: it suffices to allow $\gamma, \alpha \in C[[h]]$, $C = \mathbf{Q}[u^{(s)}]$. It is by no means obvious that such γ and α could be found, and to show that they can is the main difficulty in this approach (see [3]); this is the reason why one brings in formal power series: this is not necessary for formulations and computations, but only in proofs as a substitute for the implicit function theorem. What *is* obvious is that all the conserved densities (1.9) of the Volterra system (1.8) are left intact: since $\tilde{L} = (\cdot)L(\cdot)^{-1}$, it follows that $\text{Res}(\tilde{L}^m) \sim \text{Res}(L^m)$, where $\sim$ means equality modulo $\Im(\Delta - 1)$ (i.e., "divergencies", see [7] for details).

The plan of the paper is as follows. The next section is devoted to the equation (1.10) and its time-reversed version. Section 3 handles the equation (1.11) and its time-reversed version. In section 4 we write down the explicit motion equations, modulo $O(\Delta t)^3$, for each of the four constructed schemes. This allows us to prove that the four constructed schemes are nonisomorphic.

2. The Positive Evolution

Opening the brackets in the equation (1.10) we get

$$[1 + h(\zeta^2 + \gamma)]L = [1 + h(\zeta^2 + \gamma)](\zeta + u\zeta^{-1})$$

$$= \zeta + u\zeta^{-1} + h\zeta^3 + hu^{(2)}\zeta + h\gamma\zeta + h\gamma u\zeta^{-1}, \qquad (2.1a)$$

$$\tilde{L}[1 + h(\zeta^2 + \gamma)] = (\zeta + \tilde{u}\zeta^{-1})[1 + h(\zeta^2 + \gamma)]$$

$$= \zeta + \tilde{u}\zeta^{-1} + h\zeta^3 + h\tilde{u}\zeta + h\gamma^{(1)}\zeta + h\tilde{u}\gamma^{(-1)}\zeta^{-1}. \qquad (2.1b)$$

Equating (2.1a) to (2.1b) we find the following system of equations equivalent to the ansatz (1.10):

$$u^{(2)} + \gamma = \tilde{u} + \gamma^{(1)}, \qquad u + h\gamma u = \tilde{u} + h\gamma^{(-1)}\tilde{u}, \qquad (2.2)$$

which we can rewrite in the form

$$\tilde{u} = u^{(2)} + \gamma - \gamma^{(1)}, \qquad \tilde{u} = u(1 + h\gamma)/(1 + h\gamma^{(-1)}). (2.3)$$

The last equation, (2.3b), can be replaced by

$$[1 + h\gamma^{(-1)}][u^{(2)} + \gamma - \gamma^{(1)}] = u(1 + h\gamma). \qquad (2.4)$$

Thus, (2.3a) defines the time-step evolution in terms of the unknown γ, while (2.4) is an equation on γ itself. It is the last equation we need to solve in order to make the whole machine work. As usual, the trouble is that the equation (2.4) is nonlocal. Indeed, substituting $\gamma = \sum_{k \geq 0} \gamma_k h^k$ into (2.4) and equating terms we get

$$(1 - \Delta)(\gamma_0) = u - u^{(2)}, \qquad (2.5a)$$

$$(1 - \Delta)(\gamma_{k+1}) = u\gamma_k - u^{(2)}\gamma_k^{(-1)} - \sum_{p+q=k} \gamma_p^{(-1)}[\gamma_q - \gamma_q^{(1)}]. \qquad (2.5b)$$

So, from (2.5a) we find that

$$\gamma_0 = u + u^{(1)} + \text{const}, \qquad (2.6)$$

and from (2.5b) we see that *if γ_{k+1} can be found*, it is also defined up to a constant. However, these constants can all be absorbed in the properly redefined h. To see how it works, let us start with γ_0 (2.6). Denoting $\overline{\gamma}_0 = u + u^{(1)}$, so that $\gamma_0 = \overline{\gamma}_0 + \text{const}$, we get for the conjugation operator

$$
\begin{aligned}
1 + h(\zeta^2 + \gamma_0) &= 1 + h[\zeta^2 + \overline{\gamma}_0 + \text{const}] \\
&= (1 + h\,\text{const}) + h(\zeta^2 + \overline{\gamma}_0) \\
&= (1 + h\,\text{const})[1 + h_1(\zeta^2 + \overline{\gamma}_0)], \qquad h_1 = h/(1 + h\,\text{const}).
\end{aligned}
$$
$$(2.7)$$

The overall numerical factor $(1 + k\text{const})$ in (2.7) can be omitted, since it does not affect the conjugation of L. Having fixed γ_0 to be homogeneous in u of degree 1, with the grading $rk(u^{(s)}) = 1$, and noticing that equations (2.5) are homogeneous in the grading $rk(\gamma_k^{(s)}) = k+1$, we can perform the renormalization of h step-by-step, eliminating all constants from $\gamma_0, \gamma_1 \ldots$, until we are left with a homogeneous solution $\{\gamma_k\}$, *if it exists*. It does:

Theorem 2.8. *If γ satisfies*

$$\gamma = u + u^{(1)} - huu^{(-1)}/[1 + h\gamma^{(-2)}], \tag{2.9}$$

then γ satisfies (2.4). Moreover, the thus defined γ is homogeneous of $rk = 1$, in the grading where $rk(h) = -1$.

Proof. The grading claim is obvious from (2.9) since both parts of that equation are of grading $= 1$. To prove (2.4) we have to show that the expression

$$F := \gamma - \gamma^{(1)} + u^{(2)} - u(1 + h\gamma)/[1 + h\gamma^{(-1)}] \tag{2.10}$$

vanishes. From (2.9) we obtain

$$
\begin{aligned}
\gamma - \gamma^{(1)} + u^{(2)} &= \left\{ u + u^{(1)} - huu^{(-1)}/[1 + h\gamma^{(-2)}] \right\} \\
&\quad - \left\{ u^{(1)} + u^{(2)} - hu^{(1)}u/[1 + h\gamma^{(-1)}] \right\} + u^{(2)} \\
&= u \left\{ 1 - hu^{(-1)}/[1 + h\gamma^{(-2)}] + hu^{(1)}/[1 + h\gamma^{(-1)}] \right\},
\end{aligned}
$$

whence

$$
\begin{aligned}
F &= u \left\{ 1 - hu^{(-1)}/[1 + h\gamma^{(-2)}] + hu^{(1)}/[1 + h\gamma^{(-1)}] \right. \\
&\quad \left. -(1 + h\gamma)/[1 + h\gamma^{(-1)}] \right\} \\
&= \frac{u}{1 + h\gamma^{(-1)}} \left\{ [1 + h\gamma^{(-1)}] - (1 + h\gamma) \right. \\
&\quad \left. -hu^{(-1)}[1 + h\gamma^{(-1)}]/[1 + h\gamma^{(-2)}] + hu^{(1)} \right\} \\
&= \frac{u}{1 + h\gamma^{(-1)}} \left\{ \gamma^{(-1)} - \gamma + u^{(1)} - u^{(-1)}[1 + h\gamma^{(-1)}]/[1 + h\gamma^{(-2)}] \right\} \\
&= h \frac{u}{1 + h\gamma^{(-1)}} \Delta^{-1}(F) = huF^{(-1)}/[1 + h\gamma^{(-1)}]. \tag{2.11}
\end{aligned}
$$

Continuing on, we get

$$F = h^{k+1}uu^{(-1)} \ldots u^{(-k)} F^{(-k-1)} / \prod_{s=0}^{k} [1 + h\gamma^{(-s-1)}]. \tag{2.12}$$

Hence, $F = O(h^{k+1})$ for any k. Thus, F vanishes. $\qquad\square$

Remark 2.13. The sudden solution (2.9) of the nonlocal equation (2.4) appears as a deus ex machina. It has the following origin. From (1.10) we find that

$$[1 + h(\zeta^2 + \gamma)]L^2 = \tilde{L}^2[1 + h(\zeta^2 + \gamma)],\qquad(2.14)$$

and since

$$L^2 = (\zeta + u\zeta^{-1})(\zeta + u\zeta^{-1}) = \zeta^2 + [u + u^{(1)}] + uu^{(-1)}\zeta^{-2}\qquad(2.15)$$

is the Lax operator of the Toda lattice on the double of the original lattice $\mathbf{Z}\Delta x$, we can use the 1st integral of the discrete-time Toda lattice from [3] to conclude that

$$(\Delta^2 - 1)\left\{\gamma - u - u^{(1)} + huu^{(-1)}/[1 + h\gamma^{(-2)}]\right\} = 0.\qquad(2.16)$$

It follows that

$$\left\{\gamma - u - u^{(1)} + huu^{(-1)}/[1 + h\gamma^{(-2)}]\right\} = ?,\qquad(2.17)$$

where ? belongs to $\mathrm{Ker}(\Delta^2 - 1)$. If we demand that $rk(?) = 1$ and ? $\in$ $\mathbf{Q}[u^{(s)}][[h]]$ then ? must vanish, resulting in formula (2.9). To bypass the recourse to the Toda theory arguments, which in any case is more heuristic than rigorous, here is the direct proof that (2.4) implies (2.16):

Proof of (2.16). We have to show that

$$[\gamma^{(2)} - \gamma - u^{(2)} - u^{(3)}]+\qquad(2.18a)$$

$$+[u + u^{(1)}]+\qquad(2.18b)$$

$$+hu^{(2)}u^{(1)}/(1 + h\gamma) - huu^{(-1)}/[1 + h\gamma^{(-2)}].\qquad(2.18c)$$

vanishes. From (2.4) in the form

$$\gamma - \gamma^{(1)} + u^{(2)} = u(1 + h\gamma)/[1 + h\gamma^{(-1)}],\qquad(2.19)$$

we conclude, by applying the operator $(1 + \Delta)$, that

$$\gamma - \gamma^{(2)} + u^{(2)} + u^{(3)} = u(1 + h\gamma)/[1 + h\gamma^{(-1)}] + u^{(1)}[1 + h\gamma^{(1)}]/(1 + h\gamma).$$

Substituting this into (2.18a) and adding up (2.18b) we obtain

$$u - u(1 + h\gamma)/[1 + h\gamma^{(-1)}] + u^{(1)} - u^{(1)}[1 + h\gamma^{(1)}]/(1 + h\gamma)$$
$$= uh[\gamma^{(-1)} - \gamma]/[1 + h\gamma^{(-1)}] + u^{(1)}h[\gamma - \gamma^{(1)}]/(1 + h\gamma).$$

$$(2.20)$$

The first term in (2.20) can be transformed, by (2.19), as

$$\frac{uh}{1+h\gamma^{(-1)}}[\gamma^{(-1)} - \gamma] = \frac{uh}{1+h\gamma^{(-1)}}\Delta^{-1}\left\{u(1+h\gamma)/[1+h\gamma^{(-1)}] - u^{(2)}\right\}$$

$$= \frac{uh}{1+h\gamma^{(-1)}} \cdot \frac{u^{(-1)}[1+h\gamma^{(-1)}]}{1+h\gamma^{(-2)}} - \frac{uhu^{(1)}}{1+h\gamma^{(-1)}}$$

$$(2.21)$$

The second summand in (2.20) and the first summand in (2.18c) combine into

$$\frac{hu^{(1)}}{1+h\gamma}[u^{(2)} + \gamma - \gamma^{(1)}] \quad \text{[by (2.19)]} = \frac{hu^{(1)}u}{1+h\gamma^{(-1)}}. \qquad (2.22)$$

The first term in (2.21) and the second one in (2.18c) cancel, as do the second one in (2.21) and the RHS of (2.22). □

We conclude this section by discussing the "time-reversed" scheme. This amounts to replacing the ansatz (1.10) by the ansatz

$$L[1 + h(\zeta^2 + \gamma)] = [1 + h(\zeta^2 + \gamma)]\tilde{L}. \qquad (2.23)$$

This is equivalent to the interchange of tilded and untilded variables in (2.2) and (2.19):

$$\tilde{u}^{(2)} + \gamma = u + \gamma^{(1)}, \qquad \tilde{u}(1 + h\gamma) = u[1 + h\gamma^{(-1)}], \qquad (2.24a, b)$$

$$\gamma = \tilde{u} + \tilde{u}^{(1)} - h\tilde{u}\tilde{u}^{(-1)}/[1 + h\gamma^{(-2)}]. \qquad (2.24c)$$

The equation (2.24a) is equivalent to

$$\tilde{u} = u^{(-2)} + \gamma^{(-1)} - \gamma^{(-2)}. \qquad (2.25)$$

To bring the equation (2.24c) into the same-time, i.e., untilded, form, we use (2.25) and get

$$\tilde{u} + \tilde{u}^{(1)} = u^{(-2)} + \gamma^{(-1)} - \gamma^{(-2)} + u^{(-1)} + \gamma - \gamma^{(-1)} = u^{(-2)} + u^{(-1)} + \gamma - \gamma^{(-2)},$$

$$(2.26a)$$

while (2.24b)

$$\tilde{u}\tilde{u}^{(-1)} = \frac{u[1 + h\gamma^{(-1)}]}{1 + h\gamma} \frac{u^{(-1)}[1 + h\gamma^{(-2)}]}{1 + h\gamma^{(-1)}} = \frac{uu^{(-1)}[1 + h\gamma^{(-2)}]}{1 + h\gamma}. \qquad (2.26b)$$

Substituting (2.26) into (2.24c) we obtain

$$\gamma = u^{(-2)} + u^{(-1)} + \gamma - \gamma^{(-2)} - huu^{(-1)}/(1 + h\gamma),$$

which is the same as

$$\gamma^{(-2)} = u^{(-2)} + u^{(-1)} - huu^{(-1)}/(1 + h\gamma),$$

which is finally

$$\gamma = u + u^{(1)} - hu^{(2)}u^{(1)}/[1 + h\gamma^{(2)}]. \qquad (2.27)$$

This is different from (2.9).

Note that if u has initially a compact support,

$$u_n = 0, \quad n \leq a \text{ and } n \geq b, \qquad (2.28)$$

then so does $\tilde{u}$: from (2.3a), (2.9) we get

$$\tilde{u}_n = 0, \quad n \leq a - 2 \text{ and } n \geq b, \qquad (2.29a)$$

while from (2.25) and (2.27) we obtain

$$\tilde{u}_n = 0, \quad n \leq a - 3 \text{ and } n \geq b - 1, \qquad (2.29b)$$

3. The Negative Evolution

Opening the brackets in the defining equation (1.11) we obtain

$$(1 + ha\zeta^{-2})L = (1 + ha\zeta^{-2})(\zeta + u\zeta^{-1})$$

$$= \zeta + u\zeta^{-1} + ha\zeta^{-1} + hau^{(-2)}\zeta^{-2}h, \qquad (3.1a)$$

$$\tilde{L}(1 + ha\zeta^{-2}) = (\zeta + \tilde{u}\zeta^{-1})(1 + ha\zeta^{-2})$$

$$= \zeta + \tilde{u}\zeta^{-1} + ha^{(1)}\zeta^{-1} + h\tilde{u}a^{(-1)}\zeta^{-2}h, \qquad (3.1b)$$

which is equivalent to the system

$$au^{(-2)} = a^{(-1)}\tilde{u}, \qquad u + ha = \tilde{u} + ha^{(1)}. \qquad (3.2)$$

This system consists of the evolution of u in terms of α:

$$\tilde{u} = u + h[\alpha - \alpha^{(1)}] \ (= u^{(-2)}\alpha/\alpha^{(-1)}), \qquad (3.3)$$

and the equation defining α:

$$\alpha^{(-1)}\left\{u + h[\alpha - \alpha^{(1)}]\right\} = u^{(-2)}\alpha. \qquad (3.4)$$

Rewriting this last equation in the form

$$(u^{(-2)} - u\Delta^{(-1)})(\alpha) = h(1 - \Delta)(\alpha), \qquad (3.5)$$

we see that it is, typically, nonlocal. However, it can be integrated.

Introduce the new variable

$$\beta = \alpha/uu^{(-1)}, \qquad \alpha = \beta uu^{(-1)}. \tag{3.6}$$

Substituting this into (3.4) we get

$$\beta^{(-1)}u^{(-1)}u^{(-2)}\left\{u + h\beta uu^{(-1)} - h\beta^{(1)}u^{(1)}u\right\} = u^{(-2)}\beta uu^{(-1)}.$$

Dividing this by $uu^{(-1)}u^{(-2)}$, we obtain

$$\beta^{(-1)}\left\{1 + h\beta u^{(-1)} - h\beta^{(1)}u^{(1)}\right\} = \beta,$$

or

$$\beta^{(-1)}[1 - h\beta^{(1)}u^{(1)}] = \beta[1 - h\beta^{(-1)}u^{(-1)}],$$

or

$$\frac{1 - h\beta^{(1)}u^{(1)}}{\beta} = \frac{1 - h\beta^{(-1)}u^{(-1)}}{\beta^{(-1)}}.$$

Multiplying both sides of the last equation by $(1 - h\beta u)$, we find

$$\frac{(1 - h\beta^{(1)}u^{(1)}](1 - h\beta u)}{\beta} = \frac{(1 - h\beta u)[1 - h\beta^{(-1)}u^{(-1)}]}{\beta^{(-1)}},$$

or

$$(1 - \Delta^{(-1)})\left\{[1 - h\beta^{(1)}u^{(1)}](1 - h\beta u)/\beta\right\} = 0. \tag{3.7}$$

This is equivalent to

$$[1 - h\beta^{(1)}u^{(1)}](1 - h\beta u)/\beta = \text{const}, \tag{3.8}$$

where const is a nonzero function of h. Rescaling β and h if needed, we can make the constant into 1, so (3.8) becomes

$$\beta = [1 - h\beta^{(1)}u^{(1)}](1 - h\beta u). \tag{3.9}$$

Returning to the α-language (3.6), we finally obtain

$$\alpha = uu^{(-1)}[1 - h\alpha^{(1)}/u][1 - h\alpha/u^{(-1)}],$$

or

$$\alpha = [u - h\alpha^{(1)}][u^{(-1)} - h\alpha]. \tag{3.10}$$

Since α enters into the RHS of (3.10) only through the combination $h\alpha$, formula (3.10) defines a unique series

$$\alpha = \sum_{k \geq 0} \alpha_k h^k, \qquad \alpha_k \in \mathbf{Q}[u^{(s)}], \tag{3.11}$$

which solves (3.10) and (3.4).

Let's now turn to the ansatz "time-reversed" with respect to (1.11):

$$L(1 + h\alpha\zeta^{-2}) = (1 + h\alpha\zeta^{-2})\tilde{L}. \tag{3.12}$$

As before, this amounts to exchange of u and $\tilde{u}$ in formulae (3.2) and (3.10):

$$\tilde{u} = \alpha^{(1)}u^{(2)}/\alpha^{(2)}, \tag{3.13}$$

$$\tilde{u} = u + h[\alpha^{(1)} - \alpha], \qquad \alpha = [\tilde{u} - h\alpha^{(1)}][\tilde{u}^{(-1)} - h\alpha]. \tag{3.14}$$

Substituting (3.14a) into (3.14b) we finally get

$$\alpha = (u - h\alpha)[u^{(-1)} - h\alpha^{(-1)}]. \tag{3.15}$$

Again, formulae {(3.3) & (3.10)} and {(3.14a) & (3.15)} show that if u has a compact support (2.28), then so does $\tilde{u}$: from (3.3), (3.6), and (3.9) we find that

$$\tilde{u}_n = 0, \quad n \leq a \text{ and } n \geq b. \tag{3.16}$$

This looks very strange, but can be explained as follows:

$$
\begin{aligned}
\tilde{u} \text{ [by (3.2b)]} &= u + h\alpha - h\alpha^{(1)} \text{ [by (3.6)]} \\
&= u + huu^{(-1)}\beta - hu^{(1)}u\beta^{(1)} \\
&= u[1 + hu^{(-1)}\beta - hu^{(1)}\beta^{(1)}].
\end{aligned}
\tag{3.17}
$$

Since, by (3.9), $\beta = 1 + O(h)$, we see from (3.17) that $\text{supp}(\tilde{u}) \subset \text{supp}(u)$, which agrees with (3.16). Similarly, in case (3.12), we will have

$$\alpha = \tilde{u}\tilde{u}^{(-1)}\beta \text{ [by (3.13)]} = \frac{\alpha^{(1)}u^{(2)}}{\alpha^{(2)}}\frac{\alpha u^{(1)}}{\alpha^{(1)}}\beta = \alpha\frac{u^{(1)}u^{(2)}}{\alpha^{(2)}}\beta,$$

so that $alpha^{(2)} = u^{(1)}u^{(2)}\beta$, and finally

$$\alpha = u^{(-1)}u\beta^{(-2)}. \tag{3.18}$$

Substituting this into (3.14) we obtain

$$
\begin{aligned}
\tilde{u} &= u + h\alpha^{(1)} - h\alpha = u + huu^{(1)}\beta^{(-1)} - hu^{(-1)}u\beta^{(-2)} \\
&= u[1 + hu^{(1)}\beta^{(-1)} - hu^{(-1)}\beta^{(-2)}],
\end{aligned}
\tag{3.19}
$$

and again formula (3.16) results. This contradicts the intuition about the continuous system $u_t + uu_x = 0$. The reader is invited to resolve the apparent paradox.

4. Explicit Motion Equations

In this section we write down explicit equations of motion, modulo $O(h^3)$, for each of the 4 discretizations constructed previously.

1) The first system, (2.3a) and (2.9), is

$$\tilde{u} = u^{(2)} + \gamma - \gamma^{(1)}, \tag{4.1}$$

$$\gamma = u + u^{(1)} - huu^{(-1)}/[1 + h\gamma^{(-2)}]$$

$$= u + u^{(1)} - huu^{(-1)} + h^2 uu^{(-1)}[u^{(-2)} + u^{(-1)}] + O(h^3), \tag{4.2}$$

so that

$$\tilde{u} = u + hu[u^{(1)} - u^{(-1)}]$$

$$+ h^2 u \left\{ u^{(-1)}[u^{(-2)} + u^{(-1)}] - u^{(1)}[u^{(-1)} + u] \right\} + O(h^3). \tag{4.3}$$

2) The second system, (2.25) and (2.27), is

$$\tilde{u} = u^{(-2)} + \gamma^{(-1)} - \gamma^{(-2)}, \tag{4.4}$$

$$\gamma = u + u^{(1)} - hu^{(1)}u^{(2)}/[1 + h\gamma^{(2)}]$$

$$= u + u^{(1)} - hu^{(1)}u^{(2)} + h^2 u^{(1)}u^{(2)}[u^{(2)} + u^{(2)}h] + O(h^3), \tag{4.5}$$

so that

$$\tilde{u} = u + hu[u^{(-1)} - u^{(1)}]$$

$$+ h^2 u \left\{ u^{(1)}[u^{(1)} + u^{(2)}] - u^{(-1)}[u + u^{(1)}] \right\} + O(h^3). \tag{4.6}$$

3) The third system, (3.3) and (3.10), is

$$\tilde{u} = u + h[\alpha - \alpha^{(1)}], \tag{4.7}$$

$$\alpha = [u - h\alpha^{(1)}][u^{(-1)} - h\alpha]$$

$$= uu^{(-1)} - huu^{(-1)}[u + u^{(1)}] + O(h^2), \tag{4.8}$$

so that

$$\tilde{u} = u + hu[u^{(-1)} - u^{(1)}]$$

$$+ h^2 u \left\{ u^{(1)}[u^{(1)} + u^{(2)}] - u^{(-1)}[u + u^{(1)}] \right\} + O(h^3). \tag{4.9}$$

4) The fourth system, (3.14a) and (3.15), is

$$\tilde{u} = u + h[\alpha^{(1)} - \alpha], \tag{4.10}$$

$$\alpha = (u - h\alpha)[u^{(-1)} - h\alpha^{(-1)}]$$

$$= uu^{(-1)} - huu^{(-1)}[u^{(-1)} + u^{(-2)}] + O(h^2), \tag{4.11}$$

so that

$$\begin{aligned}
\tilde{u} &= u + hu[u^{(1)} - u^{(-1)}] \\
&\quad + h^2 u \left\{ u^{(-1)}[u^{(-1)} + u^{(-2)}] - u^{(1)}[u + u^{(-1)}] \right\} + O(h^3).
\end{aligned}$$

$$(4.12)$$

From the h^2-terms in the equations $(4.3k)$ we see that all these equations are nonisomorphic for $k = 1, 2, 3, 4$. We also see that, to have the correct continuous limit, we should have h as

$$h = \pm \Delta t / 2\Delta x + O(\Delta t / \Delta x)^2,$$

with the $+$ sign for the 2nd and 3rd cases, and $-$ sign for the 1st and 4th ones. As may be easily observed, the equations (4.3) and (4.6) are space reversals of each other, and similar for the equations (4.9) and (4.12). This is easily seen to be true for the full system $\{(4.1), (4.2)\} \leftrightarrow \{(4.4), (4.5)\}$, $\{(4.7), (4.8)\} \leftrightarrow \{(4.10), (4.11)\}$, and it is a counterpart of the symmetry $t \to -t$, $x \to -x$, of the equation (1.1).

Acknowledgement. This paper is based on a lecture delivered at the USSR Institute of Optics on the week the Soviet Union expired. I thank Valery Rupasov for the invitation. This work was partially supported by the NSF.

References

[1] Suris, Yu. B., *Leningrad Math. J.* **2** (1991), 339.

[2] Suris, Yu. B., *Phys. Lett. A* **145** (1990), 113.

[3] Gibbons, J. and Kupershmidt, B. A., *Phys. Lett. A* **165** (1992), 105.

[4] Veselov, A. P., *Funct. Anal. Appl.* **22** (2) (1988), 1; **25** (2) (1991), 38.

[5] Moser, J. and Veselov, A. P., *Comm. Math. Phys.* **139** (1991), 217.

[6] Veselov, A. P., *Uspekhi Mat. Nauk* **46** (5) (1991), 3.

[7] Kupershmidt, B. A., *Discrete Lax Equations and Differential-Difference Calculus*, Asterisque, Paris (1985).

The University of Tennessee Space Institute
Tullahoma, Tennessee 37388, USA

Received October, 1995

Trace Formulas and the Canonical 1-Form

Henry P. McKean[1]

Dedicated to the memory of Irene Dorfman.

1. Introduction

This paper studies the canonical 1-form of symplectic geometry in the context of the (defocussing) cubic Schrödinger system. The phase space is populated by pairs QP of smooth functions of period 1, equipped with the classical 1-form $QdP = \int_0^1 [Q(x)\, dP(x)]\, dx$. The introduction of canonically paired coordinates $Q_n P_n : n \in \mathbb{Z}$, as in Sections 2 and 6 below, suggests the identity $QdP = \Sigma_{\mathbb{Z}} Q_n\, dP_n$, up to an additive exact form, and this may be verified, as in Sections 5 and 6, with the help of new trace formulas, derived in what I believe to be a new way; see, especially Section 4, nos. 4 and 5. The discussion could be carried over to sine/sh-Gordon, *etc.*; compare Section 7 where this is done for KdV.

2. Preliminaries

The phase space comes equipped with the classical 2-form

$$dQ \wedge dP = \int_0^1 \left[dQ(x) \wedge dP(x) \right] dx,$$

which states that the associated bracket is the conventional

$$[AB] = \nabla A\, J\, \nabla B = \int_0^1 \left[\frac{\partial A}{\partial Q} \frac{\partial B}{\partial P} - \frac{\partial A}{\partial P} \frac{\partial B}{\partial Q} \right] dx$$

with $\nabla = \mathrm{grad} = (\partial/\partial Q,\ \partial/\partial P)$ and $J = \left(\begin{smallmatrix} 0 & 1 \\ -1 & 0 \end{smallmatrix} \right)$, as for $d < \infty$ degrees of freedom. This is the proper setting for the cubic Schrödinger system:

$$\partial Q/\partial t \quad = \quad -\partial^2 P/\partial x^2 + 2\left(Q^2 + P^2 \right) P = +\partial H_3/\partial P$$

---

[1]This work was performed at the Courant Institute of Mathematical Sciences with the partial support of the National Science Foundation, under NSF Grant No. DMS–9112664, which is gratefully acknowledged.

$$\partial P/\partial t \;=\; +\partial^2 Q/\partial x^2 - 2\left(Q^2{+}P^2\right)Q = -\partial H_3/\partial Q$$

with

$$H_3 = \frac{1}{2}\int_0^1\left[(Q')^2{+}(P')^2 + (Q^2{+}P^2)^2\right]dx\;.$$

The integration of this system is accomplished with the help of spectral information about the 2×2 Dirac operator[2] $\mathbf{D} = D\times\mathrm{id}-Z$ with $Z = \left(\begin{smallmatrix}Q & P\\ P & -Q\end{smallmatrix}\right)$. The 2×2 monodromy matrix $M = [m_{ij} : 1 \le i,\, j \le 2]$ is the solution of $\mathbf{D}M(x) = -(\lambda/2)\,JM(x)$ with $M(0) = 1$, taken at $x = 1$ and considered as a function of $\lambda \in \mathbb{C}$. The associated discriminant $\Delta = (1/2)\,sp\,M$ determines the 2-sheeted curve $\mathfrak{M}$ of points $\mathfrak{p} = \left[\lambda, \sqrt{\Delta^2(\lambda) - 1}\right]$. $\mathfrak{M}$ is ramified/double over the projection $\mathfrak{p} \to \lambda(\mathfrak{p})$ at the simple/double roots of $\Delta(\lambda) = \pm 1$. These come in (separated) pairs $\lambda_n^- \le \lambda_n^+$, indexed by $n \in \mathbb{Z}$, located at $2\pi n$ more or less. The intervening gap $[\lambda_n^-, \lambda_n^+]$ is covered by the nth real oval a_n of $\mathfrak{M}$; the latter reduces to a single double point of $\mathfrak{M}$ if the gap is closed $(\lambda_n^- = \lambda_n^+)$. To each phase QP with fixed discriminant corresponds a divisor $\mathfrak{P}$ of points $\mathfrak{p}_n \in \mathfrak{M}$; one such to each oval $a_n : n \in \mathbb{Z}$, determined by the rule

$$m_{12} = 0 \text{ at } \lambda = \lambda(\mathfrak{p}_n) \text{ and } \sqrt{\Delta^2 - 1}\,(\mathfrak{p}_n) = \tfrac{1}{2}\,(m_{11} - m_{22})\;,$$

the map of phase to divisor being $1:1$ onto the product $\times\,a_n$. McKean–Vaninsky [1996(1)] can be consulted for this and other technical information needed below. The assumption that QP is smooth in $0 \le x < 1$ is not really needed, but it simplifies life in that the gap size $\lambda_n^+ - \lambda_n^-$ vanishes rapidly at $n = \pm\infty$, obviating certain technical difficulties below.

3. Coordinates in the Large

The projection $Q_n = \lambda(\mathfrak{p}_n)$ and the associated quantities[3]

$$P_n = -\frac{1}{2}\,ch^{-1}\Delta(\mathfrak{p}_n) = -\frac{1}{2}\int_{\mathfrak{o}_n}^{\mathfrak{p}_n} \Delta^\bullet(\Delta^2 - 1)^{-1/2}\,d\lambda \;\; \text{with} \;\; \mathfrak{o}_n = [\lambda_n^-, 0]$$

are canonically paired, providing a system of coordinates in the whole phase space. The proof is routine but will be sketched for completeness. There are 5 items to check: 1) $[Q_i, Q_j] = 0$; 2) $[Q_i, P_j] = 1$ or 0 according as $i = j$ or not; 3) $[P_i, P_j] = 0$; 4) $Q_n\,P_n : n \in \mathbb{Z}$ determines the phase $QP(x) : 0 \le x < 1$; and the technical item 5) $P_n = -(1/2)\,ch^{-1}\,\Delta\,(\mathfrak{p}_n)$

---

[2] D means differentiation with regard to x.

[3] $\bullet$ means differentiation with regard to λ. The integral is provided with the signature inherited from the radical $\sqrt{\Delta^2 - 1}\,(\mathfrak{p}_n)$ so as to make it unambiguous.

is smooth despite the risky looks of $\nabla ch^{-1} \Delta (\mathfrak{p}_n) = (\Delta^2 - 1)^{-1/2} \, [\nabla \Delta + \Delta^\bullet \nabla Q_n]$. The identity [4]

$$\lambda \nabla M = D(J\nabla) M + 4 \begin{pmatrix} Q \\ P \end{pmatrix} \int_0^x XM + 2 \begin{pmatrix} Q \\ P \end{pmatrix} MJ$$

with $\mathbf{X} = P\partial/\partial Q - Q\partial/\partial P$ lies behind it all.

Item 1. The derived identity

$$\lambda \nabla m_{12} = DJ\nabla m_{12} + 4 \begin{pmatrix} Q \\ P \end{pmatrix} \int_0^x Xm_{12} + 2 \begin{pmatrix} Q \\ P \end{pmatrix} m_{11}$$

produces

$$
\begin{aligned}
(\lambda_1 - \lambda_2) \Big[m_{12}(\lambda_1), \ m_{12}(\lambda_2) \Big] \ &= \ \nabla m_{12} \bullet \nabla m_{12} \Big|_{x=0}^{x=1} \\
&\quad - \ 4 \int_0^1 Xm_{12} \times \int_0^1 Xm_{12} \\
&\quad - \ 2\, m_{11} \times \int_0^1 Xm_{12} \\
&\quad - \ 2 \int_0^1 Xm_{12} \times m_{11},
\end{aligned}
$$

left/right factors being taken at $\lambda = \lambda_1/\lambda_2$. But $\int_0^1 Xm_{12}$ is nothing but the field $\mathbf{X} = [\, \bullet, \, H_1 = \frac{1}{2} \int_0^1 (Q^2 + P^2)]$ applied to m_{12}, to wit, $\frac{1}{2}(m_{22} - m_{11})$, as follows from [5]

$$m_{12}(x) = \Pi(2\pi n)^{-1} \Big[\lambda(\mathfrak{p}_n) - \lambda \Big], \quad X\lambda(\mathfrak{p}_n) = \sqrt{\Delta^2 - 1}/m_{12}^\bullet (\mathfrak{p}_n),$$

and the fact that, $\sqrt{\Delta^2 - 1} = \frac{1}{2}(m_{11} - m_{22})$ at $\mathfrak{p}_n$:

$$
\begin{aligned}
\mathbf{X}_1 m_{12} \ &= \ m_{12} \sum \Big[\lambda(\mathfrak{p}_n) - \lambda \Big]^{-1} X_1 (\mathfrak{p}_n) \\
&= \ \sum \frac{m_{22} - m_{11}}{2m_{12}^\bullet} \text{ at } \lambda(\mathfrak{p}_n) \times \frac{m_{12}(\lambda)}{\lambda - \lambda(\mathfrak{p}_n)} \\
&= \ \frac{1}{2} (m_{22} - m_{11}),
\end{aligned}
$$

by interpolation,[6] and now the values of $\nabla m_{12} = (-m_{12}, m_{11})$ at $x = 0$ and (m_{12}, m_{22}) at $x = 1$ show that the whole display vanishes. Item 1 is plain from that.

---

[4]McKean–Vaninsky [1996(1)].

[5]$2\pi 0$ is understood as unity.

[6]This is like Shannon's sampling formula; it requires only $m_{22} - m_{11}$ to be band-limited to $[-1/2, 1/2]$ and $\int_{-\infty}^\infty (m_{22} - m_{11})^2 \, d\lambda < \infty$; compare McKean–Vaninsky [1996(1)].

Item 2. $[Q_i, P_j] = -(1/2)[Q_i, \Delta(\lambda)]$ taken at $\lambda(\mathfrak{p}_j)$ multiplied by $\Delta^\bullet(\Delta^2 - 1)^{-1/2}$ $(\mathfrak{p}_j)$ by item 1, and

$$\left[Q_i, \Delta(\lambda)\right] = -2 \frac{\sqrt{\Delta^2 - 1}}{m_{12}^\bullet} (\mathfrak{p}_i) \frac{m_{12}(\lambda)}{\lambda - \lambda(\mathfrak{p}_i)} ,$$

as can be confirmed following the plan of item 1 with the help of the derived identity

$$\lambda \nabla \Delta = DJ\nabla\Delta + 4 \binom{Q}{P} \int_0^x X\Delta + \binom{Q}{P} (m_{21} - m_{12}).$$

Now

$$\left[Q_i, P_j\right] = \frac{1}{m_{12}^\bullet \text{ at } \lambda(\mathfrak{p}_i)} \frac{m_{12}(\lambda)}{\lambda - \lambda(\mathfrak{p}_i)} \quad \text{taken at } \lambda = \lambda(\mathfrak{p}_j)$$

and this is 1 or 0, according as $i = j$ or not.

Item 3 follows from items 1 and 2 and the additional fact that the several values of $\Delta(\lambda) : \lambda \in \mathbb{R}$ commute; compare McKean–Vaninsky [1996(1)] who spell that out in the style of item 1.

Item 4. The numbers $Q_n = \lambda(\mathfrak{p}_n)$ determine $m_{12}(x) = \Pi(2\pi n)^{-1}\left[\lambda(\mathfrak{p}_n) - \lambda\right]$ and sign $P_n = \text{sign} \sqrt{\Delta^2 - 1}(\mathfrak{p}_n)$ takes away the remaining ambiguity in the divisor. Then the phase QP is fixed provided Δ is known, which is, in fact, the case: $\Delta(\lambda) - \cos(\lambda/2)$ is band-limited to $[-1/2, 1/2]$ with $\int_{-\infty}^\infty [\Delta(\lambda) - \cos(\lambda/2)]^2 \, dx < \infty$ and so maybe interpolated off its known values $\Delta(\lambda) - \cos(\lambda/2)$ at $\lambda(\mathfrak{p}_n)$, to wit $\text{ch}(-2P_n) - \cos(Q_n/2)$, as for $m_{22} - m_{11}$ in item 1.

Item 5. $\nabla P_n = -(1/2)(\Delta^2 - 1)^{-1/2}(\mathfrak{p}_n) \times [\nabla\Delta - (\Delta^\bullet/m_{12}^\bullet)\nabla m_{12}]$ at $\lambda = \lambda(\mathfrak{p}_n)$ is well behaved if $\lambda_n^- < \lambda(\mathfrak{p}_n) < \lambda_n^+$, but what happens at the ends? The determinant $m_{11}\,m_{22} - m_{12}\,m_{21} = 1$ is acted upon by $\nabla = (\partial/\partial Q, \partial/\partial P)$ and by $\bullet = \partial/\partial\lambda$ to produce

$$\Delta\nabla\Delta = m_{21}\,\nabla m_{12} + \frac{1}{2}\sqrt{\Delta^2 - 1}\,\nabla(m_{11} - m_{22})$$

and

$$\Delta\Delta^\bullet = m_{21}\,m_{12}^\bullet + \frac{1}{2}\sqrt{\Delta^2 - 1}\,(m_{11} - m_{22})^\bullet$$

at $\mathfrak{p}_n$.[7] Then you find

$$
\begin{aligned}
-2\Delta\,m_{12}^\bullet\,\nabla P_n &= \Delta m_{12}^\bullet\,\nabla\Delta - \Delta\Delta^\bullet\nabla m_{12} \quad \text{over} \quad \sqrt{\Delta^2 - 1}\,(\mathfrak{p}_n) \\
&= \frac{1}{2}\,m_{12}^\bullet\,\nabla(m_{11} - m_{22}) - \frac{1}{2}\nabla m_{12}\,(m_{11} - m_{22})^\bullet
\end{aligned}
$$

---

[7] $2\sqrt{\Delta^2 - 1}(\mathfrak{p}_n) = m_{11} - m_{22}$ is used.

at $\lambda(\mathfrak{p}_n)$, and that is fine no matter where $\lambda(\mathfrak{p}_n)$ sits in its gap since $|\Delta| \geq 1$ there and $|m_{12}^\bullet|$ is controlled from below; see McKean–Vaninsky [1996(1)]. The discussion is finished.

4. Some Trace Formulas

These involve the variables $Q_n = \lambda(\mathfrak{p}_n)$ and $P_n = -(1/2)\, ch^{-1}\, \Delta(\mathfrak{p}_n)$. The first 3 are illustrative; numbers 4 and 5 is what is needed for the identity of canonical 1-forms, to be proved in Section 5.

Sample 1. $\Delta^\bullet(\lambda) = 0$ has a simple root $\lambda_n^\bullet$ in the nth gap and $\sum_Z [\lambda(\mathfrak{p}_n) - \lambda_n^\bullet] = 2P(0)$.

Proof. The vector field $\mathbf{X} = [\bullet, \Delta(\lambda)]$ kills Δ, and the several flows so produced for $\lambda \in \mathbb{R}$ are (more or less) transitive on the sub-manifold of phases with fixed discriminant. Now

$$\mathbf{X} \sum [\lambda(\mathfrak{p}_n) - \lambda_n^\bullet] = \sum \frac{m_{22} - m_{11}}{m_{12}^\bullet} i \text{ at } \lambda(\mathfrak{p}_n) \times \frac{m_{12}(\lambda)}{\lambda - \lambda(\mathfrak{p}_n)} = m_{22} - m_{11},$$

as in item 3.1,[8] and the self-same thing is produced by

$$\mathbf{X}\, 2P(0) = -2\partial\Delta(\lambda)/\partial Q(0) = -sp\, M(1)\, M^{-1}(x) \begin{pmatrix} 1 & 0 \\ 0 & -1 \end{pmatrix} M(x)$$

taken at $x = 0$, from which you learn that $\sum [\lambda(\mathfrak{p}_n) - \lambda_n^\bullet]$ and $2P(0)$ differ by an additive constant, depending, not upon the divisor, but upon Δ alone. To see that this constant vanishes, let $\omega_n : n \in \mathbb{Z}$ be the differentials (of the third kind) with simple poles at the two points of $\mathfrak{M}$ covering ∞ (and no others), normalized by $a_i(\omega_j) = 1$ or 0 according as $i = j$ or not. Then[9]

$$\mathfrak{P} \longrightarrow \left[\theta_n = \sum_Z \int_{\mathfrak{o}_k}^{\mathfrak{p}_k} \omega_n : n \in \mathbb{Z} \right],$$

considered modulo $\mathbb{Z}^\infty$, is a faithful map of divisors $1 : 1$ onto (the real part of) the Jacobi variety of $\mathfrak{M}$, and you may introduce the associated flat volume element $d^\infty\theta$, of total mass 1, as a means of averaging over the isospectral submanifold: $\Delta(QP)$ fixed. $\int P(0)d^\infty\theta$ vanishes because $d^\infty\theta$ is invariant under the flow $\partial Q/\partial t = P$, $\partial P/\partial t = -Q$ produced by $H_1 = (1/2) \int_0^1 (Q^2 + P^2)$. The fact that the additive constant vanishes now follows from[10] $\int m_{12}\, d^\infty\theta = 2\Delta^\bullet$, as will be seen from the products

---

[8] $\mathbf{X} \lambda(\mathfrak{p}_n) = -2(\sqrt{\Delta^2 - 1}/m_{12}^\bullet)(\mathfrak{p}_n)\, m_{12}(\lambda)[\lambda - \lambda(\mathfrak{p}_n)]^{-1}$ as in item 3.2.
[9] $\mathfrak{o}_n = [\lambda_n^-, 0]$ as before.
[10] McKean–Vaninsky [1996(1)].

$m_{12}(x) = \Pi\,(2\pi n)^{-1}\,[\lambda(\mathfrak{p}_n) - \lambda]$ and $2\Delta^\bullet(\lambda) = \Pi\,(2\pi n)^{-1}\,(\lambda_n^\bullet - \lambda)$ and the associated development of the ratio $m_{12}/2\Delta^\bullet = 1 + (1/\lambda)\sum[\lambda(\mathfrak{p}_n) - \lambda_n^\bullet] + $ etc. at $\sqrt{-1}\infty$. The value $\int m_{12}\,d^\infty\theta = 2\Delta^\bullet$ is easily derived from

$$
\begin{aligned}
2\Delta^\bullet \; &= \; -\frac{1}{2}\,sp\int_0^1 M(1)\,M^{-1}(x)\begin{pmatrix} 0 & 1 \\ -1 & 0 \end{pmatrix} M(x)\,dx \\
&= \; -\frac{1}{2}\,sp\int_0^1 M(x)\,M(1)\,M^{-1}(x)\,dx\begin{pmatrix} 0 & 1 \\ -1 & 0 \end{pmatrix},
\end{aligned}
$$

in which the triple product of M's is nothing but $M(1)$ itself, updated by translation of QP in the amount x. The left side is discriminant information, so it is permitted to average the right side with regard to the translation–invariant volume element $d^\infty\theta$, with the result that

$$
\begin{aligned}
2\Delta^\bullet \; &= \; -\frac{1}{2}sp\int M\begin{pmatrix} 0 & 1 \\ -1 & 0 \end{pmatrix} d^\infty\theta \\
&= \; \frac{1}{2}\int (m_{12} - m_{21})\,d^\infty\theta = \int m_{12}\,d^\infty\theta
\end{aligned}
$$

in view of

$$
\frac{1}{2}\int (m_{12} + m_{21})\,d^\infty\theta = \int \partial\Delta(\lambda)/\partial P(0)\,d^\infty\theta = \int \mathbf{X}Q(0)\,d^\infty\theta = 0,
$$

the $\mathbf{X} = [\bullet,\,\Delta(\lambda)]$ flow being volume-preserving, as well.[11]   □

Sample 2. $-2Q(0) = \sum_Z \left(\sqrt{\Delta^2 - 1}/m_{12}^\bullet\right)(\mathfrak{p}_n)$ is produced from sample 1 by application to both sides of $\mathbf{X}_1$.

Sample 3. $\sum_Z \left[(\lambda_n^\bullet)^2 - \lambda^2(\mathfrak{p}_n)\right] = 4[Q'(0) + Q^2(0)] - \int_0^1 (Q^2 + P^2)$, and many variants of such identities maybe proved in the same style. This one is left as an exercise.

Sample 4. $\sum[m_{11}^\bullet/m_{12}^\bullet$ at $\lambda(\mathfrak{p}_n)] = P(0)$ will be helpful for the computation of the canonical 1-form.

Proof. $m_{11}^\bullet - (1/2)\,m_{12} = \Delta^\bullet + \Delta^\bullet/\partial Q(0) - (1/2)\,m_{12}$ can be interpolated off the roots of $m_{12} = 0$. This justifies

$$
\sum \frac{m_{11}^\bullet}{m_{12}^\bullet}\ \text{at}\ \lambda(\mathfrak{p}_n) = \sum \frac{m_{11}^\bullet - (1/2)m_{12}}{m_{12}^\bullet}\ \text{at}\ \lambda(\mathfrak{p}_n) \times \frac{m_{12}(\lambda)}{\lambda - \lambda(\mathfrak{p}_n)}
$$

$$
\times \frac{\lambda}{m_{12}(\lambda)}\ \text{taken at}\ \lambda = \sqrt{-1}\infty
$$

$$
= \text{the residue at}\ \sqrt{-1}\infty\ \text{of}\ m_{12}^{-1}[\Delta^\bullet + \partial\Delta^\bullet/\partial Q(0)].
$$

---

[11]McKean–Vaninsky [1996(2)].

But res $\Delta^\bullet/m_{12} = P(0)$, as in sample 1, and from the vanishing of $-m_{12}$ at $\lambda(\mathfrak{p}_n)$ and the identity $-m_{12} = \partial m_{12}/\partial Q(0) = -m_{12}^\bullet \ \partial\lambda(\mathfrak{p}_n)/\partial Q(0)$, you see that

$$\operatorname{res} \frac{1}{m_{12}} \frac{\partial\Delta^\bullet}{\partial Q(0)} = \operatorname{res} \frac{\Delta^\bullet}{m_{12}^\bullet} \sum \frac{\partial[\lambda_n^\bullet - \lambda(\mathfrak{p}_n)]/\partial Q(0)}{\lambda_n^\bullet - \lambda}$$

$$= \frac{1}{2} \frac{\partial}{\partial Q(0)} \sum [\lambda(\mathfrak{p}_n) - \lambda_n^\bullet]$$

$$= \frac{\partial P(0)}{\partial Q(0)}$$

contributes nothing.

Sample 5. $\sum_Z ch^{-1}\Delta(\mathfrak{p}_n) = Q(0) + \int_0^1 (x - \frac{1}{2})(Q^2 + P^2)\, dx$ will also be helpful with the canonical 1-form.

Proof (as for sample 1). $\mathbf{X}\Delta = 0$, so

$$\mathbf{X} \sum ch^{-1}(\mathfrak{p}_n) = \sum \frac{\Delta^\bullet}{\sqrt{\Delta^2 - 1}} (\mathfrak{p}_n) \times -2 \frac{\sqrt{\Delta^2 - 1}}{m_{12}^\bullet} (\mathfrak{p}_n) \frac{m_{12}(\lambda)}{\lambda - \lambda(\mathfrak{p}_n)}$$

$$= -2\Delta^\bullet + m_{12}$$

since this function is band-limited to $[-1/2,\ 1/2]$ and $\int(-2\Delta^\bullet + m_{12})^2\, dx < \infty$.[12] But also

$$\mathbf{X} Q(0) = \partial\Delta/\partial P(0) = \frac{1}{2}(m_{12} + m_{21})$$

and[13]

$$\mathbf{X} \int_0^1 \left(x - \frac{1}{2}\right) (Q^2 + P^2)\, dx$$

$$= 2\int_0^1 \left(x - \frac{1}{2}\right) \left[Q\frac{\partial\Delta}{\partial P} - P\frac{\partial\Delta}{\partial Q}\right] dx$$

$$= \int_0^1 \left(x - \frac{1}{2}\right) sp\, M(1)\, M^{-1}(x)$$

$$\left[\begin{pmatrix} -PQ \\ QP \end{pmatrix} = -J\begin{pmatrix} Q & P \\ P & -Q \end{pmatrix}\right] M(x)\, dx$$

$$= -\int_0^1 \left(x - \frac{1}{2}\right) sp\, M(1)\, M^{-1}(x)\, JM'(x)\, dx$$

---

[12] McKean–Vaninsky [1996(1)].

[13] $DM = \begin{pmatrix} Q & P \\ P & -Q \end{pmatrix} M - (\lambda/2)\, JM$; the last bit does not contribute to the integral.

$$= -\frac{1}{2} \, sp \, [\, JM(1) + M(1) \, J \,]$$

$$+ \int_0^1 sp \, M(1) \, M^{-1}(x) \, JM(x) \, dx$$

$$- \int_0^1 \left(x - \frac{1}{2} \right) sp \, M(1) \, M^{-1}(x) \left(\begin{array}{cc} -PQ \\ QP \end{array} \right) M(x) \, dx,$$

in which you recognize: line 1 as $m_{12} - m_{21}$, line 2 as $-4\,\Delta^\bullet$ from the proof of sample 1, and line 3 as $-\mathbf{X} \int_0^1 (x - \frac{1}{2})(Q^2 + P^2)$, the upshot being

$$\mathbf{X} \left[Q(0) + \int_0^1 \left(x - \frac{1}{2} \right) \left(Q^2 + P^2 \right) \, dx \right] = m_{12} - 2\Delta^\bullet :$$

in short, $\sum ch^{-1} \Delta(\mathfrak{p}_n)$ and $Q(0) + \frac{1}{2} \int_0^1 (x - \frac{1}{2})(Q^2 + P^2)$ differ, as in sample 1, by an additive constant depending, not upon the divisor, but upon Δ alone, and this is seen to vanish by choice of the special divisor $\mathfrak{p}_n = \mathfrak{o}_n : n \in \mathbb{Z}$. The sum vanishes by inspection; also, the divisor is invariant under the sheet map $\mathfrak{p} = [x, \sqrt{\Delta^2 - 1}] \to \mathfrak{p}^\# = [\lambda, -\sqrt{\Delta^2 - 1}]$ which is the effect, upon the divisor, of the map. $Q(\bullet) \to -Q(-\bullet)$ and $P(\bullet) = P(-\bullet)$ in phase-space — in short, $Q(0) = 0$ and $Q^2 + P^2$ is symmetrical about $x = 1/2$, being even and of period 1, so that $\int_0^1 (x - \frac{1}{2}) (Q^2 + P^2)$ vanishes as well. The proof is finished.

Sample 6. $\mathbf{X}$ applied to sample 5 produces a variant of sample 4:[14]

$$P(0) \;=\; \mathbf{X}_1 \, Q(0) = \mathbf{X}_1 \left[Q(0) + \int_0^1 \left(x - \frac{1}{2} \right) \left(Q^2 + P^2 \right) \right]$$

$$=\; \mathbf{X}_1 \sum ch^{-1} \Delta(\mathfrak{p}_n) = \sum \Delta^\bullet / m_{12}^\bullet;$$

likewise, $\mathbf{X}_2 = $ the infinitesimal translation $[\bullet, \int_0^1 Q'P]$ kills Δ, and $\mathbf{X}_2 \, \lambda(\mathfrak{p}_n) = [\, 2P(0) - \lambda(\mathfrak{p}_n) \,] \times \sqrt{\Delta^2 - 1} / m_{12}^\bullet$ at $\mathfrak{p}_n$, so

$$Q'(0) \;+\; Q^2(0) + P^2(0) - \int_0^1 (Q^2 + P^2)$$

$$=\; \mathbf{X}_2 \left[Q(0) + \int_0^1 \left(x - \frac{1}{2} \right) (Q^2 + P^2) \right]$$

$$=\; \mathbf{X}_2 \sum ch^{-1} \Delta(\mathfrak{p}_n)$$

$$=\; \sum [\, 2P(0) - \lambda(\mathfrak{p}_n) \,] \times \frac{\Delta^\bullet}{m_{12}^\bullet} \quad \text{at} \quad \lambda(\mathfrak{p}_n)$$

$$=\; 2P^2(0) - \; \underset{m_{12}}{\text{res}} \; \frac{\lambda}{m_{12}} \left[\Delta^\bullet - \frac{1}{2} m_{12} - \frac{P(0)}{\lambda} \right],$$

---

[14]This could also be deduced from sample 4 by the sheet map, which exchanges m_{11} and m_{22} while fixing m_{12} and $P(0)$.

from which you may deduce sample 3 without much difficulty.

5. The Canonical 1-Form

The identity of canonical 1-forms:

$$\sum_Z Q_n \, dP_n = \int_0^1 Q \, dP + d\left[-Q(0)\, P(0) + \int_0^1 \left(x - \frac{1}{2}\right) Q'P - \frac{1}{2}P'(0)\right]$$

will now be proved with the help of traces nos. 4 and 5. To fix ideas, let all gaps be open: if a gap shuts, then $M = id$, $\nabla\Delta$ and $\Delta^{\bullet}$ vanish, and dP_n does too, $i.e.$, the sum takes care of itself.

Proof. The identity $\lambda\nabla M = D(J\nabla)M+$ $etc.$ of Section 3 implies[15]

$$Q_n\nabla P_n = JD\nabla P_n + 4\binom{Q}{P}\int_0^x X P_n + \binom{Q}{P}\frac{m_{11}^{\bullet}}{m_{12}^{\bullet}} \quad \text{taken at } \lambda\,(\mathfrak{p}_n),$$

so

$$
\begin{aligned}
Q_n \, dP_n &= Q_n \int_0^1 \nabla P_n \cdot \left(\frac{dQ}{dP}\right)\\
&= \left. -\nabla P_n \cdot \left(\frac{dP}{-dQ}\right)\right|_0^1\\
&\quad + \int_0^1 \nabla P_n \cdot \left(\frac{dP'}{-dQ'}\right)\\
&\quad + 2\int_0^1 \nabla P_n \cdot \binom{P}{-Q} d\int_x^1 (Q^2+P^2)\\
&\quad + \frac{1}{2}\frac{m_{11}^{\bullet}}{m_{12}^{\bullet}} d\int_0^1 (Q^2+P^2)\\
&= (1)+(2)+(3)+(4) \text{ all taken at } \lambda = \lambda(\mathfrak{p}_n).
\end{aligned}
$$

This is to be summed over $n \in \mathbb{Z}$. (4) contributes $\frac{1}{2}P(0)\, d\int_0^1 (Q^2+P^2)$ by trace no. 4, and (1) is easy, too: $\nabla\Delta$ is periodic in $0 \le x < 1$ and so drops out, leaving

$$\sum -\frac{1}{2}\frac{\Delta^{\bullet}/m_{12}^{\bullet}}{\sqrt{\Delta^2-1}} \times \nabla m_{12}^{\bullet} \left.\left(\frac{dP}{-dQ}\right)\right|_0^1$$

$$= -\frac{1}{2}\sum \frac{\Delta^{\bullet}/m_{12}^{\bullet}}{\sqrt{\Delta^2-1}}\left[\binom{0}{m_{22}}-\binom{0}{m_{11}}\right]\cdot\left(\frac{dP}{-dQ}\right)(0)$$

---

[15]$X = P\partial/\partial Q - Q\partial/\partial P$. $\quad m_{11}^{\bullet}m_{22} + m_{11}m_{22}^{\bullet} = m_{21}m_{12}^{\bullet}$ and $2\sqrt{\Delta^2-1} = m_{11} - m_{22}$ are used.

$$= \sum \frac{\Delta^\bullet}{m_{12}^\bullet} \times -dQ(0)$$

$$= -P(0)\ dQ(0)\ ,$$

by trace no. 6.[16] [17] (2) is trickier: trace no. 5 states that

$$\sum P_n = -\frac{1}{2}\left[Q(0) + \int_0^1 \left(x - \frac{1}{2}\right)\left(Q^2 + P^2\right)\right],$$

so (2) should contribute

$$-\ \frac{1}{2}\int_0^1 \nabla\left[Q(0) + \int_0^1 \left(x - \frac{1}{2}\right)\left(Q^2 + P^2\right)\right] \cdot \left(\begin{array}{c} dP' \\ -dQ' \end{array}\right)$$

$$=\ -\frac{1}{2}\,dP'(0) + \int_0^1 \left(x - \frac{1}{2}\right)\left(PdQ' - QdP'\right)$$

$$=\ -\frac{1}{2}\,dP'(0) + \int_0^1 \left(x - \frac{1}{2}\right)PdQ' - \left(x - \frac{1}{2}\right)QdP'\Big|_0^1$$

$$+ \int_0^1 QdP + \int_0^1 \left(x - \frac{1}{2}\right)Q'dP$$

$$=\ -\frac{1}{2}\,dP'(0) + d\int_0^1 \left(x - \frac{1}{2}\right)Q'P - Q(0)\,dP(0) + \int_0^1 QdP\ ,$$

provided the exchange of sum and gradient finessed in line 1 is legitimate.
(3) is tricky, too: formally, it contributes

$$\sum 2\int_0^1 \nabla P_n \cdot \left(\begin{array}{c} P \\ -Q \end{array}\right) d\int_x^1 \left(Q^2 + P^2\right)$$

$$=\ -\int_0^1 \nabla\left[Q(0) + \int_0^1 \left(x - \frac{1}{2}\right)\left(Q^2 + P^2\right)\right]$$

$$\bullet \left(\begin{array}{c} P \\ -Q \end{array}\right) d\int_x^1 \left(Q^2 + P^2\right)$$

$$=\ -\frac{1}{2}P(0)\,d\int_0^1 \left(Q^2 + P^2\right),$$

cancelling the contribution of (4), but with the same proviso as to sum and
gradient and a little fast talking in line 3 to finesse the jump of $d\int_x^1 (Q^2 + P^2)$

---

[16] $\nabla m_{12} = (0, m_{11})$ at $x = 0$ and $(0, m_{22})$ at $x = 1$.
[17] $2\sqrt{\Delta^2 - 1} = m_{11} - m_{22}$.

at $x = 0$ alias $x = 1$ — in short,

$$\sum_Z Q_n \, dP_n = \int_0^1 Q \, dP + d\left[-Q(0)P(0) + \int_0^1 \left(x - \frac{1}{2} \right) Q'P - \frac{1}{2} P'(0) \right],$$

as advertised, or so you may hope.

Technicalities. The expression of $Q_n dP_n = (1) + (2) + (3) + (4)$ is used in case dQ or dP jumps at $x = 0$ to obtain enough control to justify the formal summations of (2) and (3).

Step 1. $Q_n dP_n$ is the sum

$$
\begin{aligned}
(1) \quad &= \quad -\nabla P_n \cdot \left(\begin{array}{c} dP \\ -dQ \end{array} \right) \Big|_{0+}^{1-} \\[2mm]
+(2) \quad &= \quad \int_0^1 \nabla P_n \bullet \left(\begin{array}{c} dP' \\ -dQ' \end{array} \right) \\[2mm]
+(3) \quad &= \quad 2\int_0^1 \nabla P_n \bullet \left(\begin{array}{c} P \\ -Q \end{array} \right) d\int_x^1 (Q^2 + P^2) \\[2mm]
+(4) \quad &= \quad \frac{1}{2} \frac{m_{11}^{\bullet}}{m_{12}^{\bullet}} \, d\int_0^1 (Q^2 + P^2).
\end{aligned}
$$

This is valid if dQ or dP jumps at $x = 0$ but is otherwise smooth, and the sum is controlled by $|\nabla P_n|_\infty + |m_{11}^{\bullet}/m_{12}^{\bullet}| \leq C_1[H_1 = \frac{1}{2}\int_0(Q^2 + P^2)]$ for fixed dQ and dP; compare item 3.5.[18] But $Q_n = \lambda(\mathfrak{p}_n) = 2\pi n$, up to an error controlled by H_1, independently of $n \in \mathbb{Z}$[19] so $|dP_n| \leq C_2(H_1)/n$.

Step 2. $Q_n \times (2)$ is a sum of the same type:

$$
\begin{aligned}
(1') \quad &= \quad -dP'(0) \, \Delta^{\bullet}/m_{12}^{\bullet} \\[2mm]
+(2') \quad &= \quad -\int_0^1 \nabla P_n \bullet \left(\begin{array}{c} dQ'' \\ dP'' \end{array} \right) \\[2mm]
+(3') \quad &= \quad 2\int_0^1 \nabla P_n \bullet \left(\begin{array}{c} P \\ -Q \end{array} \right) 2\int_x^1 (Q \, dP' - P \, dQ') \\[2mm]
+(4') \quad &= \quad -\frac{m_{11}^{\bullet}}{m_{12}^{\bullet}} \, d\int_0^1 Q'P,
\end{aligned}
$$

and now each command is controlled by $C_3(H_1)/n$. This is seen: for $(1')$ from the fact that $\Delta^{\bullet}(\lambda) = 0$ has a root in every gap and from the bound[20]

$$\sum n^2 \left(\lambda_n^+ - \lambda_n^- \right) \leq C_4(H_3);$$

---

[18] McKean–Vaninsky [1996(1)] has details.

[19] McKean–Vaninsky [1996(1)].

[20] McKean–Vaninsky [1996(1)].

for $(2')+(3')$ from step 1; and for $(4')$ from $m_{11}^{\bullet}+m_{22}^{\bullet} = 2\Delta^{\bullet}$, from the rapid vanishing of $m_{11} - m_{22}$ in the gaps, and from the interpolatory formula

$$(m_{11}^{\bullet} - m_{22}^{\bullet})(\mathfrak{p}_i) = \sum_{j \neq i} \frac{m_{11} - m_{22}}{m_{12}^{\bullet}}(\mathfrak{p}_j) \frac{m_{12}^{\bullet}(\mathfrak{p}_i)}{\lambda(\mathfrak{p}_i) - \lambda(\mathfrak{p}_j)}$$

$$+ \frac{1}{2} \frac{m_{11} - m_{22}}{m_{12}^{\bullet}} m_{12}^{\bullet\bullet}(\mathfrak{p}_i),$$

the upshot being $|(2)| \leq C_5(H_3)$.

Step 3 does the same for $Q_n \times (3)$. It is the sum

$$(1'') \quad = \quad \left[Q(0) - (m_{11}^{\bullet}/m_{12}^{\bullet}) P(0) \right] \times d \int_0^1 \left(Q^2 + P^2 \right)$$

$$+(2'') \quad = \quad -\int_0^1 \nabla P_n \bullet \left[\binom{Q}{P} d \int_0^1 \left(Q^2 + P^2 \right) \right]$$

$$+(3'') \quad = \quad 0$$

$$+(4'') \quad = \quad 0.$$

The only trouble comes form the first piece of $(1'')$ and that maybe fixed by pairing $+n$ to $-n$:

$$|(3)_+ + (3)_-| \quad \leq \quad C_6(H_3)/n^2 + \left| \frac{1}{Q_n} + \frac{1}{Q_{-n}} \right| \times \left| Q(0) d \int_0^1 \left(Q^2 + P^2 \right) \right|$$

$$\leq \quad C_7(H_3)/n^2 ,$$

as desired.

Step 4. The estimate of step 2 is applied to $Q^* = Q + tdP'$ and $P^* = P - tdQ'$ with the old dP' and $-dQ'$. Then $|t^{-1}(P_n^* - P_n)| \leq C_8(H_3)/n^2$, independently at $0 \leq t < 1$, so you may sum over $n \in \mathbb{Z}$ with the help of trace no. 5 and make $t \downarrow 0$ to obtain the value $\sum(2) = -\frac{1}{2} dP'(0) - \int_0^1 (x - \frac{1}{2}) (dP' - PdQ')$ found before by formal means.

Step 5. The same applies to (3), paired as above, $+n$ to $-n$. Now $Q^* = Q + tPd \int_x^1 (Q^2 + P^2)$, $P^* = P - tQd \int_x^1 (Q^2 + P^2)$, and the only novelty is that these functions jump at $x = 0$. This requires a more sophisticated version of no. 5:

$$\sum ch^{-1}\Delta(\mathfrak{p}_n) = \frac{1}{2} Q(0+) + \frac{1}{2} Q(1-) + \int_0^1 \left(x - \frac{1}{2} \right) \left(Q^2 + P^2 \right).$$

The rest is the same.

6. Actions and all That

The identity of canonical 1-forms permits the evaluation of the classical actions $I_n = a_n(QdP)$:

$$\frac{1}{2} \sum ch^{-1} \Delta(\mathfrak{p}_n)\, dQ_n = \int_0^1 QdP + \quad \text{an exact form}$$

is integrated as $\mathfrak{p}_n$ moves counterclockwise, once about its private oval a_n, to produce $I_n = a_n[\frac{1}{2} ch^{-1} \Delta(\mathfrak{p})\, d\lambda]$. You may also apply the field $X_1 = [\bullet, \frac{1}{2} \int_0^1 (Q^2 + P^2) = H_1]$ to obtain

$$\frac{1}{2} \sum ch^{-1} \Delta(\mathfrak{p}_n)\, X_1\, Q_n = \int_0^1 Q\, XP = -\int_0^1 Q^2 + \quad \text{exact stuff}$$

and integrate this over 1 period $(T = 2\pi)$ of the flow it engenders: now, too, $\mathfrak{p}_n$ is carried once clockwise about its private oval a_n,[21] so that[22][23]

$$\begin{aligned}
\sum I_n &= \int_0^{2\pi} dt\, e^{tX_1} \int_0^1 Q^2 = \int d^\infty \theta \int_0^{2\pi} dt\, e^{tX_1} \int_0^1 Q^2 \\
&= 2\pi \int d^\infty \theta \int_0^1 Q^2 \\
&= 2\pi \times \frac{1}{2} \int_0^1 (Q^2 + P^2) \\
&= 2\pi H_1
\end{aligned}$$

— in short, you have a new trace formula, no. 7: $\sum I_n = 2\pi H_1$; see McKean–Vaninsky [1996(1)].

Amplification 1. This merits some comment. In ordinary circumstances, the classical actions $a_n(QdP)$ have no real mechanical meaning, the homology basis being determined only up to a unimodular integral substitution. Not so here: the real ovals provide a God-given homology basis and $(2\pi)^{-1} \sum I_n = H_1$ is the natural constant of motion of lowest degree. This is novel. The mechanical meaning (if any) of the higher symmetric functions such as $\sum_{1<j} I_i I_j$ is unknown.

Amplification 2.[24] The identification of the action I_n as $a_n[\frac{1}{2} ch^{-1} Dd\lambda]$ can be made in another way: with the angles $\theta_n = \sum \int_{0_k}^{\mathfrak{p}_k} w_n$ of Section 4,

---

[21] McKean–Vaninsky [1996(1)].
[22] $\sum I_n$ is independent of the angles and so may be averaged without change.
[23] $\exp(tX_1)$ preserves $d^\infty \theta$.
[24] McKean–Vaninsky [1996(1)].

you find[25]

$$-\left[\theta_i; \frac{1}{2}\, a_j\,(ch^{-1}\Delta)\right] = -\frac{1}{2}\int_{a_j} \frac{d\lambda}{\sqrt{\Delta^2-1}}\left[\sum \int_{o_k}^{\mathfrak{p}_k} \omega_i,\, \Delta(\lambda)\right]$$

$$= -\frac{1}{2}\int_{a_j} \frac{d\lambda}{\sqrt{\Delta^2-1}} \sum \frac{\omega_i}{d\mathfrak{p}_k} \times -2\,\frac{\sqrt{\Delta^2-1}}{m_{12}^\bullet}\,(\mathfrak{p}_n)\,\frac{m_{12}(\lambda)}{\lambda - \lambda(\mathfrak{p}_n)}$$

$$= \int_{a_j} \omega_i = id,$$

so, with $\mathbf{X} = [\bullet, -I_n]$,[26][27]

$$I_n = -a_n\,(PdQ) \;=\; \int_0^1 dt\, e^{t\mathbf{X}} \int_0^1 P\mathbf{X}Q$$

$$= \int d^\infty\theta \int_0^1 dt\, e^{t\mathbf{X}} \int_0^1 P\mathbf{X}Q$$

$$= \int d^\infty\theta\, P(0)\,\mathbf{X}Q(0)$$

$$= \frac{1}{2}\int_{a_n} \frac{d\lambda}{\sqrt{\Delta^2-1}} \int d^\infty\theta\, P(0)\,\frac{\partial\Delta(\lambda)}{\partial P(0)}$$

$$= \frac{1}{2}\int_{a_n} \frac{d\lambda}{\sqrt{\Delta^2-1}} \int d^\infty\theta\, P(0)\,\frac{1}{2}\,(m_{12}+m_{21}).$$

The rest follows from the evaluation of the inner integral: $m_{12} - m_{21}$ is unchanged by the phase flow produced by $H_1 = \left(\frac{1}{2}\right)\int(Q^2 + P^2)$ while $P(0)$ averages to 0, so the inner integral reduces to $\int P(0)\, m_{12}$. Now examination of $d^\infty\theta$ shows that it has the form $m_{12}^\bullet\,(\Delta^2 - 1)^{-\frac{1}{2}}\,(\mathfrak{p}_n)\, d\lambda\,(\mathfrak{p}_n)$ on the nth oval $\times$ a reduced volume element indicated by $d^{\infty-1}\theta$ on the other ovals, so

$$\int P(0)\, m_{12}\, d^\infty\theta \;=\; \int\sum\left[\frac{\Delta^\bullet}{m_{12}^\bullet}\ \text{at}\ \lambda(\mathfrak{p}_n)\right] m_{12}\, d^\infty\theta$$

$$= \sum\int \frac{m_{12}}{\lambda - \lambda(\mathfrak{p}_n)}\, d^{\infty-1}\theta \int_{a_n} \Delta^\bullet\,[\lambda - \lambda(\mathfrak{p}_n)]\, d\lambda(\mathfrak{p}_n)$$

$$= \sum\int \frac{m_{12}}{\lambda - \lambda(\mathfrak{p}_n)}\, d^{\infty-1}\theta \times \int_{a_n} ch^{-1}\,\Delta\ .$$

---

[25] $\sqrt{\Delta^2-1}(\omega_i/d\mathfrak{p}_n)$ is an entire function capable of interpolation off the roots of $m_{12} = 0$.

[26] I_n is independent of the angles.

[27] $\exp(t\,\mathbf{X})$ preserves $d^\infty\theta$, as does translation.

But, by the same principle in reverse,[28]

$$\int \frac{m_{12}}{\lambda - \lambda(\mathfrak{p}_n)} d^{\infty-1}\theta = \sum \int \frac{m_{12}}{\lambda - \lambda(\mathfrak{p}_k)} d^{\infty-1}\theta \int_{a_k} \omega_n$$

$$= \int \sum \frac{\phi_n}{m_{12}^\bullet} \text{ at } \lambda(\mathfrak{p}_k) \times \frac{m_{12}}{\lambda - \lambda(\mathfrak{p}_k)} d^{\infty}\theta ,$$

$$= \int \phi_n(\lambda) d^{\infty}\theta$$

$$= \phi_n(\lambda) .$$

That does the trick:

$$I_n = \frac{1}{2} \int_{a_n} \frac{d\lambda}{\sqrt{\Delta^2 - 1}} \sum_k \phi_k(x) \, ch^{-1}\Delta \int_{a_k} = \frac{1}{2} a_n \left[ch^{-1}\Delta \right] ,$$

as advertised; compare McKean–Vaninsky [1996(2)].

Amplification 3. The actions I_n are canonically paired to the angles, up to a common signature, as per amplication 2, so these should be honest action-angle variables, as noted by Flaschka–McLaughlin [1976] in the context of KdV and by Krichever [1977] in the wider context of KP; see McKean–Vaninsky [1996(1)] for technical details in the present context. It is only the commutative character of the angles that is tricky as they are not smooth in the large. Anyhow, $[[\theta_i, \theta_j], I_k] = 0$ by Jacobi's identity and the pairing, and as the action flows are transitive on the individual angles, so $[\theta_i, \theta_j]$ is a constant of motion which is found to vanish at the special divisor $\mathfrak{p}_n = \mathfrak{o}_n : n \in \mathbb{Z}$.[29] The (canonical) transformation from the prior variables $Q_n = \lambda(\mathfrak{p}_n)$ and $P_n = -\frac{1}{2} ch^{-1} \Delta(\mathfrak{p}_n) : n \in \mathbb{Z}$ is mediated by the generating function

$$S(\mathfrak{P}, I) = -\frac{1}{2} \sum_{\mathbb{Z}} \int_{\mathfrak{o}_k}^{\mathfrak{p}_k} ch^{-1} \Delta(\mathfrak{p}) \, d\lambda :$$

in fact,

$$\partial S/\partial \mathfrak{p}_i = -\frac{1}{2} ch^{-1} \Delta(\mathfrak{p}_i) = P_i \quad \text{and} \quad \partial S/\partial I_j = \sum \int_{\mathfrak{o}_k}^{\mathfrak{p}_k} \omega_j = \theta_j$$

in view of

$$id = \frac{\partial I_i}{\partial I_j} = \int_{a_i} \frac{\partial \Delta/\partial I_j}{\sqrt{\Delta^2 - 1}} \, d\lambda,$$

so $\int_0^1 Q dP = \sum I_n \, d\theta_n$, up to a closed differential.

---

[28] $\omega_n = \phi_n (\Delta^2 - 1)^{-\frac{1}{2}} d\lambda$ and ϕ_n can be interpolated off the roots of $m_{12} = 0$.
[29] McKean–Vaninsky [1996(1)]

7. Korteweg–de Vries

Naturally, most of what has been said adapts to KdV (and other fashion-able integrable systems, as well). The phase space is now the class of *single* functions Q, of period 1, equipped with the nonclassical 2-form $dQ \wedge dP$ with $P = -(1/2)D^{-1}Q$. This makes perfect sense if $\int_0^1 Q = 0$, as will be assumed below. The corresponding bracket is

$$[A, B] = \int_0^1 \frac{\partial A}{\partial Q} \, D \frac{\partial B}{\partial Q} \, dx,$$

this being the restriction, in the style of Dirac [1935] of the classical bracket in the ambient QP space to the submanifold $P = -(1/2)D^{-1}Q$. The Dirac operator is replaced by $\mathbf{Q} = -D^2 + Q$ with monodromy $M = M(1)$ determined the 2 solutions of $\mathbf{Q}y = \lambda y : m_{11} = y_1(1, \lambda)$, $m_{12} = y_2(1, \lambda)$, $m_{21} = y_1'(1, \lambda)$, and $m_{22} = y_2'(1, \lambda)$ with $y_1 = \cos \sqrt{\lambda} x +$ etc. and $y_2 = \lambda^{-1/2} \sin \sqrt{\lambda} x +$ etc. The discriminant $\Delta = (1/2) \, sp \, M$ is now an entire function of order $1/2$ and type 1 and the curve $\mathfrak{M}$ is ram-ified/double over the simple/double roots of $\Delta(\lambda) = \pm 1$. These come in (separated) pairs $\lambda_n^- \leq \lambda_n^+$, indexed by $n \geq 1$ and located at $n^2\pi^2$ more or less, with a single root λ_0 of $\Delta(\lambda) = +1$ to the left of them all. The divisor $\mathfrak{P} = [\mathfrak{p}_n : n \geq 1]$ has a single point on each of the real ovals a_n covering the gaps $[\lambda_n^-, \lambda_n^+]$. It is determined by the self-same rule, $m_{12} = 0$ and $\sqrt{\Delta^2 - 1} = \frac{1}{2}(m_{11} - m_{22})$, and the map $Q \to \mathfrak{P}$ is $1:1$ onto the product $\times a_n$. The canonically paired variables $Q_n = \lambda(\mathfrak{p}_n)$ and $P_n = 2 \, ch^{-1}\Delta(\mathfrak{p}_n)$ are adopted. They determine the divisor and so also Q itself if Δ is known, as is, in fact, the case: $m_{12}(\lambda) = \Pi_1^\infty (n\pi)^{-2} \, [\lambda(\mathfrak{p}_n) - \lambda]$ is known, and, in the presence of $\int_0^1 Q = 0$, $\Delta(\lambda) - \cos \sqrt{\lambda}$ can be interpolated off its known values at the roots $Q_n = \lambda(\mathfrak{p}_n)$ of $m_{12} = 0$.

TRACE FORMULAS, old and new, can be obtained as in Section 4.

Sampe 1 (old).[30] $\sum_{n=1}^\infty (\lambda_n^\bullet - Q_n) = \frac{1}{2} Q(0)$.

Proof. The field $\mathbf{X} = [\bullet, \Delta(\lambda)]$ is employed as before.

$$\mathbf{X}Q_n = \frac{1}{2} \frac{\sqrt{\Delta^2 - 1}}{m_{12}^\bullet} (\mathfrak{p}_n) \frac{m_{12}(\lambda)}{\lambda - \lambda(\mathfrak{p}_n)},$$

so

$$\mathbf{X}\sum(\lambda_n^\bullet - Q_n) = -\frac{1}{2} \sum \frac{\sqrt{\Delta^2 - 1}}{m_{12}^\bullet} (\mathfrak{p}_n) \frac{m_{12}(\lambda)}{\lambda - \lambda(\mathfrak{p}_n)} = -\frac{1}{4} (m_{11} - m_{22}).$$

---

[30] $\lambda_n^\bullet$ is the root of $\Delta^\bullet(\lambda) = 0$ in the nth gap.

But also

$$\mathbf{X}Q(0) = D\,\frac{\partial\Delta(x)}{\partial Q(x)}\ \text{ taken at }\ x = 0\ \ \ \ = \frac{1}{2}\,(m_{22} - m_{11}),$$

and the discrepancy between $\sum(\lambda_n^\bullet - Q_n)$ and $(1/2)\,Q(0)$ is found to be $-\frac{1}{2}\int_0^1 Q = 0$ with the aid of $\int m_{12}\,d^\infty\theta = -2\Delta^\bullet$ and of $\int Q(0)\,d^\infty\theta = \int\!\!\int_0^1 Q(x)\,dx\,d^\infty\theta = \int_0^1 Q$, the latter being a constant of motion. This seems much simpler than the standard proofs. $\qquad\square$

Sample 2 (new). $\sum_1^\infty P_n = \int_0^1(x - \frac{1}{2})\,Q(x)\,dx.$

Proof. [31]

$$\mathbf{X}\sum P_n\ =\ \sum\frac{2\Delta^\bullet}{\sqrt{\Delta^2 - 1}}\,(\mathfrak{p}_n)\times\frac{1}{2}\,\frac{\sqrt{\Delta^2 - 1}}{m_{12}^\bullet}\,(\mathfrak{p}_n)\,\frac{m_{12}}{\lambda - \lambda(\mathfrak{p}_n)}$$

$$=\ \Delta^\bullet + \frac{1}{2}\,m_{12}$$

and

$$\mathbf{X}\int_0^1\left(x - \frac{1}{2}\right)Q\ =\ \int_0^1\left(x - \frac{1}{2}\right)D\,\frac{\partial\Delta}{\partial Q(x)}$$

$$=\ \frac{\partial\Delta}{\partial Q(0)} - \int_0^1\frac{\partial\Delta}{\partial Q}=\frac{1}{2}\,m_{12} + \Delta^\bullet.$$

Now compare both sides at the special divisor $\mathfrak{p}_n = \mathfrak{o}_n : n \ge 1$. The latter is invariant under the sheet map, which is to say that Q is even about $x = 0$ and so also about $x = 1/2$, as it is of period 1. This makes both sides vanish. $\qquad\square$

Canonical 1-Form. This is easier, technically, than before. The desired identity is

$$\sum_{n=1}^\infty Q_n\,dP_n = \int_0^1 Q\,dP + \frac{1}{4}\,d\left[\int_0^1\left(x - \frac{1}{2}\right)Q^2 - Q(0)\right].$$

for dQ with vanishing mean-value.

Step 1. P_n is smooth, with gradient

$$\partial P_n/\partial Q = 2 \times -m_{11}m_{12}(x) + (m_{11}^\bullet/m_{12}^\bullet)\,m_{12}^2\,(x),$$

much as in Section 3.

Step 2. $dP_n = \int_0^1(\partial P_n/\partial Q)\,dQ$ vanishes like n^{-4} for any smooth dQ of period 1 with $\int_0^1 dQ = 0$; in particular $\sum Q_n\,dP_n$ converges fine.

---

[31]$\Delta^\bullet + (1/2)\,m_{12} = 0$ is capable of interpolation off the roots of $m_{12} = 0$.

Proof. $\partial P_n/\partial Q = f$ solves $Kf = 2\lambda\, Df$ with $K = QD + DQ - (1/2)\, D^3$; it also vanishes at $x = 0$ and at $x = 1$ and has mean-value $\int_0^1 f = 0$. Now

$$
\begin{aligned}
2Q_n \int_0^1 f\, dQ &= -2Q_n \int_0^1 f' \int_0^x dQ \\
&= -\int_0^1 Kf\, dP \\
&= \frac{1}{2}\left[f''\, dP - f'\, dP' + f\, dP'' \right]\Big|_0^1 + \int_0^1 fK\, dP \\
&= \int_0^1 f\, K\, dP,
\end{aligned}
$$

and since $\int_0^1 f = 0$, you may subtract from $K\, dP$ its mean value $\int_0^1 K\, dP$ and repeat the computation. $Q_n \simeq n^2\pi^2$ does the rest.

Step 3 is the actual evaluation of the 1-form using the domination provided by step 2, as in step 5.4:

$$
\begin{aligned}
\sum Q_n\, dP_n &= \sum Q_n \int_0^1 \frac{\partial P_n}{\partial Q}\, dQ \\
&= -\sum \int_0^1 \frac{\partial P_n}{\partial Q} \left[K\, dP - \int_0^1 K\, dP \right] \qquad \text{as in step 2} \\
&= -\int_0^1 \left(x - \frac{1}{2} \right) \left[K\, dP - \int_0^1 K\, dP \right] \qquad \text{by trace no. 2} \\
&= \frac{1}{2} \int_0^1 \left(x - \frac{1}{2} \right) \left(QD + DQ - \frac{1}{2} D^3 \right) d\int_0^x Q \\
&= \frac{1}{2} \int_0^1 \left(x - \frac{1}{2} \right) \left[2Q\, dQ + Q'\, d\int_0^x Q - \frac{1}{2}\, dQ'' \right] \\
&= d\int_0^1 \left(x - \frac{1}{2} \right) Q^2/2 - \frac{1}{2} \int_0^1 Q \left[d\int_0^x Q + \left(x - \frac{1}{2} \right) dQ \right] \\
&\qquad - \frac{1}{4}\left[\left(x - \frac{1}{2} \right) dQ' - dQ \right]\Big|_0^1 \\
&= d\int_0^1 \left(x - \frac{1}{2} \right) Q^2/4 - \frac{1}{2} \int_0^1 Q\, d\int_0^x Q - \frac{1}{4}\, dQ'(0) \\
&= \int_0^1 Q\, dP + \frac{1}{4}\, d\left[\int_0^1 \left(x - \frac{1}{2} \right) Q^2 - Q'(0) \right],
\end{aligned}
$$

as advertised.

Actions. These are $I_n = a_n \left(\int_0^1 Q\, dP \right) = 2\, a_n \left[ch^{-1}\, \Delta\, d\lambda \right]$ much as before, only now $\sum I_n$ does not seem to have any special mechanical signifance as regards the standard constants of motion $\int_0^1 Q, (1/2) \int_0^1 Q^2$, etc. This is disappointing.

References

[1] Dirac, P.A.M.: *The Principles of Quantum Mechanics.* Oxford University Press, 1935.

[2] Flaschka, H., and D. McLaughlin: Canonically conjugate variables for KdV and Toda lattice under periodic boundary conditions. *Prog. Theor. Phys.* **55** (1976), 438–456.

[3] Krichever, I. M.: Integration of nonlinear equations by methods of algebraic geometry *Funk. Anal. Priloz.* **11** (1977), 15–31.

[4] McKean, H. P., and K. L. Vaninsky: (1) Action-angle variables for the cubic Schrödinger equation. *Comm. Pure and Applied Math.,* (2) The petit ensemble in action-angle variables. *Comm. Pure and Applied Math.,* to appear 1996.

Acknowledgement. The bulk of this paper was worked out at the Euler Institute at Sankt Petersburg, November 1992, for which hospitality thanks.

Courant Institute of Mathematical Sciences
251 Mercer Street
New York, NY 10012

Received January, 1996
Revised March, 1996

On some "Schwarzian" Equations and their Discrete Analogues

Frank Nijhoff

In remembrance of my dear friend Irene Dorfman

Abstract

Some integrable discrete analogues of the Schwarzian KdV (Krichever–Novikov) equation and of other Möbius-invariant equations, are discussed together with their Miura chains relating them to associated equations like the lattice KdV and lattice modified KdV equation. Furthermore, the similarity solutions of such lattice equations are considered, as well as reductions to discrete Painlevé equations.

1. Introduction

One of the equations that Irene Dorfman has been working on intensively is an equation introduced by Krichever and Novikov in [1]. The Schwarzian KdV (SKdV)

$$\psi_t = \psi_x S(\psi) \quad , \quad S(\psi) \equiv \frac{\psi_{xxx}}{\psi_x} - \frac{3}{2}\frac{\psi_{xx}^2}{\psi_x^2} \quad , \tag{1.1}$$

in which $S(\psi)$ denotes the Schwarzian derivative of ψ, is a specialisation of that equation which is invariant under Möbius transformations

$$\psi \;\longmapsto\; \frac{a\psi + b}{c\psi + d} \; , \tag{1.2}$$

i.e. under the group $PSL(2)$. Within the family of KdV-related equations (KdV, modified KdV, etc.) it is probably the most fundamental equation. In recent years, therefore, the Schwarzian KdV and its generalisations, cf. e.g. [2], have gained a lot of attention, cf. e.g. [3]–[5]. Whilst a complicated, probably not so workable, Lax pair was established by the inventors, cf. also [4], many of the integrability features needed to be established by separate means. Irene gave a major contribution to this by establishing in [6] the bi-symplectic structure of the equation. This was entirely natural in her train of thoughts, taking into account that one of her most famous works as a student of I.M. Gel'fand had been the very invention of the bi-Hamiltonian phenomenon in integrable systems, which she did independently and more or less simultaneously with F. Magri. These features have been at the very core of her mathematical works ever since, cf. e.g. [7, 8], culminating in her important monograph [9]. More recently she dedicated herself to the similar phenomenon in the 2+1-dimensional situation, cf. e.g. [10]. Within that general setting Irene and myself investigated not long ago the 2+1 dimensional analogue of the SKdV in [11], the Schwarzian KP equation which was first given in [12] and later investigated again in [13]. In view of the special affection that Irene had for the Krichever–Novikov system and related equations, I think it is fit to dedicate on this occasion a few thoughts to Schwarzian equations and variations of them.

One of the aspects that make integrable equations very special is that one can find discrete analogues of these equations, tending to the original continuous equation in special limits, and that are *integrable themselves*, (cf. [14] for a recent review, and references therein). These discrete analogues can be either ordinary or partial difference equations or mixed differential-difference equations like the Toda lattice equation. My experience especially with fully discrete equations (which I care to call "lattice equations") is that they have a very rich structure and could be considered to be more fundamental than the original continuous equation. In fact, the connection with the continuous counterpart goes via the Bäcklund transformations, which as an iterative transformation makes up a one-dimensional or two-dimensional lattice of solutions in view of their nice permutability properties. It is this lattice of Bäcklund transformations that has an interpretation as an integrable discrete dynamical system itself! In this contribution I would like to focus on some Schwarzian lattice equations, i.e. equations that constitute the discrete analogues of the Schwarzian KdV and of other Möbius-invariant equations.

Another aspect of the Schwarzian equations that I would like to discuss is the similarity reduction of these equations and of their lattice analogues. As is well-known the similarity solutions of integrable nonlinear partial differential equations (PDE's) give rise to the Painlevé transcendents, [15]–[18]. This connection between Painlevé equations and soliton-type equations has led to the famous ARS conjecture, [19]. There is, to date, no exhaustive treatment of classifying the various soliton equations according to their connection with Painlevé transcendents or other Painlevé-related equations, such as the Chazy equations, cf. [20]. We know only of isolated examples for which such a connection has been established. It would be certainly of great use to have a more systematic oversight of where we can expect certain transcendental equations to occur. In such an endeavour one should take into special account the reductions of the Schwarzian equations.

For example, the similarity reduction of the Schwarzian KdV is obtained very straightforwardly by considering (1.1) together with

$$\psi(x,t) = t^{\mu}\Psi(\xi) \ , \quad \xi = \frac{x}{t^{1/3}} \ , \tag{1.3}$$

leading to the third-order ODE

$$\frac{\Psi'''}{\Psi'} - \frac{3}{2}\frac{\Psi''^2}{\Psi'^2} = \mu\frac{\Psi}{\Psi'} - \frac{1}{3}\xi \ . \tag{1.4}$$

Eq. (1.4) is no longer Möbius invariant, as the scaling symmetry breaks the invariance under $PSL(2)$. However, eq. (1.4) is directly related to the Painlevé II equation, using the well-known correspondence

$$y = -\frac{1}{2}\frac{\Psi''}{\Psi'} \quad \Rightarrow \quad S(\Psi) = -2(y' + y^2) \ , \tag{1.5}$$

thus leading to the Painlevé II (PII) equation

$$y'' = 2y^3 - \frac{1}{3}\xi y + \left(\frac{1}{6} - \frac{1}{2}\mu\right) \ . \tag{1.6}$$

It is curious that the third-order equation (1.4), which we could loosely baptise as the Schwarzian PII (SPII) equation, doesn't seem to occur in the class of Chazy equations of [20], cf. also [21]. There exist other Schwarzian equations that have a similar relation with Painlevé equations and which we will encounter below.

Coming back to the issue of discrete analogues of the Schwarzian equations, also for those equations it would be of interest to investigate their connection with what will be now discrete versions of the Painlevé equations. Such lattice or difference analogues of the Painlevé equations have been a subject of growing interest, cf. [22] for a review. A substantial number of such discrete Painlevé equations have been found recently[1], e.g. in [23]–[27]. One of the methods to obtain discrete Painlevé equations is a lattice analogue of the procedure of similarity reduction that was introduced in [23]. In this paper I will apply this approach to the discrete Schwarzian equations leading to the discrete version of the procedure described above.

2. Lattice Schwarzian KdV

We shall start by considering the lattice analogue of the SKdV and its similarity reduction. This is actually a very simple equation in terms of a

---

[1]It is interesting to note that on the discrete level, there exist various different versions of Painlevé type equations that tend to one and the same transcendent in the continuum limit.

variable $z_{n,m}$ depending on the discrete (lattice) variables $n, m \in \mathbf{Z}$, and which reads, cf. [14],

$$\frac{(z_{n,m} - z_{n+1,m})(z_{n,m+1} - z_{n+1,m+1})}{(z_{n,m} - z_{n,m+1})(z_{n+1,m} - z_{n+1,m+1})} = \frac{q^2}{p^2} . \tag{2.1}$$

In eq. (2.1), p, q are arbitrary complex parameters which have the interpretation of lattice parameters measuring the grid. On the left-hand side of eq. (2.1) we recognize immediately the canonical conformally-invariant cross-ratio of four points in the complex plane. That the equation

$$\text{crossratio} = \text{constant} ,$$

is actually an integrable partial difference equation might come as a surprise. In fact, eq. (2.1) is a special case of a more general equation with free parameters α and β that was derived more than a decade ago in [28, 29] and that reads

$$\frac{1 - (p+\beta)s_{n+1,m} + (p-\alpha)s_{n,m}}{1 - (q+\beta)s_{n,m+1} + (q-\alpha)s_{n,m}}$$
$$= \frac{1 - (q+\alpha)s_{n+1,m+1} + (q-\beta)s_{n+1,m}}{1 - (p+\alpha)s_{n+1,m+1} + (p-\beta)s_{n,m+1}} . \tag{2.2}$$

The cross-ratio equation (2.1) is obtained in the special case that $\alpha = \beta = 0$, setting $s_{n,m} = z_{n,m} + \frac{n}{p} + \frac{m}{q}$.

There are a number of properties of the cross-ratio equation (2.1) that are of importance. First of all, it is obvious that the equation is invariant under Möbius-transformations

$$z \mapsto \frac{az+b}{cz+d} , \quad ad - bc \neq 0 .$$

It is well-known that the cross-ratio is a discrete analogue of the Schwarzian derivative, cf. e.g. [30], and indeed a proper continuum limit of (2.1) discussed below yields the Schwarzian KdV equation. The integrability aspect of the cross-ratio equation is manifested in the existence of a Lax pair for (2.1) of the form

$$\phi_{n+1,m}(k) = L_{n,m}(k^2)\phi_{n,m}(k) , \tag{2.3a}$$
$$\phi_{n,m+1}(k) = M_{n,m}(k^2)\phi_{n,m}(k) , \tag{2.3b}$$

in which the 2×2-matrices L and M are given by

$$L_{n,m}(k) = \begin{pmatrix} 1 & z_{n,m} - z_{n+1,m} \\ \frac{k^2/p^2}{z_{n,m} - z_{n+1,m}} & 1 \end{pmatrix} ,$$

$$M_{n,m}(k) = \begin{pmatrix} 1 & z_{n,m} - z_{n,m+1} \\ \frac{k^2/q^2}{z_{n,m} - z_{n,m+1}} & 1 \end{pmatrix} . \tag{2.4}$$

The compatibility of the two different ways of calculating $\phi_{n+1,m+1}$ from (2.3) yields eq. (2.1). The integrability of (2.1) has not yet been fully explored to date, but in view of the important role that the cross-ratio plays in complex analysis it is to be expected that here there is a seed for developing a discrete version of complex function theory. Such ideas have recently been put forward also in [31].

The more general equation (2.2) is very important too, since by tuning the free parameters α and β we obtain a number of different lattice equations related to the KdV system. In fact, in the limit $\beta \to \infty$, taking $1 - \beta s_{n,m} \to v_{n,m}$ we obtain the equation

$$(p-\alpha)\frac{v_{n,m+1}}{v_{n+1,m+1}} - (q-\alpha)\frac{v_{n+1,m}}{v_{n+1,m+1}} = (p+\alpha)\frac{v_{n+1,m}}{v_{n,m}} - (q+\alpha)\frac{v_{n,m+1}}{v_{n,m}} , \quad (2.5)$$

which for $\alpha = 0$ constitutes a lattice version of the (potential) modified KdV (MKdV) equation. Next, by taking the limit $\alpha, \beta \to \infty$ such that $\alpha\beta s_{n,m} \to u_{n,m}$, we obtain the equation

$$(p - q + u_{n,m+1} - u_{n+1,m})(p + q - u_{n+1,m+1} + u_{n,m}) = p^2 - q^2 . \quad (2.6)$$

which is the lattice (potential) KdV equation. All these variables are not unrelated. In fact, we can establish the relation

$$1 - (p + \beta)s_{n+1,m} + (p - \alpha)s_{n,m} = v_{n+1,m}w_{n,m} , \quad (2.7)$$

in which $w_{n,m}$ is the solution of eq. (2.5) for the parameter-value β instead of α, and discrete Miura-type of relations of the form

$$p - q + u_{n,m+1} - u_{n+1,m} = (p - \alpha)\frac{v_{n,m+1}}{v_{n+1,m+1}} - (q - \alpha)\frac{v_{n+1,m}}{v_{n+1,m+1}} , \quad (2.8a)$$

$$p + q + u_{n,m} - u_{n+1,m+1} = (p - \alpha)\frac{v_{n,m}}{v_{n+1,m}} + (q + \alpha)\frac{v_{n+1,m+1}}{v_{n+1,m}} . \quad (2.8b)$$

The KdV case is distinguished by the fact that the variable $s_{n,m}$ is invariant under the interchange of the parameters α and β.

Let us now consider the similarity reduction of the cross-ratio equation (the lattice Schwarzian KdV equation). In [23] we investigated ways of imposing scaling invariance on the solutions of the lattice KdV and MKdV equation. Since the scaling invariance of the lattice equation is not explicit (in contrast to the continuous case), it is not possible to find an explicit similarity variable in terms of which one can reduce the equation (like we have done in eq. (1.3)). However, it *is* possible to impose the scaling invariance as a constraint on the solution. In the case of the Schwarzian KdV equation this would amount to impose the additional differential constraint

$$3\mu\psi = x\psi_x + 3t\psi_t , \quad (2.9)$$

on the solution of (1.1). This indeed we can also do on the discrete level and impose an additional discrete *similarity constraint* on the solutions of the lattice equation. As a consequence of the implicit nature of the scaling invariance of the discrete equation, we find that the similarity constraint in the discrete case is *nonlinear* in contrast to the linear constraint (2.9) for the continuum situation. This seems to be a general feature of the dicrete similarity approach.

Let us see what the similarity constraint for the lattice Schwarzian KdV equation looks like. Without giving any details of the derivation (cf. e.g. [23] for the ingredients for the construction), we mention that the lattice similarity reduction of the cross-ratio equation is obtained by imposing on solutions of (2.1) the following nonlinear *nonautonomous* constraint

$$
\begin{aligned}
z_{n,m} \;=\; & 2n\frac{(z_{n+1,m} - z_{n,m})(z_{n,m} - z_{n-1,m})}{z_{n+1,m} - z_{n-1,m}} \\
+ \;& 2m\frac{(z_{n,m+1} - z_{n,m})(z_{n,m} - z_{n,m-1})}{z_{n,m+1} - z_{n,m-1}} \,.
\end{aligned}
\tag{2.10}
$$

which is the lattice analogue of (2.9), be it for a special value of the parameter μ. It is not so obvious that eq. (2.10) is indeed compatible with the lattice equation (2.1). The reason why it is, follows from the construction along the lines set out in [23]. We will not give a derivation of (2.10), but just mention that there is a monodromy problem which, in combination with the Lax pair (2.3),(2.4), yields the constraint (2.10). In fact, supplying the Lax pair (2.3) for the Schwarzian lattice KdV with the following differential equation

$$
\left(J_+ + k\frac{d}{dk} \right) \phi_{n,m}(k) = (n+m)J_-\phi_{n,m}(k) +
$$

$$
+2n\frac{z_{n,m} - z_{n-1,m}}{z_{n-1,m} - z_{n+1,m}} \begin{pmatrix} 0 & (z_{n+1,m} - z_{n,m}) \\ 0 & 1 \end{pmatrix} \phi_{n-1,m}(k)
$$

$$
+2m\frac{z_{n,m} - z_{n,m-1}}{z_{n,m-1} - z_{n,m+1}} \begin{pmatrix} 0 & (z_{n,m+1} - z_{n,m}) \\ 0 & 1 \end{pmatrix} \phi_{n,m-1}(k) \,,
\tag{2.11}
$$

where

$$
J_+ = \begin{pmatrix} 1 & 0 \\ 0 & 0 \end{pmatrix} \,, \quad J_- = \begin{pmatrix} 0 & 0 \\ 0 & 1 \end{pmatrix} \,,
$$

we obtain from the compatibility conditions the system consisting of the lattice equation (2.1) together with the constraint (2.10). This constitutes the discrete analogue of the system containing the original SKdV equation and the constraint (2.9), be it for a special choice of the parameter μ. Thus, we can refer to this system as constituting actually a lattice version of the Schwarzian Painlevé equation (1.4).

3. Semicontinuous Limits

Let us now investigate what happens under a continuum limit, bringing us eventually back to the original SKdV equation. Since there are two discrete variables in the lattice equation, n and m, we have to perform the continuum limit in two steps: one letting the variable m become continuous, reducing our equation to a *differential-difference equation*, i.e. an equation with one discrete and one continuous variable, and a second step in which the remaining discrete variable will become continuous. Both steps are achieved by shrinking the corresponding lattice step (encoded in the parameters p and q) to zero. The most convenient way of doing this is first, on the lattice, to do a change of discrete variables, namely $u_{n,m} =: u_{n'}(m)$, and then doing the limit by taking

$$\delta \equiv p - q \mapsto 0 \ , \quad m \mapsto \infty \ , \quad \delta m \mapsto \tau \ , \tag{3.1}$$

where $n' = n + m$ is to remain fixed. This limit is motivated from the behaviour of discrete plane-wave factors

$$\rho_k(n, m) = \left(\frac{p+k}{p-k}\right)^n \left(\frac{q+k}{q-k}\right)^m \ , \tag{3.2}$$

which govern the linear dispersion of the lattice equations, cf. [28]. Under (3.1) these plane-wave factors behave as

$$\left(\frac{p+k}{p-k}\right)^n \left(\frac{q+k}{q-k}\right)^m \mapsto \left(\frac{p+k}{p-k}\right)^{n'} e^{\frac{2k\tau}{p^2-k^2}} \ , \tag{3.3}$$

cf. [28, 29].

The continuum limit applied to the general lattice KdV (2.2) yields

$$\partial_\tau s_n \ = \ \frac{s_{n+1} - s_{n-1} + 2p(s_n^2 + s_{n+1}s_{n-1})}{2p - (p+\alpha)(p+\beta)s_{n+1} + (p-\alpha)(p-\beta)s_{n-1}}$$
$$- \ \frac{(2p+\alpha+\beta)s_n s_{n+1} - (2p-\alpha-\beta)s_n s_{n-1}}{2p - (p+\alpha)(p+\beta)s_{n+1} + (p-\alpha)(p-\beta)s_{n-1}} \ . \tag{3.4}$$

Analogous to the fully discrete case, eq. (3.4) has various interesting specifications for special choices of the variables α and β. For instance the differential-difference analogue of the lattice KdV (2.6) is given by

$$1 + \partial_\tau u_n \ = \ \frac{2p}{2p - u_{n+1} + u_{n-1}} \ , \tag{3.5}$$

which is related to the Kac–van Moerbeke–Volterra equation, whereas the continuum limit of (2.5) is given by

$$\partial_\tau \log v_n = \frac{v_{n+1} - v_{n-1}}{(p+\alpha)v_{n+1} + (p-\alpha)v_{n-1}} \ , \tag{3.6}$$

which for $\alpha = 0$ is a differential-difference version of the MKdV equation, while for $\alpha = p$ it is a potential version of the Toda lattice equation, cf. [28]. For $\alpha = \beta = 0$, eq. (3.4) reduces to the following differential-difference version of the Schwarzian KdV, cf. also [30, 32]

$$\frac{p}{2} \partial_\tau z_n = \frac{(z_{n+1} - z_n)(z_n - z_{n-1})}{z_{n+1} - z_{n-1}} \;, \tag{3.7}$$

i.e. eq. (3.4) for $\alpha = \beta = 0$.

It is interesting to look at the similarity solutions in the semicontinuum limit, because as we have shown in [23] in that limit one can actually use the similarity constraint to obtain a reduction to a single ordinary non-autonomous difference equation which can be identified to be a difference version of the Painlevé II equation. In fact, starting from the similarity constraint for the MKdV equation which in the limit (3.1) was given in [23] and which reads

$$b_n = \frac{v_{n+1} - v_{n-1}}{v_{n+1} + v_{n-1}} \left[n + 2\tau \frac{\partial_\tau (v_{n+1} v_{n-1})}{v_{n+1}^2 - v_{n-1}^2} \right] \;, \quad b_n = c_0 + c_1 (-1)^n \;, \tag{3.8}$$

c_0, c_1 being constants, one can use (3.6) to eliminate the derivatives with respect to τ to obtain the discrete painlevé II (dPII) equation

$$x_{n+1} + x_{n-1} = \frac{2p}{\tau} \frac{b_n - n x_n}{1 - x_n^2} \;, \tag{3.9}$$

in terms of the variable

$$x_n \equiv \frac{v_{n+1} - v_{n-1}}{v_{n+1} + v_{n-1}} \;. \tag{3.10}$$

Considering now the similarity reduction of the semicontinuous SKdV equation, eq. (2.10) in the limit (3.1) reduces to

$$z_n = 2n \frac{(z_{n+1} - z_n)(z_n - z_{n-1})}{z_{n+1} - z_{n-1}} + 2\tau \frac{(z_{n+1} - z_n)\dot{z}_{n-1} + (z_n - z_{n-1})\dot{z}_{n+1}}{z_{n+1} - z_{n-1}}$$
$$- 2\tau \frac{(z_{n+1} - z_n)(z_n - z_{n-1})}{(z_{n+1} - z_{n-1})^2} (\dot{z}_{n+1} + \dot{z}_{n-1}) \;. \tag{3.11}$$

Eliminating again the derivatives with respect to the variable τ from (3.11) one obtains, after one "integration", the following third-order non-autonomous nonlinear difference equation

$$\frac{4\tau}{p} \frac{(z_{n+2} - z_{n+1})(z_n - z_{n-1})}{(z_{n+2} - z_n)(z_{n+1} - z_{n-1})} = \frac{z_n}{z_{n+1} - z_n} - a_n \;, \tag{3.12}$$

in which $a_n \equiv n + (-1)^n a_0$, a_0 being an integration constant. We will refer to eq. (3.12) as the discrete Schwarzian Painlevé II (dSPII) equation, as in a continuum limit it reduces to the Schwarzian Painlevé II equation (1.4). Eq. (3.12) is related to the dPII equation (3.9) via the discrete Hopf-Cole transformation

$$x_n = \frac{z_{n+1} + z_{n-1} - 2z_n}{z_{n+1} - z_{n-1}} . \qquad (3.13)$$

In fact, eq. (3.13) follows directly from (3.10) together with the relation

$$v_{n+1} v_n = p(z_n - z_{n+1}) , \qquad (3.14)$$

which is a direct consequence of (2.7) for $\alpha = \beta = 0$. Then, using (3.12) we obtain

$$a_n + \frac{\tau}{p}(1 + x_{n+1})(1 - x_n) = \frac{z_n}{z_{n+1} - z_n} , \qquad (3.15)$$

which will subsequently yield (3.9) identifying $b_n = -a_0(-1)^n$. Thus the similarity reduction of the SKdV to the second Painlevé transcendent outlined in the introduction holds perfectly well also for the discrete case.

It would be of interest to study systematically integrable third-order difference equations with respect to their singularity behaviour. Already on the continuous level an exhaustive classification of third-order equations does not yet exist in spite of a large amount of early work in this direction, [20, 21]. Although equations containing a Schwarzian derivative form a special class in Chazy's work, eq. (1.4) doesn't seem to be among the equations he investigated. Nonetheless, as we have shown, the equation is significant both on the continuous as well as discrete level. The singularity confinement analysis that was first proposed in [33] and subsequently used in [24] to construct discrete Painlevé equations is very well applicable also to eq. (3.12), and shows that this equation *is* indeed singularity confining[2]. Since the Schwarzian KdV equation stands more or less at the bottom of the pyramid of KdV-type equations, I believe that the investigation of the discrete analogues of such equations and of their similarity reductions can prove very useful in a more systematic approach to establish a preliminary classification of the discrete Painlevé equations. In view of the nonuniqueness of the discrete analogues of the Painlevé equations, such a classification is an urgent open problem. In the following sections I will present some other discrete Schwarzian equations originating from other integrable lattice systems.

Remark 1: We note that from the equations given above one can also straightforwardly derive a Miura transformation connecting the dPII equation (3.9) to an equation called discrete P34 (dP34), which is the discrete analogue of the 34th equation in Painlevé's list, cf. [17]. In fact, going back

---

[2]The calculations to check this fact were performed recently by B. Grammaticos and A. Ramani.

to the continuum situation, introducing the KdV variable

$$w = -y' - y^2 + \frac{1}{6}\xi \,, \tag{3.16}$$

where $y = y(\xi)$ obeys the PII equation (1.6), we obtain P34 in the form

$$w''w = \frac{1}{2}w'^2 - 2w^3 + \frac{1}{3}\xi w^2 - \frac{\mu^2}{8} \,. \tag{3.17}$$

In the discrete case, interpreting (3.15) as a Miura transformation between the dPII (3.9) and an equation for the variable

$$w_n \equiv \frac{z_{n+1} + z_n}{z_{n+1} - z_n} \,, \tag{3.18}$$

we obtain from the dSPII equation (3.12) the following equation for w_n

$$(w_{n+1} + w_n)(w_n + w_{n-1}) = \frac{8\tau}{p} \frac{w_n^2 - 1}{w_n - (2a_n + 1)} \,, \tag{3.19}$$

which is the dP34 equation which in a slighlty different form was first given in [27]. The Miura transformation

$$w_n = \frac{2\tau}{p}(1 + x_{n+1})(1 - x_n) + 2a_n + 1 \,, \quad a_n = n + a_0(-1)^n \,, \tag{3.20}$$

was extablished in [34]. Another equation playing an important role in this scheme is the Painlevé type equation that one obtains from the similarity constraint for the semi-continuous KdV equation (3.5), which reads

$$0 = u_n + \frac{(np - \tau)(u_{n+1} - u_{n-1})}{2p + u_{n-1} - u_{n+1}}$$
$$+ 2p\tau \frac{u_{n+1} - u_{n-1} + p\partial_\tau(u_{n+1} + u_{n-1})}{(2p + u_{n-1} - u_{n+1})^2} \,. \tag{3.21}$$

Using the equation (3.5) itself on order to eliminate as before the derivatives with respect to τ we can obtain the following equation for the variable

$$R_n \equiv \frac{2p}{h_n h_{n-1}} \,, \quad h_n \equiv 2p + u_{n-1} - u_{n+1} \,, \tag{3.22}$$

namely,

$$1 = ((n+1)p - 2\tau) R_{n+1} - ((n-1)p - 2\tau) R_n$$
$$+ p\tau \left(R_{n+2} R_{n+1} + R_{n+1}^2 - R_n^2 - R_n R_{n-1} \right) \,. \tag{3.23}$$

Eq. (3.23) is a third-order diference equation, which is the discrete analogue of the similarity reduction of the KdV equation, which is a third-order differential equation. It is related to the dPII equation via a Miura transformation that can be derived using the relation

$$h_n = p\frac{v_{n+1} + v_{n-1}}{v_n} \, , \qquad (3.24)$$

which follows from (2.7). Using this relation and eq. (3.14) we obtain the following expression for R_n in terms of the Schwarzian KdV variable

$$R_n = \frac{2}{p} \frac{(z_{n-1} - z_n)^2}{(z_{n+1} - z_{n-1})(z_n - z_{n-2})} \, . \qquad (3.25)$$

It is interesting to note that the variable R_n itself is the variable that obeys the Kac–van Moerbeke–Volterra equation, cf. [14]. Thus we see that all these equations are interconnected on the discrete level even in the similarity reduced case.

The Miura transformation (3.20) together with the discrete Hopf-Cole transformation (3.13) constitutes on the level of the similarity reduced system a sequence of "difference substitutions"

$$\text{dSPII} \stackrel{\text{Hopf-Cole}}{\longrightarrow} \text{dPII} \stackrel{\text{Miura}}{\longrightarrow} \text{dP34} \, .$$

The existence of such a sequence of difference substitutions was already demonstrated for the classical Liouville theory in [30], cf. also [32], exhibiting a similar "cancellation phenomenon" as in the continuum situation. This cancellation phenomenon was explained group-theoretically in [35], and we expect that these arguments can be extended to the discrete case as well. The above derivation demonstrates that the entire picture can be completed also for the similarity reduced difference system, i.e. for the corresponding discrete Painlevé equations.

Remark 2: The semi-continuous limit (3.1) leading to the differential-difference Schwarzian equation (3.7), is obviously not the only way to obtain a semi-discrete Schwarzian equation. In fact, one can also perform a "straight" continuum limit of the form

$$m \mapsto \infty \, , \quad q \mapsto \infty \quad \text{such that} \quad \frac{m}{q} \mapsto \xi \, , \qquad (3.26)$$

together with the expansions, $(z_n' \equiv \partial_\xi z_n)$,

$$z_{n,m} \mapsto z_n(\xi) + \frac{1}{q} z_n'(\xi) + \frac{1}{2q^2} z_n''(\xi) + \cdots \, , \qquad (3.27)$$

which leads to various other differential-difference KdV-type equations which we will not list here. We only mention that in this case we obtain the differential-difference SKdV

$$z'_{n+1}z'_n = p^2(z_{n+1} - z_n)^2 \ , \tag{3.28}$$

which, in comparison with (3.7) is of lower order in the lattice shifts, but involves two different derivatives with respect to the continuous variable. In this case the semi-continuous similarity constraint reads

$$z_n = \xi z'_n + 2n\frac{(z_{n+1} - z_n)(z_n - z_{n-1})}{z_{n+1} - z_{n-1}} \tag{3.29}$$

from which one easily derives the following alternative Schwarzian PII equation

$$p^2\xi^2\frac{(z_{n+2} - z_n)(z_{n+1} - z_{n-1})}{(z_{n+2} - z_{n+1})(z_n - z_{n-1})} = \left(\frac{z_n}{z_{n+1} - z_n} + \frac{z_n}{z_n - z_{n-1}} - 2n\right) \times$$
$$\times \left(\frac{z_{n+1}}{z_{n+2} - z_{n+1}} + \frac{z_{n+1}}{z_{n+1} - z_n} - 2(n+1)\right) \ . \tag{3.30}$$

Eq. (3.30) is a different discrete Schwarzian PII equation. By construction both discrete equations (3.12) and (3.30) lead to the SPII equation in an appropriate subsequent continuum limit on the remaining discrete variable n (see the next remark). However, as is suggested by the treatment in [23] the two discrete SPII equations have different asymptotic behaviour, and we expect, therefore, that their solutions will be given in terms of different transcendental functions.

Similar to the previous semi-continuous limit, there is a Miura chain connecting (3.30) to a discrete version of PII itself and to another discrete P34. In fact, introducing the same variable w_n as before in (3.18) we can derive the following alternative version of dP34

$$p^2\xi^2\frac{(w_{n+1} + w_n)(w_n + w_{n-1})}{(w_{n+1} + w_n - 4(n+1))(w_n + w_{n-1} - 4n)} = \frac{1}{4}(w_n^2 - 1) \ . \tag{3.31}$$

An alternative discretisation of PII is obtained from the limit of (2.5) which reads

$$\partial_\xi \log(v_{n+1}v_n) = p\left(\frac{v_{n+1}}{v_n} - \frac{v_n}{v_{n+1}}\right) \ , \tag{3.32}$$

together with the similarity constraint

$$c = \xi\partial_\xi(\log v_n) + n\frac{v_{n+1} - v_{n-1}}{v_{n+1} + v_{n-1}} \ , \tag{3.33}$$

(c being a constant). Combining (3.32) and (3.33) we obtain for the variable $f_n \equiv v_{n+1}/v_n$ and equation of the form

$$2c = n\frac{f_n f_{n-1} - 1}{f_n f_{n-1} + 1} + (n+1)\frac{f_n f_{n+1} - 1}{f_n f_{n+1} + 1} + p\xi\left(f_n - \frac{1}{f_n}\right), \qquad (3.34)$$

which is an alternative discretisation of PII different from (3.9). The Miura transformation between the alternative dP34, (3.36), and the alternative dPII, (3.34), is nonlocal and given by

$$f_n f_{n-1} = \frac{w_{n-1} + 1}{w_n - 1}, \qquad (3.35)$$

where we have taken $c = 0$. (Again the general parameter case is obtained by including a constant factor in the right-hand side of (3.29)). We mention that the alternative dPII (3.34) was first derived in [27]. Its existence shows the abundance of discrete versions of the Painlevé transcendents.

Remark 3: After having obtained the semi-continuous limit leading to the differential-difference equations (3.4)–(3.7), a second continuum limit needs to be performed to arrive at fully continuous equations. This can be achieved for instance by taking

$$p \to \infty, \quad n \to \infty, \quad \tau \to \infty, \qquad (3.36a)$$

such that

$$2\frac{n}{p} + 2\frac{\tau}{p^2} \mapsto x, \quad \frac{2}{3}\frac{n}{p^3} + 2\frac{\tau}{p^4} \mapsto t, \qquad (3.36b)$$

in which case (3.4) goes over into the following evolution equation

$$s_t = s_{xxx} + 3\frac{(s_{xx} + \alpha s_x)(s_{xx} + \beta s_x)}{1 - (\alpha + \beta)s - 2s_x}, \qquad (3.37)$$

which contains as interesting subcases the potential KdV equation

$$u_t = u_{xxx} + 3u_x^2, \qquad (3.38)$$

and for $Q \equiv \log v$ the potential MKdV equation

$$Q_t = Q_{xxx} + 2Q_x^3. \qquad (3.39)$$

Clearly, for $\alpha = \beta = 0$ we recover the Schwarzian KdV equation (1.1). An alternative way of arriving at the same equations is by starting from the straight semicontinuum limit (3.26) and applying the limit

$$p \to \infty, \quad n \to \infty, \quad \text{such that} \quad 2\xi + 2\frac{n}{p} \mapsto x, \quad \frac{2}{3}\frac{n}{p^3} \mapsto t, \qquad (3.40)$$

to (3.28)–(3.35).

4. Modified and Schwarzian NLS Equation

The modified NLS equation is an equation related to the equation of motion of the Heisenberg ferromagnet with uniaxial anisotropy, cf. [36]. It can be written as

$$iq_t + q_{xx} = 2\frac{|q_x|^2}{q^*} , \tag{4.1}$$

cf. [37], where the asterisk denotes complex conjugation. The "real form" of this equation is obtained by going to polar coordinates, $q = \kappa \exp(i\theta)$, in terms of which we have

$$\dot{\kappa} + 2\kappa'\theta' + \kappa\theta'' = 0 , \tag{4.2a}$$

$$\frac{\kappa''}{\kappa} - 2\frac{\kappa'^2}{\kappa^2} = \dot{\theta} + 3\theta'^2 . \tag{4.2b}$$

Solving the first equation by setting

$$\frac{1}{2}\kappa^2 = \varphi' , \quad \kappa^2\theta' = -\dot{\varphi} , \tag{4.3}$$

we obtain the equation

$$\partial_t\left(\frac{\dot{\varphi}}{\varphi'}\right) + \partial_x\left(\frac{\varphi'''}{\varphi'} - \frac{3}{2}\frac{\varphi''^2 + \dot{\varphi}^2}{\varphi'^2}\right) = 0 , \tag{4.4}$$

which is the real form of the MNLS. Eq. (4.4) is a Schwarzian equation and invariant under Möbius transformation. It is well-known that after similarity reduction, using $\varphi = a\log t + \Phi(\eta)$, with $\eta = x/\sqrt{t}$, eq. (4.4) reduces to the following equation

$$-\frac{1}{2}\eta\left(\frac{a}{\Phi'} - \frac{1}{2}\eta\right) + \frac{\Phi'''}{\Phi'} - \frac{3}{2}\frac{\Phi''^2}{\Phi'^2} - \frac{3}{2}\left(\frac{a}{\Phi'} - \frac{1}{2}\eta\right)^2 = b , \tag{4.5}$$

where b is an integration constant. Eq. (4.5) leads for the variable $W = 1/\Phi'$ to the Painlevé IV equation

$$W''W = \frac{1}{2}W'^2 - \frac{3}{2}a^2W^4 + a\eta W^3 - (\frac{1}{8}\eta^2 + b)W^2 , \tag{4.6}$$

cf. [36]. A natural question to ask is whether the Schwarzian NLS has an integrable discretisation similar to the cross-ratio equation (2.1). I believe the answer is positive, but a direct discretisation of (4.1) and of (4.4) is not available yet, although in principle it could be inferred from the treatments given in refs. [29], cf. also [37]. In the next section we will, nonetheless, present a discretisation of an equation very similar to (4.4), namely the lattice version of the Schwarzian Boussinesq (SBSQ) equation.

5. Lattice Schwarzian Boussinesq Equation

The lattice BSQ equation was derived in [38] as the first higher-order member in what is the discrete analogue of the Gel'fand-Dikii hierarchy, i.e. of the integrable family of partial differential equations associated with the higher-order differential spectral problems. The lattice KdV equation is naturally embedded as the lowest member in this class of equations, which is labelled by the roots of unity, $\omega \equiv \exp(2\pi i/N)$, ($N = 2, 3, \dots$). In this section we will focus on the case $N = 3$ leading to lattice analogues of the Boussinesq (BSQ) equation and of the modified BSQ equation, among others. We expect that most of our results can with some labour be extended to the other memebers of the GD hierarchy of [38].

In [39] the following "universal" lattice BSQ equation is derived

$$\frac{G_p(s_{n+2,m+2}, s_{n+1,m+2})K_{pq}(s_{n+1,m+1}, s_{n,m+2})G_q(s_{n,m+1}, s_{n,m})}{G_q(s_{n+2,m+2}, s_{n+2,m+1})K_{qp}(s_{n+1,m+1}, s_{n+2,m})G_p(s_{n+1,m}, s_{n,m})}$$

$$= \frac{L_{pq}(s_{n+1,m+2}, s_{n+1,m+1}, s_{n,m+2}, s_{n,m+1})}{L_{qp}(s_{n+2,m+1}, s_{n+1,m+1}, s_{n+2,m}, s_{n+1,m})} \tag{5.1a}$$

in which

$$G_p(s_{n+1,m}, s_{n,m}) \equiv 1 + (p - \omega\beta)s_{n+1,m} - (p - \alpha)s_{n,m} \ ,$$

$$K_{pq}(s_{n+1,m}, s_{n,m+1}) \equiv p - q + (p - \alpha)(q - \omega\beta)s_{n,m+1}$$

$$-(q - \alpha)(p - \omega\beta)s_{n+1,m} \ , \tag{5.1c}$$

(and analogous formulas with p and q and the labels $(n+1, m)$ and $(n, m+1)$ interchanged), together with

$$\begin{aligned} L_{pq}(s_{n+1,m+1}, s_{n+1,m}, s_{n,m+1}, s_{n,m}) \equiv &(1 - (q - \alpha)s_{n+1,m})\left[(q + \alpha + \omega\beta)+ \right.\\ &\left.+(q^2 + \alpha q + \alpha^2)s_{n,m+1} - (q^2 + \omega\beta q + \omega^2\beta^2)s_{n,m}\right] - \\ &- (1 - (p - \alpha)s_{n,m+1})\left[(p + \alpha + \omega\beta) + (p^2 + \alpha p + \alpha^2)s_{n+1,m}-\right.\\ &\left.-(p^2 + \omega\beta p + \omega^2\beta^2)s_{n,m}\right] - \\ &- s_{n+1,m+1}\left[(p - q)(p + q + \alpha) + (p^2 + \alpha p + \alpha^2)(p - \omega\beta)s_{n+1,m}\right.\\ &\left. - (q^2 + \alpha q + \alpha^2)(q - \omega\beta)s_{n,m+1} - (p^3 - q^3)s_{n,m}\right] \ . \end{aligned} \tag{5.1d}$$

As before, eq. (5.1) has some interesting special parameter-subcases, namely in the limits $\alpha \to \infty$ and $\beta \to \infty$. In such limits we obtain for example the lattice modified BSQ equation

$$\frac{(p^2 + p\alpha + \alpha^2)\, v_{n+1,m+1} - (q^2 + q\alpha + \alpha^2)\, v_{n,m+2}}{(p - \alpha)\, v_{n,m+2} - (q - \alpha)\, v_{n+1,m+1}} \frac{v_{n+1,m+2}}{v_{n,m+1}}$$

$$-\frac{(p^2 + p\alpha + \alpha^2)\, v_{n+2,m} - (q^2 + q\alpha + \alpha^2)\, v_{n+1,m+1}}{(p - \alpha)\, v_{n+1,m+1} - (q - \alpha)\, v_{n+2,m}} \frac{v_{n+2,m+1}}{v_{n+1,m}}$$

$$= (p - \alpha) \left(\frac{v_{n,m}}{v_{n+1,m}} - \frac{v_{n+1,m+2}}{v_{n+2,m+2}} \right)$$

$$- (q - \alpha) \left(\frac{v_{n,m}}{v_{n,m+1}} - \frac{v_{n+2,m+1}}{v_{n+2,m+2}} \right) , \tag{5.2}$$

and the lattice BSQ equation

$$\frac{p^3 - q^3}{p - q + u_{n+1,m+1} - u_{n+2,m}} - \frac{p^3 - q^3}{p - q + u_{n,m+2} - u_{n+1,m+1}}$$

$$= (p - q + u_{n+1,m+2} - u_{n+2,m+1})(2p + q + u_{n,m+1} - u_{n+2,m+2})$$

$$- (p - q + u_{n,m+1} - u_{n+1,m})(2p + q + u_{n,m} - u_{n+2,m+1}) , \tag{5.3}$$

which were derived first in [38].

As in the KdV case, the special case $\alpha = \beta = 0$ is particularly interesting. In that case eq. (5.1) reduces to

$$\frac{(z_{n+2,m+2} - z_{n+1,m+2})(z_{n,m+2} - z_{n+1,m+1})(z_{n,m+1} - z_{n,m})}{(z_{n+2,m+2} - z_{n+2,m+1})(z_{n+2,m} - z_{n+1,m+1})(z_{n+1,m} - z_{n,m})} \tag{5.4}$$

$$= \frac{p^3 (z_{n+1,m+2} - z_{n,m+2})(z_{n+1,m+1} - z_{n,m+1}) - q^3 (z_{n+1,m+2} - z_{n+1,m+1})(z_{n,m+2} - z_{n,m+1})}{q^3 (z_{n+2,m+1} - z_{n+2,m})(z_{n+1,m+1} - z_{n+1,m}) - p^3 (z_{n+2,m+1} - z_{n+1,m+1})(z_{n+2,m} - z_{n+1,m})} ,$$

for the quantity $z_{n,m} = s_{n,m} + \frac{n}{p} + \frac{m}{q}$. Eq. (5.4) is a discrete version of the Schwarzian BSQ equation, and can as such be considered to be a higher-rank version of eq. (2.1). A Lax pair for (5.4) can be found from the construction in [38] and is given in [39].

Recently, also in [39] I investigated the similarity reductions of the lattice BSQ system, and showed that this, in fact, gives rise to discrete versions of the fourth Painlevé transcendent PIV. As in the KdV situation, the similarity reduction is obtained by imposing a non-autonomous difference constraint which in the case of the Schwarzian lattice BSQ equation reads

$$z_{n,m} = 3np \frac{(z_{n+1,m} - z_{n,m})(z_{n,m} - z_{n-1,m})}{p(z_{n+1,m} - z_{n-1,m}) + v_{n-1,m} w_{n+1,m}} +$$

$$+ 3mq \frac{(z_{n,m+1} - z_{n,m})(z_{n,m} - z_{n,m-1})}{q(z_{n,m+1} - z_{n,m-1}) + v_{n,m-1} w_{n,m+1}} , \tag{5.5}$$

in which we have to insert the expressions

$$v_{n-1,m} w_{n+1,m} = \frac{(z_{n+1,m+1} - z_{n+1,m})}{p^2 (z_{n+1,m+1} - z_{n,m+1})(z_{n-1,m+1} - z_{n,m})} \times$$

$$\times \left[p^3 (z_{n,m+1} - z_{n-1,m+1})(z_{n,m} - z_{n-1,m}) - \right.$$

$$\left. - q^3 (z_{n,m+1} - z_{n,m})(z_{n-1,m+1} - z_{n-1,m}) \right] , \tag{5.6}$$

and a similar expression for $v_{n,m-1}w_{n,m+1}$ that is obtained from (5.6) interchanging p and q and also the labels $(n \pm 1, m)$ and $(n, m \pm 1)$. The variable $w_{n,m}$ in eqs. (5.5) and (5.6) is dual to the variable $v_{n,m}$ which is the variable obeying the lattice MBSQ equation (5.3) for $\alpha = 0$. Eq. (5.5) is a rather complicatedly looking equation which can be derived from a machinery similar to the one used in the KdV case of the previous section, cf. [39]. We will not go into the details of the derivation here, but just mention that like the lattice Schwarzian KdV equation (2.6) the similarity constraint can be derived from an associated monodromy problem.

As in the KdV case, it is of interest to study special continuum limits of the lattice Schwarzian BSQ equation (5.4). A similar limit as the one for the KdV case, (3.1), leads to the following differential-difference equation

$$\partial_\tau \log \left(3 \frac{(z_{n+1} - z_n)(z_n - z_{n-1})}{\dot{z}_n} - p(z_{n+1} - z_{n-1}) \right) =$$

$$= \frac{\dot{z}_{n-1}}{z_{n-1} - z_{n-2}} - \frac{\dot{z}_{n+1}}{z_{n+2} - z_{n+1}} , \tag{5.7}$$

which is higher-order both in the derivatives with respect to the continuous variable τ as well as in the translations in terms of the lattice variable n. In the same limit the similarity constraint (5.5) together with (5.6) goes over into the equation

$$\frac{z_n}{\dot{z}_n} = np - 3\tau + p\tau \left[\frac{\dot{z}_{n+1}(z_{n+2} - z_n)}{(z_{n+2} - z_{n+1})(z_{n+1} - z_n)} + \right.$$

$$+ \frac{\dot{z}_n(z_{n+1} - z_{n-1})}{(z_{n+1} - z_n)(z_n - z_{n-1})} + \frac{\dot{z}_{n-1}(z_n - z_{n-2})}{(z_n - z_{n-1})(z_{n-1} - z_{n-2})} +$$

$$+ \frac{p}{3} \frac{\dot{z}_{n+1}\dot{z}_n(z_{n-1} - z_{n+2})}{(z_{n+2} - z_{n+1})(z_{n+1} - z_n)(z_n - z_{n-1})}$$

$$\left. + \frac{p}{3} \frac{\dot{z}_n\dot{z}_{n-1}(z_{n-2} - z_{n+1})}{(z_{n+1} - z_n)(z_n - z_{n-1})(z_{n-1} - z_{n-2})} \right] . \tag{5.8}$$

In the KdV case of section 2, we were able to use the combination of differential-difference equation and differential-difference constraint to explicitly eliminate the derivatives with respect to the continuous variable and obtained a Schwarzian version of PII. In the case of the BSQ system the elimination is more complicated and leads to a discrete version of PIV, since in the continuous case we obtain PIV from the scaling-invariant solutions of the BSQ equation, cf. [40]. Details of these calculations, which would lead us too far to include here at this point, will be published elsewhere, [39].

Remark: Similarly, as in the case of the KdV system, eq. (5.7) is not the only semicontinous limit of the lattice Schwarzian BSQ equation. Again a "straight" limit (3.26) exists in the BSQ case as well and reads

$$\partial_\xi \log \left[p^3 (z_{n+1} - z_{n-1}) + \frac{z'_{n+1} z'_n z'_{n-1}}{(z_{n+1} - z_n)(z_n - z_{n-1})} \right] =$$

$$= \frac{z'_{n+1}}{z_{n+1} - z_n} - \frac{z'_{n-1}}{z_n - z_{n-1}} , \qquad (5.9)$$

where, as before, $z'_n \equiv \partial_\xi z_n$. The associated similarity constraint in this limit is given by

$$z_n = \xi z'_n + 3np^3 \frac{(z_{n+1} - z_n)^2 (z_n - z_{n-1})^2}{z'_{n+1} z'_n z'_{n-1} + p^3 (z_n - z_{n+1})(z_{n+1} - z_{n-1})(z_{n-1} - z_n)} . \qquad (5.10)$$

Details on these equations and their consequences will be given in [39]. The full continuum limit of the lattice Schwarzian BSQ equation (5.4) has the following form

$$3\partial_y \left(\frac{z_y}{z_x} \right) + \partial_x \left(\frac{z_{xxx}}{z_x} - \frac{3}{2} \frac{z_{xx}^2 - z_y^2}{z_x^2} \right) = 0 , \qquad (5.11)$$

and was studied in [41]. Note that eq. (5.11) is very similar to (4.4), and differs only through some signs (in front of the time-derivatives) and some factors. Nevertheless, both equations have a different origin, and their solutions might have a different asymptotic behaviour. These questions are under study at present.

6. Lattice Schwarzian KP Equation

The most universal integrable system is, of course, the KP system, which brings us to the (2+1)-dimensional situation. Also here, we have Schwarzian type of equations. On the lattice, we start from the general equation

$$\frac{(1 - (p+\beta)s_{n+1,m,k+1} + (p-\alpha)s_{n,m,k+1})}{(1 - (q+\beta)s_{n,m+1,k+1} + (q-\alpha)s_{n,m,k+1})} \frac{(1 - (q+\beta)s_{n+1,m+1,k} + (q-\alpha)s_{n+1,m,k})}{(1 - (r+\beta)s_{n+1,m,k+1} + (r-\alpha)s_{n+1,m,k})}$$

$$= \frac{1 - (p+\beta)s_{n+1,m+1,k} + (p-\alpha)s_{n,m+1,k}}{1 - (r+\beta)s_{n,m+1,k+1} + (r-\alpha)s_{n,m+1,k}} \qquad (6.1)$$

where now we have a dependent variable $s_{n,m,k}$ depending on three discrete variables $n, m, k \in \mathbf{Z}$, and where p, q and r are the three lattice parameters measuring the three-dimensional grid. Eq. (6.1) which was given for the first time in [42] can be derived from the same relation (2.7) as in the KdV case together with two similar relations for the other lattice directions. In contrast to the KdV situation, there is now no direct relation between the variables v and w, whereas in the KdV case w is obtained from v by replacing α by β. The variables v and w themselves obey an equation of the form

$$
(p - \alpha) \left(\frac{v_{n,m,k+1}}{v_{n+1,m,k+1}} - \frac{v_{n,m+1,k}}{v_{n+1,m+1,k}} \right)
$$
$$
+ (q - \alpha) \left(\frac{v_{n+1,m,k}}{v_{n+1,m+1,k}} - \frac{v_{n,m,k+1}}{v_{n,m+1,k+1}} \right)
$$
$$
+ (r - \alpha) \left(\frac{v_{n,m+1,k}}{v_{n,m+1,k+1}} - \frac{v_{n+1,m,k}}{v_{n+1,m,k+1}} \right) = 0 , \qquad (6.2)
$$

which for $\alpha = 0$ is a lattice version of the modified KP equation. Eq. (6.2) is Miura-related to the lattice KP equation

$$
\frac{p - r + u_{n,m+1,k+1} - u_{n+1,m+1,k}}{p - r + u_{n,m,k+1} - u_{n+1,m,k}} = \frac{q - r + u_{n+1,m,k+1} - u_{n+1,m+1,k}}{q - r + u_{n,m,k+1} - u_{n,m+1,k}} ,
$$
$$
(6.3)
$$

via the Miura-transformation consisting of

$$
p - q + u_{n,m+1,k} - u_{n+1,m,k} = (p - \alpha) \frac{v_{n,m+1,k}}{v_{n+1,m+1,k}} - (q - \alpha) \frac{v_{n+1,m,k}}{v_{n+1,m+1,k}} , \qquad (6.4)
$$

and two other equations involving the other lattice directions.

For the special choice of the parameters $\alpha = \beta = 0$ eq. (6.1) reduces again to a Möbius-invariant equation, which we call the lattice Schwarzian KP equation, i.e. the equation for the variable $s_{n,m,k} = z_{n,m,k} + \frac{n}{p} + \frac{m}{q} + \frac{k}{r}$ reading

$$
\frac{(z_{n+1,m,k+1} - z_{n+1,m,k})(z_{n,m+1,k+1} - z_{n,m,k+1})}{(z_{n,m+1,k+1} - z_{n,m+1,k})(z_{n+1,m,k+1} - z_{n,m,k+1})} = \frac{z_{n+1,m+1,k} - z_{n+1,m,k}}{z_{n+1,m+1,k} - z_{n,m+1,k}} .
$$
$$
(6.5)
$$

Continuum limits of eq. (6.1) can be obtained in several ways, leading to partial difference-differential equations and partial differential-difference equations, depending on the number of discrete variables we retain, cf. [42]. The full continuum limit, letting all three lattice variables become continuous, was worked out in [42] and yields the equation

$$
\frac{3}{4} \partial_y \left(\frac{(\alpha - \beta)s_x + s_y}{1 - (\alpha + \beta)s - s_x} \right) = \partial_x \left\{ \frac{s_t - \frac{1}{4}s_{xxx} + \frac{3}{4}(\alpha - \beta)s_y}{1 - (\alpha + \beta)s - s_x} \right.
$$
$$
\left. - \frac{3}{2} \frac{\left[\alpha s_x + \frac{1}{2}(s_y + s_{xx}) \right] \left[\beta s_x - \frac{1}{2}(s_y - s_{xx}) \right]}{(1 - (\alpha + \beta)s - s_x)^2} \right\} . \qquad (6.6)
$$

In the case of the Schwarzian KP (i.e. taking $\alpha = \beta = 0$) we obtain an equation that was studied in a number of papers, [12, 13], and investigated by Irene Dorfman and myself in [11]. It reads

$$\partial_x \left(\frac{z_t}{z_x} - \frac{1}{4} \frac{z_{xxx}}{z_x} + \frac{3}{8} \frac{z_{xx}^2 - z_y^2}{z_x^2} \right) = \frac{3}{4} \partial_y \left(\frac{z_y}{z_x} \right) . \qquad (6.7)$$

We note easily that the Schwarzian BSQ equation is a dimensional reduction of eq. (6.7) by taking the solution z to be independent of the t-variable. This reduction is trivial on the level of the continuous KP equation, but it is nontrivial for the lattice KP. In fact, a more careful analysis must be done to obtain (5.4) as a dimensional reduction of (6.5).

Remark: The similarity reduction of the lattice KP equation is being investigated in [43], constructing the nonlinear similarity constraints related to scaling symmetry. We hope that a systematic investigation of the similarity reductions of lattice matrix KP systems, the general form of which was derived in [44], will yield a connection with discrete versions of the sixth Painlevé transcendent PVI.

7. Discussion

In this paper I have presented a number of Schwarzian, i.e. Möbius-invariant, integrable equations, and their discrete analogues. When dealing with their scaling-invariant similarity solutions, the invariance under the group $PSL(2)$ is broken. Nonetheless, due to the integrability the chain of differential substitutions consisting of a Hopf-Cole and a Miura transformation remains valid, and leads to a connection with the Painlevé equations. This, we demonstrated explicitly for the KdV-family, but similar features will no doubt hold also in the other cases we investigated. That this is true on the continuous level could have been expected. That it remains true on a discrete level with analogous "difference substitutions" demonstrates the beauty of the structure underlying these integrable systems. What we still need is a more profound understanding on a group-theoretical level of the mechanisms behind these discrete structures. For the continuum case, G. Wilson has provided an explanation of the so-called cancellation phenomenon in [35], but whether those arguments are applicable also to the discrete systems remains to be seen. From my personal point of view, I consider the discrete situation to be the right setting from which these and similar questions need to be investigated. In fact, the Möbius transformation itself could possibly be viewed as a discrete dynamical operation in its own right, corresponding to a discrete version of the Ricatti equation, which consequently generates a discrete flow corresponding to the group orbits of $PSL(2)$. At this point in time, I have not yet a full understanding of the role of the scale-invariance that leads to the reduction to the Painlevé-type of systems that we have encountered. Imposing the scale-invariance breaks the invariance under the Möbius group, but nevertheless it manifests yet in another way the conformal invariance. Possibly, an in-depth study of this "cross-road" of symmetries that we observe in the similarity reduction of the Schwarzian KdV and of its discrete counterparts, could lead to a better understanding of the singularity behaviour of integrable equations including the Painlevé property.

As for the discrete Painlevé equations, there we are still a long way from understanding the singularity behaviour of the solutions. The singularity confinement phenomenon that was discovered in [33], and used in [24] for the construction of new discrete Painlevé equations, is still far from understood. Furthermore, as noted before and demonstrated again in the present paper, on the discrete level there seems to be an abundance of discrete Painlevé equations, all integrable and all going to the same transcendents in the continuum limit but without apparent links on the discrete level. To arrive at something remotely like a classification of these equations it is necessary to have a better understanding of the underlying structures. The Schwarzian versions of the discrete Painlevé eqations might serve as guiding animals in this relatively new world.

Acknowledgement. Discussions with M.J. Ablowitz, L.D. Faddeev, A.P. Fordy, B. Grammaticos, M. Mañas and A. Ramani are gratefullly acknowledged.

References

[1] I.M. Krichever and S.P. Novikov, Holomorphic Bundles over Algebraic Curves and Nonlinear Equations, *Russ. Math. Surv.* **35** (1980), 53–79.

[2] M. Antonowicz and A.P. Fordy, Multicomponent Schwarzian KdV Hierarchies, *Rep. Mod. Phys.* **32** (1993), 223–233.

[3] G. Wilson, On the Quasi-Hamiltonian Formalism of the KdV Equation, *Phys. Lett.* **A132** (1988), 445–450.

[4] F. Guil and M. Mañas, Loop Algebras and the Krichever–Novikov Equation, *Phys. Lett.* **153A** (1991), 90–94.

[5] A.I. Mokhov, Symplectic and Poisson Geometry on Loop Spaces of Manifolds and Nonlinear Equations, preprint `solv-int/9504076`.

[6] I.Ya. Dorfman, Krichever–Novikov Equations and Local Symplectic Structures, *Sov. Math. Dokl.* **38** (1989), 340–343.

[7] I.Ya. Dorfman, Dirac Structures of Integrable Evolution Equations, *Phys. Lett.* **125A** (1987), 240–246.

[8] I.Ya. Dorfman and O.I. Mokhov, Local Symplectic Operators and Structures Related to Them, *J. Math. Phys.* **32** (1991), 3288–3296.

[9] I.Ya. Dorfman, *Dirac Structures and Integrability of Nonlinear Evolution Equations*, (Wiley, England, 1993).

[10] I.Ya. Dorfman and A.S. Fokas, Hamiltonian theory over noncommutative rings and integrability in multidimensions, *J. Math. Phys.* **33** (1992), 2504–2514.

[11] I. Dorfman and F.W. Nijhoff, On a (2+1)-dimensional version of the Krichever–Novikov equation, *Phys. Lett.* **A157** (1991), 107–112.

[12] J. Weiss, The Painlevé Property for Partial Differential Equations II, *J. Math. Phys.* **24** (1983), 1405.

[13] B.G. Konopel'chenko and W. Strampp, On the structure and properties of the singularity manifold equations of the KP hierarchy, *J. Math. Phys.* **32** (1991), 40–49.

[14] F.W. Nijhoff and H.W. Capel, The Discrete Korteweg-de Vries Equation, *Acta Applicandae Mathematicae* **39** (1995), 133–158.

[15] P. Painlevé, Memoire sur les équations différentielles dont l'intégrale générale est uniforme, *Bull. Soc. Math. France* **28** (1900), 201–261; Sur les équations différentielles du second ordre et d'ordre supérieur dont l'intégrale générale est uniforme, *Acta Math.* **25** (1902), 1–85.

[16] B. Gambier, Sur les équations différentielles du second ordre et du premier degré 'dont l'intégrale générale est à points critiques fixés, *Acta Math.* **33** (1909), 1–55.

[17] E.L. Ince, *Ordinary Differential Equations* (Dover Publ., New York, 1956).

[18] M.J. Ablowitz and P.A. Clarkson, Solitons, Nonlinear Evolution Equations and Inverse Scattering, *LMS Lect. Notes* **149**, Cambridge University Press, 1991.

[19] M.J. Ablowitz, A. Ramani and H. Segur, A connection between nonlinear evolution equations and ordinary differential equations of P-type, I,II, *J. Math. Phys.* **21** (1980), 715–721;1006–1015.

[20] J. Chazy, Sur les équations différentielles du troisième ordre et d'ordre supérieur dont l'intégrale générale a ses points critiques fixes, *Acta Math.* **34** (1911), 317–385.

[21] M.R. Garnier, Sur des équations différentielles du troisième ordre dont l'intégrale est uniforme et sur une classe d'équations nouvelles d'ordere supérieur dont l'intégrale a ses points critiques fixes, *Ann. Sci. de l'ENS* vol. XXIX, # 3, (1912), 1–126.

[22] B. Grammaticos and A. Ramani, Discrete Painlevé Equations: Derivation and Properties, in *Applications of Analytic and Geometric Methods to Nonlinear Differential Equations*, P.A. Clarkson (Ed.), NATO ASI Series C, vol. 413, (Kluwer Acad. Publ., Dordrecht, 1993), pp. 299–314.

[23] F.W. Nijhoff and V.G. Papageorgiou, Similarity Reductions of Integrable Lattices and Discrete Analogues of the Painlevé II Equation, *Phys. Lett.* **153A** (1991), 337–344.

[24] A. Ramani, B. Grammaticos and J. Hietarinta, Discrete Versions of the Painlevé Equations, *Phys. Rev. Lett.* **67** (1991), 1829–1832.

[25] A.S. Fokas, A.R. Its and A.V. Kitaev, Discrete Painlevé Equations and their Appearance in Quantum Gravity, *Comm. Math. Phys.* **142** (1991), 313–344.

[26] V.G. Papageorgiou, F.W. Nijhoff, B. Grammaticos and A. Ramani, Isomonodromic deformation problems for discrete analogues of Painlevé equations, *Phys. Lett.* **A164** (1992), 57–64.

[27] A.S. Fokas, B. Grammaticos and A. Ramani, From Continuous to Discrete Painlevé Equations, *J. Math. Anal. Appl.* **180** (1993), 342–360.

[28] F.W. Nijhoff, G.R.W. Quispel and H.W. Capel, Direct Linearization of Nonlinear Difference-Difference Equations, *Phys. Lett.* **97A** (1983), 125–128.

[29] G.R.W. Quispel, F.W. Nijhoff, H.W. Capel and J. van der Linden, Linear Integral Equations and Nonlinear Difference-Difference Equations, *Physica* **125A** (1984), 344–380.

[30] L.D. Faddeev and L.A. Takhtajan, Liouville Model on the Lattice, *Springer Lect. Notes Phys.* **246** (1986), 166–179.

[31] A. Bobenko and U. Pinkall, Discrete Isothermic Surfaces, preprint SFB 288, (November 1994).

[32] A.Yu. Volkov, Miura Transformation on the Lattice, *Theor. Math. Phys.* **74** (1988), 96–99.

[33] B. Grammaticos, A. Ramani and V.G. Papageorgiou, Do integrable mappings have the Painlevé property?, *Phys. Rev. Lett.* **67** (1991), 1825–1828.

[34] A. Ramani and B. Grammaticos, Miura Transforms for Discrete Painlevé Equations, *J. Phys. A: Math. Gen.* **25** (1992), L633–637.

[35] G. Wilson, On the Adler-Gel'fand-Dikii Bracket, *Proc. of the CRM Workshop on "Hamiltonian Systems, Transformation Groups and Spectral Transform Methods"*, Eds. J. Harnad and J.E. Marsden, *CRM Lecture Notes Series 1990*, pp. 77–85.

[36] G.R.W. Quispel and H.W. Capel, The Nonlinear Schrödinger Equation and the Anisotropic Heisenberg Spin Chain, *Phys. Lett.* **88A** (1982), 371–374; *Physica* **117A** (1983), 76–102.

[37] F.W. Nijhoff, G.R.W. Quispel, J. van der Linden and H.W. Capel, On some linear integral equations generating solutions of nonlinear partial differential equations, *Physica* **119A** (1983), 101–142.

[38] F.W. Nijhoff, V.G. Papageorgiou, H.W. Capel and G.R.W. Quispel, The Lattice Gel'fand-Dikii Hierarchy, *Inv. Probl.* **8** (1992), 597–621.

[39] F.W. Nijhoff, The Lattice Boussinesq Equation and the Discrete Painlevé IV Equation, in preparation.

[40] G.R.W. Quispel, F.W. Nijhoff and H.W. Capel, Linearization of the Boussinesq Equation and of the Modified Boussinesq Equation, *Phys. Lett.* **91A** (1982), 143–145.

[41] J. Weiss, The Painlevé property and Bäcklund transformations for the sequence of Boussinesq equations, *J. Math. Phys.* **26** (1985), 258-269.

[42] F.W. Nijhoff, H.W. Capel, G.L. Wiersma and G.R.W. Quispel, Bäcklund Transformations and Three-Dimensional Lattice Equations, *Phys. Lett.* **105A** (1984), 267–272.

[43] F.W. Nijhoff, Similarity Reduction of the Lattice KP Equation, in preparation.

[44] F.W. Nijhoff and H.W. Capel, The Direct Linearisation Approach to Hierarchies of Intgerable PDE's in 2+1 Dimensions, *Inv. Probl.* **8** (1990), 567–590.

Department of Applied Mathematical Studies
The University of Leeds, Leeds LS2 9JT, UK

Received November, 1995

Poisson Brackets
for Integrable Lattice Systems

W. Oevel

In remembrance of our dear colleague Irene Dorfman.

Abstract

Poisson brackets associated with Lax operators of lattice systems are considered. Linear brackets originate from various r-matrices on the algebra of (pseudo-) shift symbols. Quadratic brackets are investigated which provide Hamiltonian formulations for various reductions of the (modified) Lattice KP hierarchy.

1. Introduction

The investigation of (multi-)Hamiltonian formulations probably is the simplest and most direct approach to integrable systems. The Hamiltonian version of Noether's theorem states that the conserved 1-forms are in unique correspondence with the generators of symmetries. The map between the invariant co-vectorfields (in particular, the gradients of conservation laws) and invariant vectorfields (symmetries) is given by the Poisson tensor associated with the Poisson bracket. If two distinct tensors $\mathcal{P}_{1,2}$ are known for an equation, then the simple recursive scheme $\mathcal{P}_1 \nabla H_i = \mathcal{P}_2 \nabla H_{i-1}$ may be used to construct a sequence of Hamiltonian functions H_i in involution. These constitute conservation laws for that equation, the associated Hamiltonian vectorfields are the symmetry generators. The bi-Hamiltonian structure was invented and investigated in the fundamental papers [1] by Irene Dorfman and Gelfand and independently by Magri [2] and Fuchssteiner and Fokas [3]. Subsequently much work by many authors has been devoted to this machinery: we refer to Irene's important monograph [4] for an overview of the theory and for references.

Nowadays one thinks of a Poisson bracket as a geometrical quantity associated with the underlying phase space rather than as a structure associated with individual equations. In the context of an integrable system it is natural to connect the bracket with its Lax representation $L_t = [A, L]$, where L lies in a suitable operator algebra. Indeed, for the celebrated KdV hierarchy and its generalizations the fundamental papers [5] by Gelfand

and Dikii provided a construction based on the Lax operators which are to be regarded as elements of the algebra of pseudo–differential symbols. This approach also works in a context where bi-Hamiltonian formulations and recursion operators do not exist in the usual sense [6].

Generalizations of these results culminated in a Lie algebraic setting now referred to as the Adler-Kostant-Symes scheme [7], where the construction of an integrable hierarchy and its (first) Hamiltonian formulation originates solely from the properties of the underlying algebra. An interpretation in terms of classical r-matrices and a simple representation of a second ("quadratic") Poisson bracket was given by Semenov [8]. However, certain technical restrictions apply in the construction of the quadratic bracket. Semenov's original formula used "unitary" (skew-symmetric) r-matrices. This restriction was partially resolved in Refs. [9] and [10], where formulas for quadratic Poisson brackets were given for more general r-matrices:

$$\mathcal{P}_1(L): \quad \nabla H \quad \rightarrow \quad \tfrac{1}{2}\left([r(\nabla H), L] + r^*([\nabla H, L]) \right),$$
$$\mathcal{P}_2(L): \quad \nabla H \quad \rightarrow \quad \tfrac{1}{4}\left([r(L\nabla H + \nabla HL), L] + Lr^*([\nabla H, L]) \right. \qquad (1.1)$$
$$\left. + r^*([\nabla H, L])L \right)$$

are two compatible Poisson tensors that can be associated with Lax operators. Here L may be an element of an arbitrary associative algebra g endowed with a bi-invariant metric. The linear tensor $\mathcal{P}_1$ is a Poisson tensor for arbitrary r-matrices. Severe technical assumptions still apply to grant Poisson properties of the quadratic tensor: both r and its skew-symmetric part need to solve modified Yang–Baxter equations (for details see Refs. [9, 10], where also a cubic bracket was given).

These restrictions implied serious technical difficulties in the systematic construction of the quadratic brackets associated with such simple models as the modified KdV equation or the relativistic Toda equation [10].

A more general construction of quadratic brackets was recently proposed by Suris:

Theorem 1 [11] On any associative algebra $g = g^*$ equipped with a nondegenerate symmetric invariant metric $\langle a, bc \rangle = \langle ab, c \rangle = \langle c, ab \rangle$ the quadratic tensor

$$\mathcal{P}_S(L): \quad \nabla H \rightarrow \tfrac{1}{2}\left(A_1(L\nabla H)\, L - LA_2(\nabla HL) + S(\nabla HL)L - LS^*(L\nabla H)\right) \tag{1.2}$$

defines a Poisson structure, if the linear maps $A_{1,2}: g \rightarrow g$ are skew-symmetric solutions of the modified Yang–Baxter equation

$$[A(\xi), A(\eta)] + [\xi, \eta] = A([A(\xi), \eta] + [\xi, A(\eta)]) \tag{1.3}$$

and the linear map $S : g \to g$ with adjoint S^* satisfies

$$
\begin{aligned}
S\left(\,[A_2(\xi), \eta] + [\xi, A_2(\eta)]\,\right) &= [\,S(\xi), S(\eta)\,], \\
S^*\left(\,[A_1(\xi), \eta] + [\xi, A_1(\eta)]\,\right) &= [S^*(\xi), S^*(\eta)]
\end{aligned}
\tag{1.4}
$$

for all $\xi, \eta \in g$.

Since two r-matrices $A_{1,2}$ are involved, this result provides more flexibility in constructing suitable quadratic brackets on the underlying algebra of Lax operators. For many important examples the connection to the r-matrix giving rise to the first (linear) Poisson bracket is given by

$$
r = A_1 + S = A_2 + S^* ,
$$

i.e., the maps $A_1, \ldots, S^*$ originate from decompositions of the r-matrix. In this case the Poisson bracket

$$
\begin{aligned}
\{H_1, H_2\}(L) &= \langle \nabla H_2, \mathcal{P}_S(L) \nabla H_1 \rangle \\
&= \langle L \nabla H_2, (A_1 + S)(L \nabla H_1) \rangle + \langle L \nabla H_2, S([\nabla H_1, L]) \rangle \\
&\quad - \langle \nabla H_2 L, (A_2 + S^*)(\nabla H_1 L) \rangle + \langle \nabla H_2 L, S^*([\nabla H_1, L]) \rangle
\end{aligned}
$$

vanishes for all pairs of Casimir functions H_i, (i.e., $[\nabla H_i, L] = 0$). The Casimirs form the set of commuting integrals for the integrable system under consideration. The Hamiltonian system associated with a Casimir C is the Lax equation

$$
L_t = \mathcal{P}_S \nabla C = \tfrac{1}{2} [r(L \nabla C), L] .
$$

Further, the following technical result is useful[1]:

Theorem 2 *Let $r = A_1 + S = A_2 + S^*$ satisfy the modified Yang–Baxter equation (1.3). If $A_{1,2}$ are skew symmetric w.r.t. the underlying bi-invariant metric, then (1.4) imply that both A_1 and A_2 satisfy (1.3).*

Proof. From the modified Yang–Baxter equation for $r = A_1 + S$ and using (1.4) one derives

$$
\begin{aligned}
\Delta(\xi, \eta) &:= [A_1(\xi), A_1(\eta)] + [\xi, \eta] - A_1([A_1(\xi), \eta] + [\xi, A_1(\eta)]) \\
&= S([S^*(\xi), \eta] + [\xi, S^*(\eta)]) + A_1([S(\xi), \eta] \\
&\quad + [\xi, S(\eta)]) - [S(\xi), A_1(\eta)] - [A_1(\xi), S(\eta)] .
\end{aligned}
$$

The invariance of the metric and skew symmetry of A_1 leads to the identity

$$
\begin{aligned}
\langle \Delta(\xi, \eta), \theta \rangle &= \langle [S^*(\eta), S^*(\theta)] - S^*([A_1(\eta), \theta] + [\eta, A_1(\theta)]), \xi \rangle \\
&\quad + \langle [S^*(\theta), S^*(\xi)] - S^*([A_1(\theta), \xi] + [\theta, A_1(\xi)]), \eta \rangle
\end{aligned}
$$

with arbitrary $\xi, \eta, \theta \in g$. The corresponding identity holds for A_2 and S. $\qquad\square$

---

[1] A related statement is given in Theorem 3 of Ref. [11].

This observation reduces the verification of the technical condition (1.3) for $A_{1,2}$ to (1.4). The latter relations are often more easily checked.

In Section 2 we review the construction of Poisson brackets for reductions of the KP hierarchy and the modified KP hierarchy. Suris' bracket is crucial in these considerations.

In Section 3 discrete versions of these structures are presented generalizing Suris' original results which he applied in a discrete context. The underlying set of dynamical equations is the (modified) Lattice KP hierarchy, a systematic construction of the Poisson brackets associated with various constraints is given. Several Poisson brackets are naturally associated with some simple r-matrices. Results concerning the reduction properties of these brackets to finite Lax operators are presented. In particular, the Hamiltonian structures of the Toda model and the relativistic Toda model are described as restrictions of the same Poisson structure to different integral manifolds.

2. Review of Poisson Brackets for Continuous Soliton Equations

2.1. Poisson Brackets for Constrained KP Flows

In their fundamental papers [5] Gelfand and Dikii considered the Hamiltonian structures of the soliton equations associated with (scalar) Lax operators of the type

$$L = \partial^N + u_{N-2}\partial^{N-2} + u_{N-1}\partial^{N-1} + \cdots + u_0 . \qquad (2.1)$$

The corresponding hierarchies of generalized KdV equations for the fields $u_0, \ldots, u_{N-2}$ are encoded in the Lax equations

$$L_{t_q} = [(L^q)_{\geq 0}, L] , \qquad (2.2)$$

where the subscript ≥ 0 denotes the projection of the pseudo-differential fractional powers L^q onto their nonnegative differential orders. In particular, the case $N = 2$ represents the KdV hierarchy proper, $N = 3$ yields the Boussinesq hierarchy etc. The bi-Hamiltonian formulations for these equations arise from two abstract Poisson brackets associated with the Lax operator which are given by the following compatible Poisson tensors

$$\begin{aligned} \mathcal{P}_1(L): \quad \nabla H \quad &\to \quad [L, (\nabla H)_{<0}]_{\geq 0} , \\ \mathcal{P}_2(L): \quad \nabla H \quad &\to \quad (L\nabla H)_{\geq 0}L - L(\nabla H\,L)_{\geq 0} . \end{aligned} \qquad (2.3)$$

The linear tensor is the Lie Poisson structure of the Adler-Kostant-Symes scheme [7], whereas $\mathcal{P}_2$ corresponds to Semenov's quadratic bracket. Indeed, consider the algebra of pseudo-differential symbols

$$g = g_{\geq 0} \oplus g_{<0} = \left\{ \sum_{n\geq 0} u_n\partial^n \right\} \oplus \left\{ \sum_{n<0} u_n\partial^n \right\}$$

with the r-matrix $r(\xi) = \xi_{\geq 0} - \xi_{<0}$. Due to the unitarity (skew-symmetry) of r with respect to Adler's trace duality [7]

$$\langle \xi^*, \xi \rangle = \mathrm{tr}(\xi^*\xi), \quad \mathrm{tr}(\xi) = \mathrm{tr}\left(\sum_n u_n \partial^n\right) := \int u_{-1}(x)\, dx$$

both tensors (1.1) represent compatible Poisson brackets on g. These maps are nothing but (2.3). For the Casimir functions $H = \mathrm{tr}(L^{q+1})/(q+1)$ (i.e., $\nabla H = L^q$) the associated Hamiltonian vector fields are the Lax equations (2.2), which give rise to the KP hierarchy for a general pseudo-differential $L \in g$. The restriction to Lax operators of the form (2.1) represents a reduction to the integrable equations of the Gelfand-Dikii hierarchy in 1+1 dimensions for which the bi-Hamiltonian formulations are given by (1.1)/(2.3). The linear Poisson tensor admits a proper restriction to operators of the form (2.1), where (1.1.$\mathcal{P}_1$) yields (2.3.$\mathcal{P}_1$). The quadratic tensor will in general produce a term of order ∂^{N-1}, so that a term $u_{N-1}\partial^{N-1}$ has to be included in (2.1). Dirac reduction has to be used to generate a bracket preserving the constraint $u_{N-1} = 0$. Usually such a reduction procedure gives rise to nonlocal terms (integrations) in the Poisson structure. In the present case, however, this does not happen. This is explained by the following representation

$$\mathcal{P}_2^{(red)}(L): \nabla H \;\rightarrow\; (L\nabla H)_{\geq 0}L - L(\nabla HL)_{\geq 0} + \tfrac{1}{N}\,[D^{-1}(\mathrm{res}([\nabla H, L])), L] \tag{2.4}$$

of the reduced tensor, where $\mathrm{res}(\xi) = \mathrm{res}(\sum_n u_n \partial^n) = u_{-1}$ is the usual residue and $D^{-1} = \tfrac{1}{2}(\int^x - \int_x)$ is an integration stemming from the reduction. As the residue of a commutator is in the image of the differential operator [7], the formal integration yields some purely differential expression of the coordinates, wence the reduced bracket is again local. It provides the "second" Hamiltonian operator of the Gelfand-Dikii hierarchies. We note the explicit dependence of $\mathcal{P}_2^{(red)}$ on N, through which these tensors can be restricted properly to Nth order operators with $u_N = 1$ and $u_{N-1} = 0$ (a Poisson submanifold).

Suris' result as presented in Theorem 1 leads to a direct construction of (2.4). With

$$A_1 = A_2 = r: \xi \;\rightarrow\; \xi_{\geq 0} - \xi_{<0}, \quad S = S^* = 0$$

the tensor $\mathcal{P}_S(L)$ coincides with (1.1.$\mathcal{P}_2$)/(2.3.$\mathcal{P}_2$). Alternatively, with

$$A_1: \xi \;\rightarrow\; \xi_{\geq 0} - \xi_{<0} - \tfrac{1}{N} D^{-1}(\mathrm{res}(\xi)), \quad S = \tfrac{1}{N} D^{-1}(\mathrm{res}(\xi)),$$

$$A_2: \xi \;\rightarrow\; \xi_{\geq 0} - \xi_{<0} + \tfrac{1}{N} D^{-1}(\mathrm{res}(\xi)), \quad S^* = -\tfrac{1}{N} D^{-1}(\mathrm{res}(\xi))$$

the tensor $\mathcal{P}_S$ produces the reduced Poisson structure (2.4). In all cases one has

$$r = A_1 + S = A_2 + S^*: \quad \xi \;\rightarrow\; \xi_{\geq 0} - \xi_{<0}, \tag{2.5}$$

where the r-matrix r trivially satisfies the modified Yang–Baxter equation, since it stems from a Lie algebra decomposition. The technical assumptions (1.4) are easily checked.

Recently there has been much interest in reductions of the KP hierarchy via the symmetry generated by squared eigenfunctions (e.g., [12]–[17]). It was found that the resulting constrained KP flows (2.2) are characterized by pseudo-differential Lax operators of the form

$$L = \partial^N + u_{N-2}\partial^{N-2} + \cdots + u_0 + \sum_{i=1}^{K} \phi_i \partial^{-1} \psi_i, \qquad (2.6)$$

which constitute an invariant operator manifold for the KP dynamics. The evolution (2.2) implies that ϕ_i and ψ_i are eigenfunctions and adjoint eigenfunctions of the KP hierarchy satisfying

$$\phi_{it_q} = ((L^q)_{\geq 0}\phi_i), \quad \psi_{it_q} = -(((L^q)_{\geq 0})^* \psi_i), \quad i = 1, \ldots, K. \qquad (2.7)$$

These, however, are not linear equations, since the fields ϕ_i, ψ_i enter non-linearly via L^q. The corresponding $1 + 1$-dimensional hierarchies for the fields $u_0, \ldots, u_{N-2}$, $\phi_1, \ldots, \phi_K$, $\psi_1, \ldots, \psi_K$ are bi-Hamiltonian. Indeed, the operators (2.6) form a proper Poisson submanifold for both the linear tensor $(1.1.\mathcal{P}_1)$ and the quadratic tensor (2.4) associated with the standard r-matrix (2.5). A simple restriction to (2.6) provides the Hamiltonian structures. Explicit formulas can be found in Ref. [18].

2.2. Poissonbracket for Constrained Modified KP Flows

In Ref. [19] the bi-Hamiltonian construction was extended to reductions of the "nonstandard" Lax equations [20] given by

$$L_{t_q} = [(L^q)_{\geq 1}, L], \qquad (2.8)$$

where — in contrast to the KP case — zero order terms are excluded from $(L^q)_{\geq 1}$. In the remainder of this section we review the results of Ref. [19].

For a generic pseudo-differential $L = \partial + v_0 + v_{-1}\partial^{-1} + \cdots$ the resulting equations for the field v_0 constitute the $2 + 1$-dimensional hierarchy of modified KP flows [21, 22, 23], so that (2.8) shall be referred to as the modified KP hierarchy. These equations have been studied extensively in Refs. [20]–[24], where also a further class of integrable Dym-type equations is considered.

It is easily verified that the modified dynamics (2.8) may be restricted to operator manifolds of the form

$$L = \partial^N + v_{N-1}\partial^{N-1} + v_{N-2}\partial^{N-2} + \cdots + v_1\partial \qquad (2.9)$$

or

$$L = \partial^N + v_{N-1}\partial^{N-1} + v_{N-2}\partial^{N-2} + \cdots + v_1\partial + v_0 + \partial^{-1}\Psi \qquad (2.10)$$

or

$$L = \partial^N + v_{N-1}\partial^{N-1} + v_{N-2}\partial^{N-2} + \cdots + v_1\partial + v_0 + \partial^{-1}\Psi + \sum_{i=1}^{K} \phi_i\partial^{-1}\psi_i .$$

$$(2.11)$$

Operators of the form (2.9) arise from gauge transformations $\tilde{L} = \Phi^{-1}L\circ\Phi$, where L is given by (2.1). Indeed, if Φ is an eigenfunction satisfying $L\Phi = 0$, then $\tilde{L}$ is of the form (2.9). Further, if Φ satisfies $\Phi_{t_q} = ((L^q)_{\geq 0}\Phi)$, then the dynamics (2.2) yields (2.8) for $\tilde{L}$ (see e.g. Ref. [23]). Similarly, operators of the form (2.10)/(2.11) arise from gauge transformations of the operators (2.6)[18].

The underlying r-matrix is again built on the algebra g of pseudo-differential symbols, this time with a decomposition $g = g_{\geq 1} \oplus g_{<1}$ into strictly positive differential orders versus nonpositive orders. The resulting linear map $\tilde{r} : \xi \to \xi_{\geq 1} - \xi_{<1}$ is again an r-matrix, since both $g_{\geq 1}$ as well as $g_{<1}$ are sub-(Lie)algebras of g. The relevance of this "nonstandard" decomposition in the context of soliton systems had already been observed by Reiman [25].

The linear tensor $(1.1.\mathcal{P}_1)$ with $\tilde{r}$ gives rise to the Poisson tensor

$$\tilde{\mathcal{P}}_1(L) : \nabla H \quad \to \quad [\nabla H, L]_{<-1} - [(\nabla H)_{<1}, L]$$
$$= [(\nabla H)_{\geq 1}, L]_{<-1} - [(\nabla H)_{<1}, L]_{\geq -1} \qquad (2.12)$$

which reduces to $\tilde{\mathcal{P}}_1(L) = [L, (\nabla H)_{<1}]_{\geq -1}$ for $L \in g_{\geq -1} = (g_{<1})^*$, i.e., for operators of the form (2.10). The tensor (2.12) provides a first Hamiltonian formulation for the flows (2.8) with the Hamiltonians $H = \text{tr}(L^{q+1})/(q+1)$. The fact that the modified KP dynamics can be restricted to operators of the form (2.10) and (2.11), respectively, is explained by its Hamiltonian nature: these operators constitute integral manifolds of the linear bracket.

Theorem 3 *[19] The linear Poisson tensor (2.12) admits a proper restriction to Lax operators of the form (2.10) and (2.11).*

However, no proper Hamiltonian restriction to operators of the form (2.9) is possible. Dirac reduction has to be invoked to obtain a Poisson bracket for the integrable hierarchies associated with the Lax operators (2.9).

With these considerations a first Hamiltonian structure for the equations (2.8) has been established via the linear tensor (2.12).

In the construction of a quadratic Poisson bracket a serious problem arises. As discussed in Ref. [10], Semenov's quadratic tensor $(1.1.\mathcal{P}_2)$ does *not* provide a Poisson bracket for the modified r-matrix $\tilde{r}$, since it is neither skew-symmetric nor does its skew-symmetric part satisfy the Yang–Baxter equation. However, the more general construction of Theorem 1 is capable of producing suitable brackets. It can be verified that the maps

$$
\begin{aligned}
A_1 &: \xi \;\rightarrow\; \xi_{\geq 1} - \xi_0 + \partial^{-1}\mathrm{res}(\xi) - \xi_{<-1} - 2\,D^{-1}(\mathrm{res}(\xi))\,, \\
A_2 &: \xi \;\rightarrow\; \xi_{\geq 0} - \xi_{<0} + 2\,D^{-1}(\mathrm{res}(\xi))\,, \\
S &: \xi \;\rightarrow\; -2\,\partial^{-1}\mathrm{res}(\xi) + 2\,D^{-1}(\mathrm{res}(\xi))\,, \\
S^* &: \xi \;\rightarrow\; -2\,\xi_0 - 2\,D^{-1}(\mathrm{res}(\xi))
\end{aligned}
$$

$$(2.13)$$

formally satisfy the assumptions of Theorem 1. Here $\xi_0, \xi_{<-1}$ etc. are the terms of differential order 0, less than -1 etc. We note

$$
\tilde{r} = A_1 + S = A_2 + S^* : \quad \xi \rightarrow \xi_{\geq 1} - \xi_{<1}\,,
$$

so that according to Theorem 2 only the technical conditions (1.4) need to be checked. The tensor (1.2) results in

$$
\begin{aligned}
\tilde{\mathcal{P}}_S(L) \;:\; \nabla H \;\rightarrow\;\; &(L\nabla H)_{\geq 1} L - L(\nabla H\,L)_{\geq 0} + L(L\,\nabla H)_0 \\
&-\partial^{-1}\mathrm{res}([\nabla H, L])\,L + [D^{-1}(\mathrm{res}([\nabla H, L]), L]
\end{aligned}
$$

$$(2.14)$$

which was found in Ref. [18]. Swapping A_1 and A_2 and simultaneously replacing S by S^* the tensor (1.2) results in

$$
\begin{aligned}
\tilde{\mathcal{P}}_S'(L) \;:\; \nabla H \;\rightarrow\;\; &(L\,\nabla H)_{\geq 0} L - L\,(\nabla H\,L)_{\geq 1} - (\nabla H\,L)_0\,L \\
&-L\,\partial^{-1}\mathrm{res}([\nabla H, L]) - [D^{-1}(\mathrm{res}([\nabla H, L]), L]\,.
\end{aligned}
$$

$$(2.15)$$

Both $\tilde{\mathcal{P}}_S$ and $\tilde{\mathcal{P}}_S'$ are compatible with (2.12). This follows from the usual argument[2] that the deformation $L \rightarrow L + \epsilon 1$ of both quadratic tensors produces the linear tensor as the term linear in ϵ. Both $\tilde{\mathcal{P}}_S$ and $\tilde{\mathcal{P}}_S'$ play a role as quadratic Hamiltonian structures for the dynamics (2.8):

$$
L_{t_q} = [(L^q)_{\geq 1}, L] = \tilde{\mathcal{P}}_S(L)\,\nabla H = \tilde{\mathcal{P}}_S'(L)\,\nabla H\,, \quad H(L) = \mathrm{tr}(L^q)/q\,.
$$

However, when restricting these equations to Lax operators of a specified form, the two tensors behave differently:

Theorem 4 *[19] The tensor (2.14) admits a proper restriction to Lax operators of the form (2.10) and (2.11).*

---

[2]See also Theorem 3 of Ref. [11]: with $\tilde{r} = A_1 + S = A_2 + S^*$ Suris' quadratic bracket is automatically compatible with the linear bracket generated by $\tilde{r}$.

No proper restriction to (2.9) is possible. Dirac reduction has to be invoked to obtain the quadratic bracket associated with the constraint $v_0 = \Psi = 0$. It turns out that the reduction of (2.14) to the submanifold (2.9) is identical with (2.15):

Theorem 5 *[19] The tensor (2.15) admits a proper restriction to Lax operators of the form (2.9), where it coincides with the Dirac reduction of (2.14).*

We note that the integration D^{-1} is not well-defined in (2.13), so that the argument remains formal. However, in (2.14) as well as in (2.15) this problem does not arise, as the residues of the commutators can be integrated properly. The resulting Poisson brackets are local.

3. Lattice Systems

It was shown in Ref. [11] that a unified description of the Hamiltonian structures of the Toda model as well as the relativistic Toda model is provided by the construction of Theorem 1. We return to Suris' original considerations in a slightly more general context and generalize the results of Ref. [11]. Each of the results presented in the previous section has a discrete analog.

3.1. Notation

We consider the algebra of shift operators

$$g = \left\{ u_N(n)T^N + u_{N-1}(n)T^{N-1} + \cdots + u_0(n) + u_{-1}(n)T^{-1} + \cdots \right\}$$

with commuting coefficients u_j from some commutative algebra g_0. They depend on a discrete lattice index n. The action of the shift symbols on functions (of n) is given by $(T^k u)(n) = u^{(k)}(n) := u(n+k)$. Projections to various shift orders are denoted by

$$\xi_j = u_j T^j, \quad \xi_{\geq k} = \sum_{j \geq k} u_j T^j, \quad \xi_{<k} = \sum_{j<k} u_j T^j$$

for $\xi = \sum_j u_j T^j \in g$. The (pseudo-) difference symbol

$$\Delta^{-1} = T^{-1} + T^{-2} + \cdots$$

is introduced which may be regarded as the formal inverse of the difference operator $\Delta = T - 1$.

We use the trace form $\operatorname{tr}(\xi) = \sum_n \xi_0(n)$ to identify g and its dual g^* via $\langle \xi^*, \xi \rangle = \operatorname{tr}(\xi^* \xi)$. This metric is bi-invariant due to the symmetry of the trace. Indeed, the zero order term of a commutator is in the image of the difference operator:

$$
\begin{aligned}
[\xi, \eta]_0 &= \sum_j \left(u_j v_{-j}^{(j)} - v_j u_{-j}^{(j)} \right) \\
&= \sum_{j>0} \left(u_j v_{-j}^{(j)} + u_{-j} v_j^{(-j)} - v_j u_{-j}^{(j)} - v_{-j} u_j^{(-j)} \right) \\
&= \sum_{j>0} \left((T^j - 1)(v_{-j} u_j^{(-j)} - u_{-j} v_j^{(-j)}) \right) = \left((T-1)s \right)
\end{aligned}
$$

with $\xi = \sum_j u_j T^j$, $\eta = \sum_j v_j T^j$ and

$$
s = \sum_{j>0} \left((1 + T + \cdots + T^{j-1})(v_{-j} u_j^{(-j)} - u_{-j} v_j^{(-j)}) \right).
$$

Consequently, $\operatorname{tr}(\xi \eta - \eta \xi) = \sum_n [\xi, \eta]_0 = \sum_n \left((T-1)s \right) = 0$.

3.2. The Lattice KP Hierarchy and its Reductions

We consider the hierarchy of commuting Lax equations [26]

$$
\begin{aligned}
L_{t_q} &= \left[(L^q)_{\geq 1} + \tfrac{\alpha+1}{2} (L^q)_0, L \right] \\
&= -\left[(L^q)_{<0} + \tfrac{1-\alpha}{2} (L^q)_0, L \right]
\end{aligned}
\tag{3.1}
$$

with $L \in g$ and some scalar parameter α. The corresponding set of nonlinear differential-difference equations shall be called the Lattice KP hierarchy. The underlying r-matrix is

$$
r_\alpha(\xi) = \xi_{\geq 1} + \alpha \xi_0 - \xi_{<0}
$$

which originates from a "triangular" decomposition $g = g_{\geq 1} \oplus g_0 \oplus g_{<0}$ of the operators into "strictly upper", diagonal and "strictly lower" parts. We note that simple r-matrices arise from such a decomposition, defined as (minus) the identity on the strictly upper (lower) subalgebra. The action on the diagonal may be given by an arbitrary r-matrix on g_0 which, in the present case, is just multiplication by α.

One obtains a family of linear Poisson brackets $(1.1.\mathcal{P}_1)$:

$$
\mathcal{P}_1(L): \nabla H \rightarrow
$$

$$
\begin{aligned}
&[\nabla H_{\geq 1}, L] - [\nabla H, L]_{\geq 1} + \tfrac{\alpha+1}{2} [\nabla H_0, L] + \tfrac{\alpha-1}{2} [\nabla H, L]_0 \\
&= -[\nabla H_{<0}, L] + [\nabla H, L]_{<0} + \tfrac{\alpha-1}{2} [\nabla H_0, L] + \tfrac{\alpha+1}{2} [\nabla H, L]_0.
\end{aligned}
\tag{3.2}
$$

The Casimir function $H = \text{tr}(L^{q+1})/(q+1)$ with $\nabla H = L^q$ produces (3.1) as associated Hamiltonian system, so that $\mathcal{P}_1$ provides the first Hamiltonian formulation for these equations. The dynamics admits a Hamiltonian restriction to finite Lax operators:

Theorem 6 *The linear tensor (3.2) admits a proper restriction to Lax operators of the form*

$$L = u_N T^N + \cdots + u_M T^M + \sum_{i=1}^{K} \phi_i T^M \Delta^{-1} \psi_i, \quad N \geq 0 \geq M. \quad (3.3)$$

Proof. From the second representation of (3.2) it is clear that the highest shift order of $\mathcal{P}_1(L)\nabla H$ coincides with the highest shift order of L, provided it is nonnegative. We note the identities

$$(A\phi T^M \Delta^{-1})_{<M} = (A\phi) T^M \Delta^{-1}, \quad (T^M \Delta^{-1} \psi A)_{<M} = T^M \Delta^{-1} (A^* \psi) \quad (3.4)$$

which hold for arbitrary functions ϕ, ψ and symbols $A \in g_{\geq 0}$. From the first representation of (3.2) one computes for $M \leq 0$

$$
\begin{aligned}
(\mathcal{P}_1(L)\nabla H)_{<M} &= [B, \textstyle\sum_i \phi_i T^M \Delta^{-1} \psi_i]_{<M} \\
&\overset{(3.4)}{=} \textstyle\sum_i (B\phi_i) T^M \Delta^{-1} \psi_i - \sum_i \phi_i T^M \Delta^{-1} (B^* \psi_i)
\end{aligned}
$$

with $B = \nabla H_{\geq 1} + \frac{\alpha+1}{2} \nabla H_0 \in g_{\geq 0}$, so that $\mathcal{P}_1(L)\nabla H$ is tangent to L. □

From the proof it is clear that the functions ϕ_i, ψ_i inherit the dynamics

$$\phi_{it_q} = (B\phi_i), \quad \psi_{it_q} = -(B^* \psi_i), \quad B = \nabla H_{\geq 1} + \frac{\alpha+1}{2} \nabla H_0.$$

The parameter values $\alpha = \pm 1$ are special in that the corresponding r-matrices stem from simple Lie algebra decompositions $g = g_{\geq 0} \oplus g_{<0}$ $(\alpha = 1)$ and $g = g_{\geq 1} \oplus g_{<1}$ $(\alpha = -1)$, respectively. Both values lead to the same nonlinear equations since transposition of the Lax operators provides the link between the cases $\alpha = 1$ and $\alpha = -1$:

$$L_{t_q} = [(L^q)_{\geq 0}, L] \implies$$

$$L^*_{t_q} = -[((L^q)_{\geq 0})^*, L^*] = -[(L^{*q})_{\leq 0}, L^*] = [(L^{*q})_{\geq 1}, L^*].$$

For $\alpha = 1$ the highest coefficient in the Lax operator may be constrained. By transposition, the same is true for the lowest coefficient (if existent) for $\alpha = -1$:

Theorem 7   a) *For $\alpha = 1$ and $N \geq 1 \geq M$ the linear tensor (3.2) admits a proper restriction to Lax operators of the form*

$$L = T^N + u_{N-1}T^{N-1} + \cdots + u_M T^M + \sum_{i=1}^{K} \phi_i T^M \Delta^{-1} \psi_i . \quad (3.5)$$

b) *For $\alpha = -1$, $N > M$ and $0 > M$ the linear tensor (3.2) admits a proper restriction to Lax operators of the form*

$$L = u_N T^N + \cdots + u_{M+1} T^{M+1} + T^M .$$

These statements are easily verified with the first and second representation of (3.2).

We now turn to a discussion of the quadratic Poisson brackets that may be associated with the Lax equations (3.1). We note that the skew-symmetric part

$$\tfrac{1}{2}\left(r_\alpha - r_\alpha^*\right) = r_0 : \ \xi \ \to \ \xi_{\geq 1} - \xi_{<0}$$

of the r-matrix solves the modified Yang–Baxter equation (1.3), so that Semenov's quadratic tensor $(1.1.\mathcal{P}_2)$

$$\mathcal{P}_2(L): \ \nabla H \ \to \ (L\nabla H)_{\geq 1} L - L(\nabla H L)_{\geq 1} + \tfrac{1}{2}(L\nabla H)_0 L - \tfrac{1}{2}L(\nabla H L)_0$$
$$+ \tfrac{\alpha}{2}(\nabla H L)_0 L - \tfrac{\alpha}{2}L(L\nabla H)_0$$
$$= -(L\nabla H)_{<0}L + L(\nabla H L)_{<0} - \tfrac{1}{2}(L\nabla H)_0 L + \tfrac{1}{2}L(\nabla H L)_0$$
$$+ \tfrac{\alpha}{2}(\nabla H L)_0 L - \tfrac{\alpha}{2}L(L\nabla H)_0$$
$$(3.6)$$

is indeed a Poisson tensor on g. The Casimir function $H = \mathrm{tr}(L^q)/q$ yields (3.1) as the associated Hamiltonian equation. The compatibility with the linear bracket is provided by the usual deformation $L \to L + \epsilon 1$. For any value of α a proper restriction to finite Lax operators of the form (3.3) is possible:

Theorem 8 *Semenov's quadratic Poisson tensor (3.6) admits a proper restriction to Lax operators of the form (3.3).*

The proof is straightforward using the two representations (3.6) and an argument similar to that in the proof of Theorem 6.

For the case $\alpha = 1$ the constraint $u_N = 1$ is not left invariant by the tensor (3.6), whence Dirac reduction has to be invoked to obtain a quadratic Poisson bracket for Lax operators of the form (3.5). Despite the reduction process the result is a local Hamiltonian structure. For a direct approach to the reduced bracket we use Theorem 1 to generalize (3.6) for arbitrary α.

Theorem 9 *Let $\rho : g_0 \to g_0$ be an arbitrary skew-symmetric linear map on the algebra g_0 of zero order terms of g. Then*

$$A_1 : \xi \to \xi_{\geq 1} - \xi_{<0} - \rho(\xi_0), \quad S : \xi \to \alpha\xi_0 + \rho(\xi_0),$$

$$A_2 : \xi \to \xi_{\geq 1} - \xi_{<0} + \rho(\xi_0), \quad S^* : \xi \to \alpha\xi_0 - \rho(\xi_0)$$

satisfy all conditions of Theorem 1. The resulting Poisson tensor is

$$\mathcal{P}_S(L)\nabla H = \mathcal{P}_2(L)\nabla H + \tfrac{1}{2}\left[\rho([\nabla H, L]_0), L\right],$$

where $\mathcal{P}_2(L)$ is Semenov's quadratic tensor (3.6). It is compatible with the linear tensor (3.2) and admits a proper restriction to Lax operators of the form (3.3).

The technical conditions (1.4) are easily verified. With

$$A_1 + S = A_2 + S^* = r_\alpha : \xi \to \xi_{\geq 1} + \alpha\xi_0 - \xi_{<0}$$

this grants the Poisson properties of $\mathcal{P}_S$ according to Theorem 2. We note that the additonal term generated by ρ drops out for Casimir functions, so that $\mathcal{P}_S$ generates the same integrable hierarchies as $\mathcal{P}_2$.

In the following we shall concentrate on the case $\alpha = 1$. The two Poisson tensors under consideration are

$$\mathcal{P}_1(L) : \nabla H \quad \to \quad [\nabla H_{\geq 0}, L] - [\nabla H, L]_{\geq 1} \tag{3.7}$$
$$= -[\nabla H_{<0}, L] + [\nabla H, L]_{<1}$$

and

$$\mathcal{P}_S(L) : \nabla H \quad \to \quad (L\nabla H)_{\geq 1}L - L(\nabla H L)_{\geq 1} \tag{3.8}$$
$$+ \frac{1}{2}\left([(L\nabla H + \nabla H L)_0 + \rho([\nabla H, L]_0), L]\right)$$
$$= -(L\nabla H)_{<0}L + L(\nabla H L)_{<0}$$
$$+ \frac{1}{2}\left(L[\nabla H, L]_0 + [\nabla H, L]_0 L + [\rho([\nabla H, L]_0), L]\right)$$

which are compatible for any skew ρ. Since the associated integrable hierarchies $L_{t_q} = [(L^q)_{\geq 0}, L]$ (i.e., (3.1) with $\alpha = 1$) can be restricted to (3.5), one would like both $\mathcal{P}_1$ and $\mathcal{P}_S$ to admit a proper restriction to such Lax operators. For $\mathcal{P}_1$ this is granted by Theorem 7.a). For $\mathcal{P}_S$ the invariance of the constraint $u_N = 1$ can be achieved by an appropriate choice of ρ:

Theorem 10 *With $\rho(\xi_0) = ((T^N + 1)(T^N - 1)^{-1}\xi_0)$ the Poisson tensor (3.8) can be restricted properly to Lax operators of the form (3.5), where it coincides with the Dirac reduction of Semenov's tensor (3.6, $\alpha = 1$).*

Proof. According to Theorem 9 operators of the form (3.3) constitute an integral manifold of $\mathcal{P}_S$, whence it suffices to show that the constraint $u_N = 1$ is left invariant. From the second representation (3.8) the Nth shift order of $\mathcal{P}_S(L)\nabla H$ vanishes ($\xi_0 = [\nabla H, L]_0$):

$$T^N \xi_0 + \xi_0 T^N + [\rho(\xi_0), T^N] = (\xi_0^{(N)} + \xi_0 + \rho(\xi_0) - \rho^{(N)}(\xi_0)) T^N = 0.$$

For the Dirac reduction of (3.6) induced by the constraint $u_N = 1$ one computes

$$((\mathcal{P}_2(L)\nabla H)_N = S(h^{(-N)} - h + [\nabla H_{>-N}, L]_0) T^N$$

with $S : \xi_0 \to \frac{1}{2}(\xi_0 + \xi_0^{(N)})$ and $\nabla H = \nabla H_{>-N} + T^{-N} h$ (i.e., $h = \frac{\delta H}{\delta u_N}$). Further,

$$\begin{aligned}
(\mathcal{P}_2(L)\nabla H)_{<N} &= (\mathcal{P}_2(L)\nabla H_{>-N})_{<N} + (\mathcal{P}_2(L)T^{-N}h)_{<N} \\
&= (\mathcal{P}_2(L)\nabla H_{>-N})_{<N} + \frac{1}{2}[h^{(-N)} + h, L]_{<N}.
\end{aligned}$$

The reduced tensor is obtained from elimination of h, when the constraint $\dot{u}_N = (\mathcal{P}_2(L)\nabla H)_N = 0$ is imposed. Insertion of $h = ((1 - T^{-N})^{-1}[\nabla H_{>-N}, L]_0)$ into $(\mathcal{P}_2(L)\nabla H)_{<N}$ yields the reduced tensor

$$\begin{aligned}
\mathcal{P}_2^{(red)}(L)\nabla H_{>-N} &= ((\mathcal{P}_2(L)\nabla H)_{<N} \\
&= (\mathcal{P}_2(L)\nabla H_{>-N})_{<N} \\
&\quad + \frac{1}{2}[\left(\frac{1+T^{-N}}{1-T^{-N}}[\nabla H_{>-N}, L]_0\right), L]_{<N} \\
&= (\mathcal{P}_S(L)\nabla H_{>-N})_{<N}.
\end{aligned}$$

Since $\mathcal{P}_S(L)$ does not produce any terms of order $\geq N$, the final projection to orders $< N$ may be omitted: $\mathcal{P}_2^{(red)} = \mathcal{P}_S$. $\qquad\square$

Example 11. We consider the Lax operator $L = T + b + T^{-1}a$ of the celebrated Toda lattice hierarchy. The first equation $L_{t_1} = [L_{\geq 0}, L]$ is the Toda system itself:

$$a_{t_1} = a(b^{(1)} - b), \quad b_{t_1} = a - a^{(-1)}. \tag{3.9}$$

In terms of the coordinates a and b the dual fibers are parametrized by $\nabla H = \frac{\delta H}{\delta b} + \frac{\delta H}{\delta a} T$, so that

$$\text{tr}(\nabla H \dot{L}) = \text{tr}\left(\left(\frac{\delta H}{\delta b} + \frac{\delta H}{\delta a} T\right)(\dot{b} + T^{-1}\dot{a})\right) = \sum_n \frac{\delta H}{\delta b}\dot{b} + \sum_n \frac{\delta H}{\delta a}\dot{a}.$$

Insertion of ∇H into (3.7) and (3.8) yields the two Hamiltonian structures associated with (3.9):

$$\mathcal{P}_1(L)\nabla H = a\frac{\delta H}{\delta a} - \left(a\frac{\delta H}{\delta a}\right)^{(-1)} + T^{-1}a\left(\left(\frac{\delta H}{\delta b}\right)^{(1)} - \frac{\delta H}{\delta b}\right),$$

$$\mathcal{P}_S(L)\nabla H = a\left(\frac{\delta H}{\delta b}\right)^{(1)} - \left(a\frac{\delta H}{\delta b}\right)^{(-1)} + ba\frac{\delta H}{\delta a} - b\left(a\frac{\delta H}{\delta a}\right)^{(-1)}$$

$$+ T^{-1}a\left(\left(a\frac{\delta H}{\delta a}\right)^{(1)} - \left(a\frac{\delta H}{\delta a}\right)^{(-1)} + \left(b\frac{\delta H}{\delta b}\right)^{(1)} - b\frac{\delta H}{\delta b}\right),$$

i.e.,

$$\mathcal{P}_1(L):\quad \begin{pmatrix} \frac{\delta H}{\delta a} \\ \frac{\delta H}{\delta b} \end{pmatrix} \rightarrow \begin{pmatrix} 0 & a\,(T-1) \\ (1-T^{-1})\,a & 0 \end{pmatrix} \begin{pmatrix} \frac{\delta H}{\delta a} \\ \frac{\delta H}{\delta b} \end{pmatrix},$$

$$\mathcal{P}_S(L):\quad \begin{pmatrix} \frac{\delta H}{\delta a} \\ \frac{\delta H}{\delta b} \end{pmatrix} \rightarrow \begin{pmatrix} a\,(T-T^{-1})\,a & a\,(T-1)\,b \\ b\,(1-T^{-1})\,a & aT - T^{-1}\,a \end{pmatrix} \begin{pmatrix} \frac{\delta H}{\delta a} \\ \frac{\delta H}{\delta b} \end{pmatrix}.$$

The Hamiltonians of (3.9) are

$$H_1 = \operatorname{tr}(L) = \sum_n b \quad \text{and} \quad H_2 = \tfrac{1}{2}\operatorname{tr}(L^2) = \sum_n(a + \tfrac{1}{2}b^2).$$

Example 12. We consider Lax operators of the form (3.5), $N = M = K = 1$:

$$L = T + \phi T\Delta^{-1}\psi. \tag{3.10}$$

The associated hierarchy of Lax equations $L_{t_q} = [(L^q)_{\geq 0}, L]$ is equivalent to

$$\phi_{t_q} = ((L^q)_{\geq 0}\phi), \quad \psi_{t_q} = -(((L^q)_{\geq 0})^*\psi).$$

The basic system

$$\phi_{t_1} = \phi^{(1)} + \psi\phi^2, \quad \psi_{t_1} = -\psi^{(-1)} - \phi\psi^2 \tag{3.11}$$

is related to the relativistic Toda model (see Example 18). The dual fibers of the manifold (3.10) with coordinates ϕ, ψ may be parametrized by operators $\nabla H \in g$ with

$$((\nabla H_{\geq 0})^*\psi) = \frac{\delta H}{\delta\phi}, \quad (\nabla H_{\geq 0}\phi) = \frac{\delta H}{\delta\psi},$$

so that via (3.4)

$$(T\Delta^{-1}\psi\nabla H)_0 = (T\Delta^{-1}((\nabla H_{\geq 0})^*\psi))_0 = \frac{\delta H}{\delta\phi},$$

$$(\nabla H\phi T\Delta^{-1})_0 = ((\nabla H\phi)T\Delta^{-1})_0 = \frac{\delta H}{\delta\psi}$$

and

$$\begin{aligned}
\operatorname{tr}(\nabla H\,\dot{L}) &= \operatorname{tr}(\nabla H\,(\dot\phi T\Delta^{-1}\psi + \phi T\Delta^{-1}\dot\psi)) \\
&= \operatorname{tr}(\dot\phi T\Delta^{-1}\psi\nabla H \\
&= \sum_n \dot\phi\,(T\Delta^{-1}\psi\nabla H)_0 + \sum_n(\nabla H\phi T\Delta^{-1})_0\,\dot\psi \\
&= \sum_n \dot\phi\,\frac{\delta H}{\delta\phi} + \sum_n \frac{\delta H}{\delta\psi}\,\dot\psi.
\end{aligned}$$

Insertion of ∇H into (3.7) and (3.8) yields the Hamiltonian structures of (3.11):

$$P_1(L)\nabla H = [\phi T\Delta^{-1}\psi, \nabla H_{\geq 0}]_{<1} = \phi T\Delta^{-1}\tfrac{\delta H}{\delta\phi} - \tfrac{\delta H}{\delta\psi}T\Delta^{-1}\psi,$$

and (after some computation)

$$P_S(L)\nabla H = \left(\left(\tfrac{\delta H}{\delta\psi}\right)^{(1)} + \phi\psi\,\tfrac{\delta H}{\delta\psi} + \phi\rho(\psi\tfrac{\delta H}{\delta\psi}) - \phi\rho(\phi\tfrac{\delta H}{\delta\phi})\right)T\Delta^{-1}\psi$$

$$-\phi T\Delta^{-1}\left(\left(\tfrac{\delta H}{\delta\phi}\right)^{(-1)} + \psi\phi\,\tfrac{\delta H}{\delta\phi} - \psi\rho(\psi\tfrac{\delta H}{\delta\phi}) + \psi\rho(\psi\tfrac{\delta H}{\delta\psi})\right),$$

i.e.,

$$P_1(L) \; : \; \begin{pmatrix} \tfrac{\delta H}{\delta\phi} \\ \tfrac{\delta H}{\delta\psi} \end{pmatrix} \rightarrow \begin{pmatrix} 0 & -1 \\ 1 & 0 \end{pmatrix} \begin{pmatrix} \tfrac{\delta H}{\delta\phi} \\ \tfrac{\delta H}{\delta\psi} \end{pmatrix},$$

$$P_S(L) \; : \; \begin{pmatrix} \tfrac{\delta H}{\delta\phi} \\ \tfrac{\delta H}{\delta\psi} \end{pmatrix} \rightarrow$$

$$\begin{pmatrix} -\phi\,\rho\circ\phi & T + \phi\psi + \phi\,\rho\circ\psi \\ -T^{-1} - \psi\phi + \psi\,\rho\circ\phi & -\psi\,\rho\circ\psi \end{pmatrix} \begin{pmatrix} \tfrac{\delta H}{\delta\phi} \\ \tfrac{\delta H}{\delta\psi} \end{pmatrix}$$

with

$$\rho = (T+1)(T-1)^{-1} : \; u(n) \; \rightarrow \; \sum_{j<n} u(j) - \sum_{j>n} u(j).$$

The Hamiltonians of (3.11) are

$$H_1 = \operatorname{tr}(L) = \sum_n \phi\psi \quad \text{and} \quad H_2 = \tfrac{1}{2}\operatorname{tr}(L^2) = \sum_n(\phi^{(1)}\psi + \tfrac{1}{2}(\phi\psi)^2).$$

Remark 13. The connection with the version of the relativistic Toda lattice of Ref. [11] is as follows. A LU factorization (or, actually, a UL factorization) yields

$$L = T + \phi T\Delta^{-1}\psi = (d+T)(1 - T^{-1}c)^{-1}$$

with $\psi = c\psi^{(1)}$ and $\phi\psi\psi^{(1)} = d\psi^{(1)} + \psi$, where we note

$$(1 - T^{-1}c)^{-1} = (\psi^{-1}(1 - T^{-1})\psi)^{-1} = \psi^{-1}T\Delta^{-1}\psi.$$

In Ref. [11] the adjoint of L was used, parametrized by the coordinates c and d. This explains why the underlying r-matrices for the (c,d) version of the relativistic Toda lattice equation and the usual Toda lattice equation (3.9) were slightly different, since Suris' adjoint of L is related to the class $\alpha = -1$ instead of $\alpha = 1$. In the present situation both the classical as well as the relativistic Toda model are representatives of the same abstract bi-Hamiltonian system $L_{t_1} = [L_{\geq 0}, L]$ restricted to different operator submanifolds.

3.3. The Modified Lattice KP Hierarchy and its Reductions

We consider

$$\tilde{r}(\xi) = \xi_{\geq 1} - \xi_{<1} - 2(\xi\Delta^{-1})_0 \qquad (3.12)$$

with adjoint

$$\tilde{r}^*(\xi) = \xi_{<0} - \xi_{\geq 0} - 2\Delta^{-1}\xi_0 .$$

This r-matrix solves the modified Yang–Baxter equation (1.3) which follows easily from the following interpretation. One can rewrite the operators in g as being ordered w.r.t. powers of $\Delta = T - 1$ instead of powers of T:

$$\xi = u_N T^N + \cdots + u_1 T + u_0 + u_{-1}T^{-1} + \cdots$$
$$= v_N \Delta^N + \cdots + v_1 \Delta + v_0 + v_{-1}\Delta^{-1} + \cdots$$

with

$$v_N = u_N , \quad v_{N-1} = u_{N-1} + N u_N , \quad \cdots \quad , \quad v_0 = u_0 + u_1 + \cdots + u_N , \quad \ldots .$$

There are two obvious decompositions of g into sub-(Lie)algebras. The first consists of powers ≥ 0 of Δ versus powers < 0. This is the same as splitting the operators into powers ≥ 0 and < 0 of T, so that the corresponding integrable systems are covered by the considerations of Section 3.2. The second decomposition $g = g_{\widetilde{\geq 1}} \oplus g_{\widetilde{<0}}$ into powers ≥ 1 of Δ versus powers < 1 is new. The r-matrix (3.12) is the identity on $g_{\widetilde{\geq 1}}$ and minus the identity on $g_{\widetilde{<1}}$, since $(\xi\Delta^{-1})_0 = u_N + \cdots + u_1 = v_0 - u_0$. We emphasize that we do not change the notation in this section: $\xi_{\geq 1}, \xi_0, \xi_{<1}$ etc. still denote the projection to various shift orders **w.r.t. powers of** T.

The linear Poisson tensor $(1.1.\mathcal{P}_1)$ with the r-matrix (3.12)

$$\tilde{\mathcal{P}}_1(L)\nabla H = [\nabla H_{\geq 1} - (\nabla H \Delta^{-1})_0, L] - [\nabla H, L]_{\geq 0} - \Delta^{-1}[\nabla H, L]_0$$
$$= -[\nabla H_{<1} + (\nabla H \Delta^{-1})_0, L] + [\nabla H, L]_{<0} - \Delta^{-1}[\nabla H, L]_0 \qquad (3.13)$$

produces the integrable hierarchy [26]

$$L_{t_q} = [(L^q)_{\geq 1} - (L^q \Delta^{-1})_0, L]$$
$$= -[(L^q)_{<1} + (L^q \Delta^{-1})_0, L] \qquad (3.14)$$

with the commuting Hamiltonians $H = \mathrm{tr}(L^{(q+1)})/(q+1)$. These equations are Miura related to the Lattice KP hierarchy (3.1, $\alpha = 1$) and shall be referred to as the modified Lattice KP hierarchy. For the Lax operators the link is given by a simple gauge transformation:

Theorem 14 If $L_{t_q} = [(L^q)_{\geq 0}, L]$ and $\Phi_{t_q} = ((L^q)_{\geq 0}\Phi)$, then $\tilde{L} = \Phi^{-1}L \circ \Phi$ satisfies (3.14).

Proof. One computes

$$\tilde{L}_{t_q} = [\Phi^{-1}(L^q)_{\geq 0} \circ \Phi - \Phi^{-1}\Phi_{t_q}, \tilde{L}] = [(\tilde{L}^q)_{\geq 1} - \Phi^{-1}((L^q)_{\geq 1}\Phi), \tilde{L}],$$

where

$$\Phi^{-1}((L^q)_{\geq 1}\Phi) = \Phi^{-1}((L^q)_{\geq 1} \circ \Phi\Delta^{-1})_0 = (\Phi^{-1}L^q \circ \Phi\Delta^{-1})_0 = (\tilde{L}^q\Delta^{-1})_0.$$

$\square$

It is easily checked that the dynamics (3.14) can be restricted to Lax operators of the form

$$L = u_N T^N + u_{N-1}T^{N-1} + \cdots + u_M T^M + T^M\Delta^{-1}\Psi + \sum_{i=1}^{K}\phi_i T^M\Delta^{-1}\psi_i$$

$$(3.15)$$

or

$$L = u_N T^N + u_{N-1}T^{N-1} + \cdots + u_M T^M + T^M\Delta^{-1}\Psi \qquad (3.16)$$

or

$$L = u_N T^N + u_{N-1}T^{N-1} + \cdots + u_M T^M \qquad (3.17)$$

or [3]

$$L = u_N T^N + u_{N-1}T^{N-1} + \cdots + u_M T^M, \quad u_N + u_{N-1} + \cdots + u_M = 0. \quad (3.18)$$

Indeed, from the second representation of (3.14) it is clear that the highest shift order of the commutator coincides with the highest order of L. The invariance of the "low end" of the operators is verified with the first representation of (3.14) and (3.4). We note that the functions ϕ_i and ψ_i, Ψ, respectively, inherit the dynamics

$$\phi_{it_q} = ((L^q)_{\geq 1} - (L^q\Delta^{-1})_0)\phi_i), \quad \psi_{it_q} = (-((L^q)_{\geq 1})^* + (L^q\Delta^{-1})_0)\psi_i)$$

of eigenfunctions and adjoint eigenfunctions, respectively. The equations associated with the Lax operators (3.17) are in general non-Hamiltonian. However, the operators (3.15) and (3.16) constitute integral manifolds of the linear Poisson tensor, so that the restriction to these operators is Hamiltonian:

---

[3]The constraint on the coefficients implies that such L may be written in the form

$$L = (v_{N-1}T^{N-1} + \cdots + v_M T^M)\Delta.$$

With $(L^q)_{\geq 1} - (L^q\Delta^{-1})_0 = (\cdots)\Delta$ it is clear that (3.18) is invariant under the flows (3.14).

Theorem 15 *The linear tensor (3.13) admits a proper restriction to Lax operators of the form (3.15) and (3.16) with $N \geq 1 \geq M$.*

This is verified by a straightforward computation using the two representations (3.13) and (3.4) to check the highest and lowest orders of $\tilde{\mathcal{P}}_1(L)\nabla H$.

We now turn to a discussion of the quadratic Poisson bracket associated with (3.14). As noted in Ref. [10], the skew-symmetric part of the r-matrix

$$\tfrac{1}{2}(\tilde{r} - \tilde{r}^*) : \quad \xi \;\to\; \xi_{\geq 1} - \xi_{<0} - (\xi\Delta^{-1})_0 + \Delta^{-1}\xi_0$$

does *not* satisfy the modified Yang–Baxter equation, so that Semenov's quadratic tensor $(1.1.\mathcal{P}_2)$ does *not* enjoy any Poisson properties. However, Suris' construction of Theorem 1 is capable of producing a suitable quadratic Poisson bracket:

Theorem 16 *Let* $\rho = \frac{T+1}{T-1} : \; u(n) \;\to\; \sum_{j<n} u(j) - \sum_{j>n} u(j)$. *Then*

$$
\begin{aligned}
A_1 : \quad & \xi \;\to\; \xi_{\geq 1} - \xi_{<0} - 2(\xi\Delta^{-1})_0 - \rho(\xi_0) + 2\Delta^{-1}\xi_0\,, \\
A_2 : \quad & \xi \;\to\; \xi_{\geq 1} - \xi_{<0} + \rho(\xi_0)\,, \\
S : \quad & \xi \;\to\; \rho(\xi_0) - \xi_0 - 2\Delta^{-1}\xi_0\,, \\
S^* : \quad & \xi \;\to\; -\rho(\xi_0) - \xi_0 - 2(\xi\Delta^{-1})_0
\end{aligned}
$$

satisfy all conditions of Theorem 1. The resulting Poisson tensor

$$
\begin{aligned}
\tilde{\mathcal{P}}_S(L)\nabla H &= (L\nabla H)_{\geq 1}L - L(\nabla H L)_{\geq 1} + \tfrac{1}{2}[L, \nabla H]_0 L + \tfrac{1}{2}L[L, \nabla H]_0 \\
&\quad + \Delta^{-1}[L, \nabla H]_0 L + [L, (L\nabla H\Delta^{-1})_0] + \tfrac{1}{2}[\rho([\nabla H, L]_0), L] \\[4pt]
&= -(L\nabla H)_{<0}L + L(\nabla H L)_{<0} + \tfrac{1}{2}[L, (L\nabla H + \nabla H L)_0] \\
&\quad + \Delta^{-1}[L, \nabla H]_0 L + [L, (L\nabla H\Delta^{-1})_0] + \tfrac{1}{2}[\rho([\nabla H, L]_0), L]
\end{aligned}
$$
$$(3.19)$$

is compatible with the linear tensor (3.13) and admits a proper restriction to Lax operators of the form (3.15) and (3.16), $N \geq 1 \geq M$.

With $A_1 + S = A_2 + S^* = \tilde{r}$ one has to verify (1.4) in Theorem 2:

$$
\begin{aligned}
S(\,[A_2(\xi), \eta] + [\xi, A_2(\eta)]\,) &\overset{(i)}{=} 0 \overset{(ii)}{=} [\,S(\xi), S(\eta)\,]\,, \\
S^*(\,[A_1(\xi), \eta] + [\xi, A_1(\eta)]\,) &\overset{(iii)}{=} 0 \overset{(iv)}{=} [S^*(\xi), S^*(\eta)]\,.
\end{aligned}
$$

Here (i), (ii) and (iv) are (almost) trivial, whereas the verification of (iii) requires a substantial computational effort. We omit all details. The reduction properties of $\tilde{\mathcal{P}}_S$ are easily checked using the two representations (3.19) and (3.4).

Remark 17. After the exchange $A_1 \leftrightarrow A_2$ and $S \leftrightarrow S^*$ Theorem 1 yields the Poisson tensor

$$
\begin{aligned}
\tilde{P}'_S(L)\nabla H &= (L\nabla H)_{\geq 1}L - L(\nabla HL)_{\geq 1} + \frac{1}{2}[L, \nabla H]_0 L + \frac{1}{2}L[L, \nabla H]_0 \\
&\quad + L\Delta^{-1}[L, \nabla H]_0 + [L, (\nabla HL\Delta^{-1})_0] - \frac{1}{2}[\rho([\nabla H, L]_0), L] \\
&= -(L\nabla H)_{<0}L + L(\nabla HL)_{<0} + \frac{1}{2}[L, (L\nabla H + \nabla HL)_0] \\
&\quad + L\Delta^{-1}[L, \nabla H]_0 + [L, (\nabla HL\Delta^{-1})_0] - \frac{1}{2}[\rho([\nabla H, L]_0), L]
\end{aligned}
$$

$$(3.20)$$

which is again compatible with the linear tensor (3.13) and admits a proper restriction to Lax operators of the form (3.18). This yields *the* Hamiltonian formulation of the integrable hierarchies arising from Lax operators (3.18). We note that the linear tensor (3.13) cannot be restricted properly to such operators. Dirac reduction results in a nonlocal bracket stemming from $\tilde{P}_1$.

Example 18. The gauge transformation $L \to \tilde{L} = \phi^{-1} L\phi$ of Theorem 14 applied to the Lax operator of Example 12 yields

$$
L = T + \phi T\Delta^{-1}\psi \quad \to \quad \tilde{L} = uT + T\Delta^{-1}\Psi
$$

with

$$
u = \phi^{-1}\phi^{(1)}, \quad \Psi = \psi\phi. \tag{3.21}
$$

The basic equation of the integrable hierarchy (3.14) for $\tilde{L}$ is the relativistic Toda lattice equation [27, 28]

$$
u_{t_1} = u\left(u^{(1)} - u + \Psi^{(1)} - \Psi\right), \quad \Psi_{t_1} = u\Psi - u^{(-1)}\Psi^{(-1)} \tag{3.22}
$$

which is related to (3.11) by the "Miura transformation" (3.21). The first two Hamiltonian structures found in Ref. [28] can be retrieved from (3.13) and (3.19) by inserting $\nabla H = T^{-1}\frac{\delta H}{\delta u} + \frac{\delta H}{\delta \Psi}$. From

$$
\tilde{P}_1(\tilde{L})\nabla H = u\left(\left(\tfrac{\delta H}{\delta \Psi}\right)^{(1)} - \tfrac{\delta H}{\delta \Psi}\right)T + T\Delta^{-1}\left(u\tfrac{\delta H}{\delta u} - \left(u\tfrac{\delta H}{\delta u}\right)^{(-1)}\right)
$$

and a corresponding expression for $\tilde{P}_S(L)\nabla H$ one obtains

$$
\tilde{P}_1(L) \quad : \quad \begin{pmatrix} \frac{\delta H}{\delta u} \\[4pt] \frac{\delta H}{\delta \Psi} \end{pmatrix} \to \begin{pmatrix} 0 & u(T-1) \\[4pt] (1-T^{-1})u & 0 \end{pmatrix} \begin{pmatrix} \frac{\delta H}{\delta u} \\[4pt] \frac{\delta H}{\delta \Psi} \end{pmatrix},
$$

$$
\tilde{P}_S(L) \quad : \quad \begin{pmatrix} \frac{\delta H}{\delta u} \\[4pt] \frac{\delta H}{\delta \Psi} \end{pmatrix} \to
$$

$$
\begin{pmatrix} u(T-T^{-1})u & u(T-1)(uT+\Psi) \\[4pt] (T^{-1}u+\Psi)(1-T^{-1})u & u\Psi T - T^{-1}u\Psi \end{pmatrix} \begin{pmatrix} \frac{\delta H}{\delta u} \\[4pt] \frac{\delta H}{\delta u} \end{pmatrix}.
$$

The Hamiltonians of (3.22) are

$$H_1 = \text{tr}(L) = \sum_n \Psi \quad \text{and} \quad H_2 = \tfrac{1}{2}\text{tr}(L^2) = \sum_n \left(u\Psi + \tfrac{1}{2}\Psi^2\right).$$

Example 19. We discuss an example of the class (3.18). The basic equation associated with the Lax operator $L = (u + vT^{-1})\Delta$ is

$$u_{t_1} = u\left(v^{(1)} - v\right), \quad v_{t_1} = v\left(u^{(-1)} - u\right).$$

Insertion of $\nabla H = \Delta^{-1}(\frac{\delta H}{\delta u} + T\frac{\delta H}{\delta v})$ into (3.20) yields the Poisson structure of this equation:

$$\mathcal{P}'_S(L)\nabla H = \left(u\left((v\tfrac{\delta H}{\delta v})^{(1)} - v\tfrac{\delta H}{\delta v}\right) + v\left(u\tfrac{\delta H}{\delta u} - (u\tfrac{\delta H}{\delta u})^{(-1)}\right)T^{-1}\right)\Delta,$$

i.e.,

$$\tilde{\mathcal{P}}'_S(L): \quad \begin{pmatrix} \frac{\delta H}{\delta u} \\ \frac{\delta H}{\delta v} \end{pmatrix} \rightarrow \begin{pmatrix} 0 & u\,(T-1)\,v \\ v\,(1-T^{-1})\,u & 0 \end{pmatrix} \begin{pmatrix} \frac{\delta H}{\delta u} \\ \frac{\delta H}{\delta v} \end{pmatrix}.$$

The Hamiltonian is $H = \text{tr}(L) = \sum_n(v - u)$.

4. Conclusions

We have derived the Poisson brackets associated with scalar Lax equations $L_t = [A, L]$, where L is a shift operator and A involves only shifts into one lattice direction. The main examples of such systems are the Lattice KP hierarchy and the modified Lattice KP hierarchy, in which A is generated via (fractional) powers of L. They include the Toda model and the relativistic Toda model as the most prominent equations. As already observed by Suris, both systems may be regarded as the same dynamical system (the Lattice KP hierarchy) restricted to different operator submanifolds. The unifying structure is provided by simple r-matrices on the algebra of shift symbols. By restriction or reduction to finite Lax operators they generate Poisson brackets which yield the multi-Hamiltonian formulations for a variety of integrable lattice hierarchies. Reductions of the Lattice KP hierarchy include all Lax operators of the form

$$L = T^N + u_{N-1}T^{N-1} + \cdots + u_M T^M + \sum_{i=1}^K \phi_i T^M \Delta^{-1}\psi_i$$

with arbitrary K and $N \geq 1 \geq M$. Reductions of the modified Lattice KP include the Lax operators

$$L = u_N T^N + \cdots + u_M T^M + T^M \Delta^{-1}\Psi + \sum_{i=1}^K \phi_i T^M \Delta^{-1}\psi_i$$

with arbitrary K and $N \geq 1 \geq M$.

Acknowledgement.   The author gratefully acknowledges the kind hospitality of Prof. S. Carillo and the Department MeMoMat, University "La Sapienza", Rome, where parts of this research were undertaken.

References

[1] I.M. Gelfand and I.Y. Dorfman, *Funct. Anal. Appl.* **13** (1979) 248; *Funct. Anal. Appl.* **14** (1980) 223.

[2] F. Magri, *J. Math. Phys.* **19** (1978) 1156.

[3] B. Fuchssteiner and A.S. Fokas, *Physica* **4D** (1981) 47.

[4] I. Y. Dorfman, *Dirac Structures and Integrability of Nonlinear Evolution Equations*, Wiley, Chichester 1993.

[5] I.M. Gelfand and L.A. Dikii, *Funct. Anal. Appl.* **10** (1976) 259; *Funct. Anal. Appl.* **11** (1977) 93.

[6] I.Y. Dorfman and A.S. Fokas, *J. Math. Phys.* **33** (1992) 2504.

[7] M. Adler, *Invent. Math.* **50** (1979) 219.

[8] M.A. Semenov-Tian-Shansky, *Funct. Anal. Appl.* **17** (1983) 259.

[9] L.C. Li and S. Parmentier, *Comm. Math. Phys.* **125** (1989) 545.

[10] W. Oevel and O. Ragnisco, *Physica A* **161** (1990) 181.

[11] Y.B. Suris, *Phys. Lett. A* **180** (1993) 419.

[12] Y. Cheng, *J. Math. Phys.* **33** (1992) 3774.

[13] Y. Cheng and Y. Li, *Phys. Lett. A* **157** (1991) 22; *J. Phys. A* **25** (1992) 419.

[14] B.G. Konopelchenko, J. Sidorenko and W. Strampp, *Phys. Lett. A* **157** (1991) 17.

[15] B.G. Konopelchenko and W. Strampp, *Inverse Problems* **7** (1991) L17.

[16] J. Sidorenko and W. Strampp, *Inverse Problems* **7** (1991) L37.

[17] Y. Zeng, *J. Phys. A* **24** (1991) L1065.

[18] W. Oevel and W. Strampp, *Comm. Math. Phys.* **157** (1993) 51.

[19] W. Oevel, *Phys. Lett. A* **186** (1994) 79.

[20] B.A. Kupershmidt, *Comm. Math. Phys.* **99** (1985) 51.

[21] K. Kiso, *Progr. Theor. Phys.* **83** (1990) 1108.

[22] H. Aratyn, E. Nissimov, S. Pacheva and I. Vaysburd, *Phys. Lett. B* **294** (1992) 167.

[23] W. Oevel and C. Rogers, *Rev. Math. Phys.* **5** (1993) 299.

[24] B.G. Konopelchenko and W. Oevel, *Publ. RIMS, Kyoto University* **29** (1993) 1.

[25] A.G. Reiman, *J. Soviet Math.* **19** (1982) 1507.

[26] B.A. Kupershmidt, *Astérisque* **123** (1985) 1.

[27] S.N.M. Ruijsenaars, *Comm. Math. Phys.* **133** (1990) 217.

[28] W. Oevel, H. Zhang, B. Fuchssteiner and O. Ragnisco, *J. Math. Phys.* **30** (1989) 2664.

FB 17–Mathematik, Universität Paderborn,
D 33095 Paderborn, Germany
Email: **walter@uni-paderborn.de**

Received November, 1995

On the r-Matrix Structure of the Neumann System and its Discretizations

Orlando Ragnisco and Yuri B. Suris

0. Introduction

A novel impetus to the construction of integrable discretisations of given integrable continuous-time hamiltonian systems has been given in recent years by a number of relevant findings: we mention the successful application to differential-difference integrable hierarchies of the *stationary flow* or *restricted flow* approach [1], the results obtained toward the identification of integrable mappings of the standard type [2], the discovery of Backlund transformations for Calogero-Moser and Rujsenaars systems [3], including relativistic Toda [4], and finally the construction of non-autonomous mappings which are endowed with a proper discrete analog of the Painleve' property [5]. Of course, time-discretisation is a highly non-unique procedure, even if we restrict considerations to integrability-preserving difference schemes. One may just ask to get an integrable Poisson map such that the discrete dynamics it describes goes into the continuous one in a suitable asymptotic limit, together with integrals of motion and Poisson structure, or require that Poisson structure and integrals of motion be exactly preserved by the discretisation. Stationary or restricted flow technique typically lead to discretisation of the former type, while Backlund transformations provide an example of the latter one. In the present paper, we compare two integrable discretisations of the Neumann system, both belonging to the first family; indeed, one of them is obtained by applying the restricted flow technique to the Toda lattice hierarchy. It turns out that they bear quite different features: namely, the difference scheme found by Ragnisco leads to a Poisson map whose associated r-matrix is essentially constant, and whose invariant functions commute both for the unconstrained and for the constrained Poisson brackets, while the Veselov scheme leads to a Poisson map with dynamical r-matrix and whose invariant functions commute only for the constrained Poisson brackets.

1. Continuous–time Neumann system

The famous Neumann system [6-10] is described by a system of differential equations

$$\ddot{x}_k = -\omega_k x_k - u x_k, \quad 1 \le k \le N,$$

where the Lagrange multiplier $u = u(x, \dot{x})$ is defined by the condition that the trajectory $x(t)$ remains for all $t \in \mathbf{R}$ on the sphere $S = \{x : \langle x, x \rangle = 1\}$ in the configurational space. It is easy to see that

$$u(x, \dot{x}) = \langle \dot{x}, \dot{x} \rangle - \langle \Omega x, x \rangle,$$

where $\Omega = \mathrm{diag}(\omega_1, \ldots, \omega_N)$, and $\langle \cdot, \cdot \rangle$ stands for the standard Euclidean scalar product in $\mathbf{R}^N$.

Rewritten as

$$\dot{x}_k = p_k, \quad \dot{p}_k = -\omega_k x_k - u(x, p) x_k, \tag{1.1}$$

$$u(x, p) = \langle p, p \rangle - \langle \Omega x, x \rangle, \tag{1.2}$$

this system may be presented as a constrained Hamiltonian one:

$$\dot{x}_k = \{x_k, H\}_{\mathrm{Dirac}}, \quad \dot{p}_k = \{p_k, H\}_{\mathrm{Dirac}}.$$

Here Hamiltonian function reads

$$H(x, p) = \frac{1}{2} \langle p, p \rangle + \frac{1}{2} \langle \Omega x, x \rangle, \tag{1.3}$$

and Poisson bracket $\{\cdot, \cdot\}_{\mathrm{Dirac}}$ arises from the standard one on $\mathbf{R}^{2N}\{x, p\}$:

$$\{x_k, p_j\} = \delta_{kj}, \tag{1.4}$$

constrained on the tangent bundle TS to the unit sphere S, i.e. on the set

$$\varphi_1(x, p) = \langle x, x \rangle = 1, \quad \varphi_2(x, p) = \langle x, p \rangle = 0. \tag{1.5}$$

The general theory due to Dirac assures that the modified Poisson structure is defined by

$$\{F, G\}_{\mathrm{Dirac}} = \{F, G\} + \frac{\{F, \varphi_2\}\{G, \varphi_1\} - \{F, \varphi_1\}\{G, \varphi_2\}}{\{\varphi_1, \varphi_2\}}. \tag{1.6}$$

The Poisson bracket $\{\cdot, \cdot\}_{\mathrm{Dirac}}$, when compared with the initial Poisson structure $\{\cdot, \cdot\}$, has two additional Casimir functions φ_1, φ_2, which serve as integrals of motion for *every* Hamiltonian system.

In our case the non-vanishing constrained Poisson brackets of the co-ordinate functions are:

$$\{x_k, p_j\}_{\text{Dirac}} = \delta_{kj} - \frac{x_k x_j}{\langle x, x \rangle}, \quad \{p_k, p_j\}_{\text{Dirac}} = \frac{p_k x_j - x_k p_j}{\langle x, x \rangle}. \tag{1.7}$$

It can be immediately checked that these brackets and Hamiltonian (1.3) result in the system (1.1), (1.2).

The Neumann system possesses N integrals of motion

$$F_k(x, p) = x_k^2 + \sum_{j \neq k} \frac{(x_k p_j - p_k x_j)^2}{\omega_k - \omega_j}, \quad 1 \leq k \leq N. \tag{1.8}$$

Only $N - 1$ of them are independent on TS, since they satisfy the identity $\sum_{k=1}^{N} F_k = \langle x, x \rangle = 1$. They commute in the constrained Poisson structure $\{\cdot, \cdot\}_{\text{Dirac}}$ (as well as in the unconstrained one), which assures the complete integrability of the system.

The integrals (1.8) may be derived systematically from the Lax representation for the Neumann system, and their involutivity may be proved systematically with the help of the r–matrix theory. While the unconstrained version of the Neumann system was given an r–matrix interpretation in [11], [12], such an interpretation of the constrained one seems to lack.

It may be checked by direct computation that the system (1.1), (1.2) may be presented in the Lax form

$$\dot{L} = [M, L] \tag{1.9}$$

with the matrices $L = L(x, p, \lambda)$, $M = M(x, p, \lambda)$ depending on the phase space coordinates and on an additional spectral parameter:

$$L = \begin{pmatrix} A(\lambda) & B(\lambda) \\ C(\lambda) & -A(\lambda) \end{pmatrix}, \quad M = \begin{pmatrix} 0 & \lambda + u \\ -1 & 0 \end{pmatrix}. \tag{1.10}$$

Here

$$A(\lambda) = \sum_{k=1}^{N} \frac{x_k p_k}{\lambda - \omega_k}, \quad B(\lambda) = 1 + \sum_{k=1}^{N} \frac{p_k^2}{\lambda - \omega_k}, \quad C(\lambda) = -\sum_{k=1}^{N} \frac{x_k^2}{\lambda - \omega_k}. \tag{1.11}$$

The determinant of the Lax matrix $L(\lambda)$ serves as a generating function for the integrals of motion (1.8):

$$\Delta(\lambda) = -A^2(\lambda) - B(\lambda)C(\lambda) = \sum_{k=1}^{N} \frac{F_k}{\lambda - \omega_k}.$$

By direct computation one obtains the following pairwise Poisson brackets:

$$\{A(\lambda), A(\mu)\}_{\text{Dirac}} = \{C(\lambda), C(\mu)\}_{\text{Dirac}} = 0,$$

$$\{B(\lambda), B(\mu)\}_{\text{Dirac}} = \frac{4}{\langle x, x\rangle}(A(\lambda) - A(\mu)) + \frac{4}{\langle x, x\rangle}(B(\lambda)A(\mu) - A(\lambda)B(\mu)),$$

$$\{A(\lambda), B(\mu)\}_{\text{Dirac}} = -2\frac{B(\lambda) - B(\mu)}{\lambda - \mu} - \frac{2}{\langle x, x\rangle}C(\lambda) + \frac{2}{\langle x, x\rangle}C(\lambda)B(\mu),$$

$$\{A(\lambda), C(\mu)\}_{\text{Dirac}} = 2\frac{C(\lambda) - C(\mu)}{\lambda - \mu} - \frac{2}{\langle x, x\rangle}C(\lambda)C(\mu),$$

$$\{B(\lambda), C(\mu)\}_{\text{Dirac}} = -4\frac{A(\lambda) - A(\mu)}{\lambda - \mu} + \frac{4}{\langle x, x\rangle}A(\lambda)C(\mu).$$

These relations may be presented in the following compact form:

$$\{L(\lambda) \overset{\otimes}{,} L(\mu)\}_{\text{Dirac}} = [r(\lambda, \mu), I \otimes L(\mu)] - [r^*(\lambda, \mu), L(\lambda) \otimes I] +$$

$$+ \rho\,(L(\lambda) \otimes L(\mu)) + (L(\lambda) \otimes L(\mu))\,\rho -$$

$$- (I \otimes L(\mu))\,\rho\,(L(\lambda) \otimes I) - (L(\lambda) \otimes I)\,\rho\,(I \otimes L(\mu)), \tag{1.12}$$

where

$$r(\lambda, \mu) = \frac{2\Pi}{\lambda - \mu} + \frac{2}{\langle x, x\rangle}E_{12} \otimes E_{12}, \tag{1.13}$$

$$r^*(\lambda, \mu) = \Pi r(\mu, \lambda)\Pi = -\frac{2\Pi}{\lambda - \mu} + \frac{2}{\langle x, x\rangle}E_{12} \otimes E_{12}, \tag{1.14}$$

and

$$\rho = \frac{1}{\langle x, x\rangle}\Big(E_{12} \otimes (E_{11} - E_{22}) - (E_{11} - E_{22}) \otimes E_{12}\Big). \tag{1.15}$$

Here Π is the 2×2 permutation matrix, namely $\Pi = \sum_{k,j=1}^{2} E_{kj} \otimes E_{jk}$.

It is instructive to compare these formulas with the ones which hold in the unconstrained Poisson structure and which were found and used in [11], [12]:

$$\{A(\lambda), A(\mu)\} = \{B(\lambda), B(\mu)\} = \{C(\lambda), C(\mu)\} = 0,$$

$$\{A(\lambda), B(\mu)\} = -2\frac{B(\lambda) - B(\mu)}{\lambda - \mu}, \quad \{A(\lambda), C(\mu)\} = 2\frac{C(\lambda) - C(\mu)}{\lambda - \mu},$$

$$\{B(\lambda), C(\mu)\} = -4\frac{A(\lambda) - A(\mu)}{\lambda - \mu},$$

or in a compact form:

$$\{L(\lambda) \overset{\otimes}{,} L(\mu)\} = \left[\frac{2\Pi}{\lambda - \mu}, I \otimes L(\mu) + L(\lambda) \otimes I\right]. \qquad (1.16)$$

Let us recall the geometric meaning of the formula (1.16). Consider an assosiative algebra $sl(2)[\lambda, \lambda^{-1}]$ of (semi–infinite) Laurent series over $sl(2)$. The generic element of this algebra may be written as

$$U(\lambda) = \sum_{j,k=1}^{2} \sum_{n} u_{jk}^{(n)} \lambda^n E_{jk}.$$

This algebra may be equipped with an invariant non–degenerate scalar product in an infinite number of ways, for example,

$$(U(\lambda), V(\lambda)) = \text{res}\left(\text{tr } U(\lambda)V(\lambda)\right), \qquad (1.17)$$

where res stands for the coefficient of λ^{-1} term in the Laurent series. Given a linear operator R on $sl(2)[\lambda, \lambda^{-1}]$, satisfying the so called modified Yang–Baxter equation [13], one can define the following Poisson bracket on this algebra:

$$\{F, G\}(U) = \left(U, [R(\nabla F), \nabla G] + [\nabla F, R(\nabla G)]\right). \qquad (1.18)$$

Here the gradient is defined according to the formula

$$(\nabla H(U), X) = \frac{d}{d\epsilon} H(U + \epsilon X)\Big|_{\epsilon=0} \quad \forall X \in sl(2)[\lambda, \lambda^{-1}].$$

Hamiltonian equations of motion defined by this Poisson bracket and a conjugation–invariant Hamiltonian function $H(U)$, have the Lax form

$$\dot{U} = [U, R(\nabla H(U))]. \qquad (1.19)$$

Now the formula (1.16) means that the set of matrices

$$\{L(x, p, \lambda) : (x, p) \in \mathbf{R}^{2N}\}$$

forms a Poisson subspace in the space $sl(2)[\lambda, \lambda^{-1}]$ equipped with a scalar product (1.17) and a Poisson bracket (1.18) with the operator

$$R(U(\lambda)) = \mathcal{P}_+(U(\lambda)) - \mathcal{P}_-(U(\lambda)),$$

where $\mathcal{P}_+$ ($\mathcal{P}_-$) stands for the non–negative (resp. negative) part of the Laurent series.

Analogously, our main formula (1.12) means that the set of matrices $\{L(x, p, \lambda) : (x, p) \in TS\}$ also forms a Poisson subspace in the space

$sl(2)[\lambda, \lambda^{-1}]$ equipped with a scalar product (1.17) and another Poisson bracket, namely the sum of a linear Poisson bracket (1.18) with the operator

$$R(U(\lambda)) = \mathcal{P}_+(U(\lambda)) - \mathcal{P}_-(U(\lambda)) + 2u_{21}^{(-1)}E_{12}, \qquad (1.20)$$

and a particular case of the general quadratic Poisson bracket, introduced and studied in [14].

Let us mention some of the features of this remarkable r–matrix Poisson bracket.

- The correction to (1.16) caused by the Dirac constraints, consists not only in the appearance of the quadratic terms in (1.12), but also in an additional linear term. Both sorts of corrections seem to be *dynamical*, i.e. to depend on the point of the phase space. However, such dependence appears only through the Casimir function $\langle x, x \rangle$, hence these matrices are *constant* on each symplectic leaf of the bracket $\{\cdot, \cdot\}_{\text{Dirac}}$, in particular on TS.

- Both the linear and the quadratic parts of the Poisson bracket (1.12) are themselves Poisson brackets, which are, henceforth, compatible.

- The quadratic part of the bracket (1.12) does not influence the equations of motion corresponding to conjugation-*invariant* Hamiltonian functions $H = H(L)$, which, according to the general theory [13, 14], have the Lax form (1.19) with the operator (1.20). (It is easy to check that this operator satisfies the modified Yang–Baxter equation).

This explains the Lax equation (1.9) with the matrices (1.10). Indeed, it is easy to see that the matrix M from (1.10) may be presented as

$$M(\lambda) = \mathcal{P}_+(\lambda L(\lambda)) + (\lambda L(\lambda))_{21}^{(-1)}E_{21} == -(R + I)(\nabla H(L));$$

the last equality holds since the Hamiltonian (1.3) is equal to

$$H(L) = -\frac{1}{4}\Big(L(\lambda), \lambda L(\lambda)\Big),$$

so that $\nabla H(L) = -\frac{1}{2}\lambda L(\lambda)$.

2. Ragnisco's discretization of the Neumann system

The Ragnisco's discrete Neumann system [15] reads:

$$\frac{x_k(t+h)}{2\langle x(t+h), x(t)\rangle} + \frac{x_k(t-h)}{2\langle x(t), x(t-h)\rangle} = \omega_k x_k(t) - u(t)x_k(t), \quad 1 \le k \le N,$$

$$(2.1)$$

where the Lagrange multiplier $u(t)$ is chosen to assure that the vector $x(t)$ lies on $S = \{x \in \mathbf{R}^N : \langle x, x \rangle = 1\}$ for all $t \in h\mathbf{Z}$, so that

$$u(t) = \langle x(t), \Omega x(t) \rangle - 1. \tag{2.2}$$

Equations (2.1), (2.2) may be presented as Lagrangian equations on $S \times S$:

$$\frac{\partial \mathcal{L}(x(t+h), x(t))}{\partial x_k} + \frac{\partial \mathcal{L}(x(t), x(t-h))}{\partial x_k} = 0,$$

with a Lagrange function

$$\mathcal{L}(x(t+h), x(t)) = \frac{1}{2} \log \left(\langle x(t+h), x(t) \rangle \right) - \frac{1}{2} \langle x(t), \Omega x(t) \rangle.$$

The unconstrained variables p canonically conjugated to x, are given by [16]

$$p_k(t) = \frac{\partial \mathcal{L}(x(t), x(t-h))}{\partial x_k(t)} = \frac{x_k(t-h)}{2\langle x(t), x(t-h) \rangle}. \tag{2.3}$$

It is easy to see that the two constraints which have to be imposed on these variables read:

$$\varphi_1(x, p) = \langle x, x \rangle = 1, \quad \varphi_2(x, p) = \langle x, p \rangle = \frac{1}{2}. \tag{2.4}$$

Hence the constrained Poisson brackets take exactly the same form (1.7) as in the continuous–time case.

In terms of canonically conjugated variables the map generated by (2.1), (2.2) reads:

$$x(t+h) = a^{-1}(t)\Big(\Omega x(t) - u(t)x(t) - p(t)\Big), \quad p(t+h) = a(t)x(t), \tag{2.5}$$

where

$$u = \langle x, \Omega x \rangle - 1 \tag{2.6}$$

$$\begin{aligned} a^2 &= \langle \Omega x - ux - p, \Omega x - ux - p \rangle \\ &= \langle \Omega x - p, \Omega x - p \rangle - \langle \Omega x, x \rangle^2 + \langle \Omega x, x \rangle. \end{aligned} \tag{2.7}$$

It may be checked that this map can be represented in the following discrete Lax form:

$$L(t+h) = M^{-1}(t)L(t)M(t), \tag{2.8}$$

where $L = L(x, p, \lambda)$, $M = M(x, p, \lambda)$ have the form

$$L = \begin{pmatrix} A(\lambda) & B(\lambda) \\ C(\lambda) & -A(\lambda) \end{pmatrix}, \quad M = \begin{pmatrix} \lambda - u & a^2 \\ -1 & 0 \end{pmatrix}, \tag{2.9}$$

with u, a^2 given by (2.6), (2.7), and

$$A(\lambda) = \frac{1}{2} + \sum_{k=1}^{N} \frac{x_k p_k}{\lambda - \omega_k}, \tag{2.10}$$

$$B(\lambda) = \sum_{k=1}^{N} \frac{p_k^2}{\lambda - \omega_k}, \quad C(\lambda) = -\sum_{k=1}^{N} \frac{x_k^2}{\lambda - \omega_k}. \tag{2.11}$$

As in the continuous case, $\det(L(\lambda))$ serves as a generating function for N integrals:

$$\Delta(\lambda) = -A^2(\lambda) - B(\lambda)C(\lambda) = \sum_{k=1}^{N} \frac{F_k}{\lambda - \omega_k},$$

where

$$F_k = -x_k p_k + \sum_{j \neq k} \frac{(x_k p_j - p_k x_j)^2}{\omega_k - \omega_j}.$$

Only $N-1$ of them are independent on our phase space, because $\sum_{k=1}^{N} F_k = -\varphi_2(x,p) = -\frac{1}{2}$. In order to prove their involutivity in the constrained Poisson bracket we will obtain the r-matrix structure for the Lax matrix $L(\lambda)$. A slight modification of the Lax pair (2.9) as compared with [15] (a gauge transformation) simplifies the resulting formulas considerably.

It is straightforward to calculate the pairwise Poisson brackets $\{\cdot,\cdot\}_{\text{Dirac}}$ of the matrix elements of the Lax matrix $L(x,p,\lambda)$, starting from the analogous expressions for the continuous–time case:

$$\{A(\lambda), A(\mu)\}_{\text{Dirac}} = \{C(\lambda), C(\mu)\}_{\text{Dirac}} = 0,$$

$$\{B(\lambda), B(\mu)\}_{\text{Dirac}} = -\frac{2}{\langle x, x \rangle}\Big(B(\lambda) - B(\mu)\Big)$$
$$+ \frac{4}{\langle x, x \rangle}\Big(B(\lambda)A(\mu) - A(\lambda)B(\mu)\Big),$$

$$\{A(\lambda), B(\mu)\}_{\text{Dirac}} = -2\frac{B(\lambda) - B(\mu)}{\lambda - \mu} + \frac{2}{\langle x, x \rangle}C(\lambda)B(\mu),$$

$$\{A(\lambda), C(\mu)\}_{\text{Dirac}} = 2\frac{C(\lambda) - C(\mu)}{\lambda - \mu} - \frac{2}{\langle x, x \rangle}C(\lambda)C(\mu),$$

$$\{B(\lambda), C(\mu)\}_{\text{Dirac}} = -4\frac{A(\lambda) - A(\mu)}{\lambda - \mu}$$
$$- \frac{2}{\langle x, x \rangle}C(\mu) + \frac{4}{\langle x, x \rangle}A(\lambda)C(\mu).$$

These relations may be again put in a compact form:

$$\{L(\lambda) \overset{\otimes}{,} L(\mu)\}_{\text{Dirac}} = [r(\lambda, \mu), I \otimes L(\mu)] - [r^*(\lambda, \mu), L(\lambda) \otimes I] +$$

$$+\rho\left(L(\lambda)\otimes L(\mu)\right)+\left(L(\lambda)\otimes L(\mu)\right)\rho-$$

$$-\left(I\otimes L(\mu)\right)\rho\left(L(\lambda)\otimes I\right)-\left(L(\lambda)\otimes I\right)\rho\left(I\otimes L(\mu)\right),\qquad(2.12)$$

where the matrix ρ is just the same as in continuous–time case (1.15), and

$$r(\lambda,\mu)=\frac{2\Pi}{\lambda-\mu}+\frac{1}{\langle x,x\rangle}E_{12}\otimes(E_{11}-E_{22}),\qquad(2.13)$$

$$r^*(\lambda,\mu)=\Pi r(\mu,\lambda)\Pi=-\frac{2\Pi}{\lambda-\mu}+\frac{1}{\langle x,x\rangle}(E_{11}-E_{22})\otimes E_{12}.\qquad(2.14)$$

The geometrical meaning of these formulas is just the same as in the previous section, except for the operator R which enters the basic Poisson bracket (1.18): this time it is equal to

$$R(U(\lambda))=\mathcal{P}_+(U(\lambda))-\mathcal{P}_-(U(\lambda))+u_{21}^{(-1)}(E_{11}-E_{22}).\qquad(2.15)$$

The problem of finding the Hamiltonian flow (1.19) interpolating the integrable map (2.8) in $sl(2)[\lambda,\lambda^{-1}]$ is meaningful and waits for its solution.

3. Veselov's discretization of the Neumann system

The Veselov's discrete Neumann system [16] with the frequency matrix $\Omega=\mathrm{diag}(\omega_1,\ldots,\omega_N)$ is the system of difference equations

$$x_k(t+h)+x_k(t-h)=u(t)\omega_k x_k(t),\quad 1\le k\le N,\quad t\in h\mathbf{Z},\qquad(3.1)$$

where the Lagrange multiplier $u(t)$ is choosen to assure that the vector $x(t)=(x_1(t),\ldots,x_N(t))$ lies on $S=\{x\in\mathbf{R}^N:\langle x,x\rangle=1\}$ for all $t\in h\mathbf{Z}$, so that

$$u(t)=\frac{2\langle x(t),\Omega x(t-h)\rangle}{\langle\Omega x(t),\Omega x(t)\rangle}.\qquad(3.2)$$

Equations (3.1), (3.2) define a map $(x(t),y(t))\mapsto(x(t+h),y(t+h))$, $y(t)=x(t-h)$ from $S\times S$ onto itself. It is easy to see that (3.1), (3.2) imply the equality

$$\langle x(t+h),\Omega x(t)\rangle=\langle x(t),\Omega x(t-h)\rangle,\qquad(3.3)$$

which immediately gives one integral $\langle x,\Omega y\rangle$ for the above–mentioned map.

In fact it is a completely integrable Hamiltonian system. In order to demonstrate this, one needs first of all the invariant symplectic structure. According to the procedure outlined in [17], [16], this can be derived from the representation of (3.1) in a Lagrangian form:

$$\frac{\partial\mathcal{L}(x(t+h),x(t))}{\partial x_k}+\frac{\partial\mathcal{L}(x(t),x(t-h))}{\partial x_k}=0$$

with the Lagrangian function $\mathcal{L}(x,y) = \langle x, \Omega^{-1}y\rangle$ on $S \times S$. The invariant symplectic structure is then the restriction to $S \times S$ of the 2-form $\sum_{j,k=1}^{N}(\partial^2\mathcal{L}/\partial x_j\partial y_k)dx_j \wedge dy_k = \sum_{k=1}^{N}\omega_k^{-1}dx_k \wedge dy_k$. Hence, the invariant Poisson bracket is a Dirac reduction of the bracket

$$\{x_k, y_j\} = \omega_k\delta_{kj} \tag{3.4}$$

to the set $S \times S$ described by the equations

$$\varphi_1(x,y) = \langle x, x\rangle = 1, \quad \varphi_2(x,y) = \langle y, y\rangle = 1. \tag{3.5}$$

The explicit formulas for the invariant Poisson structure follow from (1.6) and read:

$$\{x_k, x_j\}_{\text{Dirac}} = \{y_k, y_j\}_{\text{Dirac}} = 0, \tag{3.6}$$

$$\{x_k, y_j\}_{\text{Dirac}} = \omega_k\delta_{kj} - \frac{\omega_k\omega_j x_j y_k}{\langle x, \Omega y\rangle}. \tag{3.7}$$

A systematic way to derive the integrals of motion for the discrete Neumann system is to represent it in a (discrete–time) Lax form. An $N \times N$ variant of a Lax representation was given in [17], we suggest here an alternative 2×2 variant:

$$L(t + h) = M^{-1}(t)L(t)M(t), \tag{3.8}$$

where $L = L(x,y,\lambda)$, $M = M(x,y,\lambda)$ have the form

$$L = \begin{pmatrix} A(\lambda) & B(\lambda) \\ C(\lambda) & -A(\lambda) \end{pmatrix}, \quad M = \begin{pmatrix} u & \lambda \\ -\lambda & 0 \end{pmatrix}, \tag{3.9}$$

with $u = \dfrac{2\langle x, \Omega y\rangle}{\langle \Omega x, \Omega x\rangle}$ from (3.2),

$$A(\lambda) = \sum_{k=1}^{N}\frac{\omega_k^{-1}x_k y_k}{\lambda^2 - \omega_k^{-2}}, \quad B(\lambda) = \lambda\sum_{k=1}^{N}\frac{y_k^2}{\lambda^2 - \omega_k^{-2}}, \quad C(\lambda) = -\lambda\sum_{k=1}^{N}\frac{x_k^2}{\lambda^2 - \omega_k^{-2}}. \tag{3.10}$$

The $\det(L(\lambda))$ serves again as a generating function for the integrals of motion:

$$\Delta(\lambda) = -A^2(\lambda) - B(\lambda)C(\lambda) = \sum_{k=1}^{N}\frac{F_k}{\lambda^2 - \omega_k^{-2}}$$

with

$$F_k(x,y) = x_k^2 y_k^2 + \sum_{j\neq k}\frac{\omega_j^2(x_k^2 y_j^2 + y_k^2 x_j^2) - 2\omega_k\omega_j x_k y_k x_j y_j}{\omega_j^2 - \omega_k^2}. \tag{3.11}$$

Note that $\sum_{k=1}^{N} F_k = \langle x, x\rangle\langle y, y\rangle$ (so that only $N-1$ of F_k's are independent on $S \times S$), and that $\sum_{k=1}^{N} \omega_k^2 F_k = \langle x, \Omega y\rangle^2$. Integrals F_k commute in the constrained Poisson brackets (3.6), (3.7). A systematic way to demonstrate this is to find an r-matrix structure for the Lax matrix $L(x, y, \lambda)$ from (3.9). A direct calculation gives:

$$\{B(\lambda), B(\mu)\}_{\text{Dirac}} = \{C(\lambda), C(\mu)\}_{\text{Dirac}} = 0,$$

$$\{A(\lambda), A(\mu)\}_{\text{Dirac}} = \frac{1}{\lambda\mu\langle x, \Omega y\rangle}(B(\lambda)C(\mu) - C(\lambda)B(\mu)),$$

$$\{A(\lambda), B(\mu)\}_{\text{Dirac}} = -2\frac{\lambda\mu^{-1}B(\lambda) - B(\mu)}{\lambda^2 - \mu^2} - \frac{2}{\lambda\mu\langle x, \Omega y\rangle}B(\lambda)A(\mu),$$

$$\{A(\lambda), C(\mu)\}_{\text{Dirac}} = 2\frac{\lambda\mu^{-1}C(\lambda) - C(\mu)}{\lambda^2 - \mu^2} + \frac{2}{\lambda\mu\langle x, \Omega y\rangle}C(\lambda)A(\mu),$$

$$\{B(\lambda), C(\mu)\}_{\text{Dirac}} = -4\frac{\lambda\mu^{-1}A(\lambda) - \lambda^{-1}\mu A(\mu)}{\lambda^2 - \mu^2} - \frac{4}{\lambda\mu\langle x, \Omega y\rangle}A(\lambda)A(\mu).$$

The compact form of these relations reads:

$$\{L(\lambda) \overset{\otimes}{,} L(\mu)\}_{\text{Dirac}} = [r(\lambda, \mu), I \otimes L(\mu)] - [r^*(\lambda, \mu), L(\lambda) \otimes I] +$$

$$+ \rho(\lambda, \mu)\,(L(\lambda) \otimes L(\mu)) + (L(\lambda) \otimes L(\mu))\,\rho(\lambda, \mu) -$$

$$- (I \otimes L(\mu))\,\rho(\lambda, \mu)\,(L(\lambda) \otimes I) - (L(\lambda) \otimes I)\,\rho(\lambda, \mu)\,(I \otimes L(\mu)), \quad (3.12)$$

where

$$r(\lambda, \mu) = \frac{2}{\lambda^2 - \mu^2}(E_{11} \otimes E_{11} + E_{22} \otimes E_{22}) + \frac{2\lambda^{-1}\mu}{\lambda^2 - \mu^2}(E_{12} \otimes E_{21} + E_{21} \otimes E_{12}),$$
$$(3.13)$$

$$-r^*(\lambda, \mu) = \frac{2}{\lambda^2 - \mu^2}(E_{11} \otimes E_{11} + E_{22} \otimes E_{22}) + \frac{2\lambda\mu^{-1}}{\lambda^2 - \mu^2}(E_{12} \otimes E_{21} + E_{21} \otimes E_{12}),$$

and

$$\rho(\lambda, \mu) = \frac{1}{\lambda\mu\langle x, \Omega y\rangle}(E_{12} \otimes E_{21} - E_{21} \otimes E_{12}). \quad (3.14)$$

Several remarks are here in turn.

- The relevant algebra is now no longer $sl(2)[\lambda, \lambda^{-1}]$, but its (twisted) subalgebra consisting of matrices with even diagonal entries $A(\lambda)$, $-A(\lambda)$, and odd off-diagonal entries $B(\lambda)$, $C(\lambda)$. If the scalar product in this subalgebra is chosen as

$$(U(\lambda), V(\lambda)) = \text{coeff. of } \lambda^{-2} \text{ in tr } U(\lambda)V(\lambda),$$

then the matrix (3.13) corresponds to the "standard" operator

$$R(U(\lambda)) = \mathcal{P}_+(U(\lambda)) - \mathcal{P}_-(U(\lambda)),$$

where $\mathcal{P}_+$ ($\mathcal{P}_-$) is the non–negative (resp. negative) part of the Laurent series, however on the *twisted subalgebra*.

- The matrix $\rho(\lambda, \mu)$ this time is essentially *dynamical*, i.e. it depends on the point of the phase space. This dependence appears through the function $\langle x, \Omega y \rangle$. Although this function is a spectral invariant of the Lax matrix and hence is constant on each trajectory of our dynamical system, it is however *not a Casimir function* of the bracket (3.6),(3.7), i.e. it is *not constant* on symplectic leaves of the bracket $\{\cdot, \cdot\}_{\text{Dirac}}$, in particular on $S \times S$.

- The quadratic part ot this Poisson bracket, as before, does not influence the equations of motion corresponding to *invariant* Hamiltonian functions $H = H(L)$.

- In contrast with the continuous–time case, in the unconstrained Poisson structure (3.4) the integrals F_k *do not commute*, and hence the pairwise Poisson brackets of the entries of the Lax matrix (3.9) *cannot* be put into an r–matrix form. The simplest way to see it is to compute the Poisson bracket ot the two above–mentioned linear combinations of F_k's:

$$\{\langle x, x \rangle \langle y, y \rangle, \langle x, \Omega y \rangle\} = \langle \Omega x, \Omega x \rangle \langle y, y \rangle - \langle \Omega y, \Omega y \rangle \langle x, x \rangle,$$

which does not vanish identically on $\mathbf{R}^{2N}\{x, y\}$ and even on $S \times S$. The formula analogous to (3.12) looks like

$$\{L(\lambda) \overset{\otimes}{,} L(\mu)\} = -[r^*(\lambda, \mu), I \otimes L(\mu)] + [r(\lambda, \mu), L(\lambda) \otimes I] +$$

$$+ \frac{4\langle x, \Omega y \rangle}{\lambda \mu}(E_{12} \otimes E_{21} - E_{21} \otimes E_{12}).$$

It is the third term on the right–hand side that serves as an obstacle for commutativity of integrals.

As in the previous section, it would be highly desirable to find the Hamiltonian flow interpolating the integrable map (3.8).

4. Separation of variables

As a simple application of the results above we give a construction of separation variables for all the three models discussed above. This construction is due to Sklyanin (see, e.g., [18]) and became standard in the recent years. We define the variables ξ_k, π_k ($1 \leq k \leq N - 1$) for all three models by the relations

$$C(\xi_k) = 0, \quad \pi_k = A(\xi_k) \tag{4.1}$$

(for the Veselov model only non–trivial zeros of $C(\lambda)$ must be taken into account, i.e. $\lambda \neq 0$).

Obviously, for all models it holds:

$$\pi_k^2 + \Delta(\xi_k) = 0. \tag{4.2}$$

One has the following Poisson brackets (we omit in this section the subscript "Dirac" for the sake of brevity):

$$\{\xi_j, \xi_k\} = 0, \quad \{\pi_j, \pi_k\} = 0 \tag{4.3}$$

for all models, and

$$\{\xi_j, \pi_k\} = 2\delta_{jk} \tag{4.4}$$

for the continuous–time Neumann system and the Ragnisco's discretization,

$$\{\xi_j, \pi_k\} = \xi_j^{-1}\delta_{jk} \tag{4.5}$$

for the Veselov's one.

These results imply, according to the standard definition (see [18]) that $(\xi_j, \frac{1}{2}\pi_k)$ are the separation variables for the continuous–time model and the Ragnisco's discretization, and $(\xi_j^2, \frac{1}{2}\pi_k)$ serve as the separation variables for the Veselov's discretization. (Recall that the coefficients of $\Delta(\lambda)$ are always integrals of motion, and that in Veselov's case this function depends only on λ^2).

We restrict ourselves here with the construction of the separation variables. Some results on the discrete–time dynamics in these variables can be found in the second paper of Ref. [15], but much remains to be done.

Proof of (4.3)–(4.5). The first equality in (4.3) follows from the identity

$$\{C(\lambda), C(\mu)\} = 0$$

which holds for all the models. To prove (4.4), (4.5) note first that from the definition (4.1) it follows:

$$0 = \{C(\xi_j), \pi_k\} = \{C(\lambda), \pi_k\}_{\lambda=\xi_j} + C'(\xi_j)\{\xi_j, \pi_k\},$$

so that

$$\{\xi_j, \pi_k\} = -\frac{\{C(\lambda), \pi_k\}_{\lambda=\xi_j}}{C'(\xi_j)}.$$

Here the numerator can be computed with the help of the formulas

$$\{C(\lambda), A(\mu)\} = -2\frac{C(\lambda) - C(\mu)}{\lambda - \mu} + \frac{2}{\langle x, x \rangle}C(\lambda)C(\mu)$$

for the continuous model and its Ragnisco's discretization, and

$$\{C(\lambda), A(\mu)\} = -2\frac{C(\lambda) - \lambda^{-1}\mu C(\mu)}{\lambda^2 - \mu^2} - \frac{2}{\lambda\mu\langle x, \Omega y \rangle}A(\lambda)C(\mu)$$

for the Veselov's model. Substituting $\mu = \xi_k$ in these expression, we are left with:

$$\{C(\lambda), \pi_k\} = -2\frac{C(\lambda)}{\lambda - \xi_k} \quad \text{or} \quad \{C(\lambda), \pi_k\} = -2\frac{C(\lambda)}{\lambda^2 - \xi_k^2},$$

respectively. (It is remarkable that the quadratic part of the Poisson bracket does not contribute to the resulting formulas). Now (4.4), (4.5) are a plain consequence of the l'Hopital rule.

Last, to prove the second equality in (4.3), we deduce:

$$\{\pi_j, \pi_k\} = \{A(\lambda), A(\mu)\}_{\lambda=\xi_j, \mu=\xi_k} + A'(\xi_j)\{\xi_j, \pi_k\} + A'(\xi_k)\{\pi_j, \xi_k\}.$$

Here the sum of the second and the third terms on the right–hand side vanishes due to (4.4), (4.5), and the first term vanishes because

$$\{A(\lambda), A(\mu)\} = 0$$

for the continuous–time Neumann system and the Ragnisco's discretization, and

$$\{A(\lambda), A(\mu)\} = \frac{1}{\lambda\mu\langle x, \Omega y\rangle}\Big(B(\lambda)C(\mu) - C(\lambda)B(\mu)\Big)$$

for the Veselov's one (the contribution of the quadratic part of the Poisson bracket vanishing once more!).

5. Concluding Remarks

Our main goal has been to show, on a classical prototype example, the essential "non-uniqueness" of the (time-)discretisation procedure, which cannot be cured by the "preserving integrability" requirement. The schemes we have considered are moreover interesting by themselves, being related to different r-matrix structures. As we have mentioned, a question that remains open is the discovery of the interpolating hamiltonian flows or, what is the same, the characterisation of the M matrix in terms of r-matrix and invariant functions of the Lax matrix, via a suitable group factorisation.

References

[1] G.R.W.Quispel, J.A.Roberts, C.J.Thompson. – *Physica D, 1990, v.34*, p.183–192
M.Bruschi, O.Ragnisco, P.Santini, and G.Tu. Integrable symplectic maps. – *Physica D, 1991, v. 49, p.273–294.*
O.Ragnisco. A simple method to generate integrable symplectic maps – in: *Solitons and chaos, eds. I.Antoniou, F.J.Lambert (Springer, 1991),* p.227–231;

[2] Yu.B.Suris.–*Funct. Anal. Appl., 1989, v.23, p.84–85*

[3] F.Nijhoff, G.D. Pang. Discrete-time Calogero-Moser model and lattice KP equation.– in: *Symmetries and Integrability of Difference Equations, D. Levi, P. Winternitz and L. Vinet eds., CRM Montreal 1995;* F.Nijhoff, O.Ragnisco, V.Kuzsnetsov. Integrable time-discretization of the Rujisenaars-Schneider model.–*Comm. Math. Phys. (to appear);*

[4] Yu.B.Suris. A discrete-time relativistic Toda lattice.– *Preprint University of Bremen, Centre for Complex Systems and Visualization, 1995.*

[5] B.Grammaticos, A.Ramani. Discrete Painleve equations: coalescences, limits and degeneracies. –*Preprint Universite' de Paris VII, October 1995, solv-int/9510011*

[6] C.Neumann. De problemate quodam mechanica, quod ad primam integralium ultraellipticorum classem revocatur. – *J. Reine Angew. Math., 1859, v.56, p.46–69.*

[7] M.Adler, P. van Moerbecke. Completely integrable systems, Euclidean Lie algebras, and curves. – *Adv. Math., 1980, v.38, p.267–317.*

[8] J.Moser. Geometry of quadrics and spectral theory, – *In: Chern Symposium 1979. Springer, 1981, p.147–188.*

[9] T.Ratiu. The C.Neumann problem as a completely integrable system on an adjoint orbit. – *Trans. Amer. Math. Soc., 1981, v.264, p.321–329.*

[10] H.Flashka. Towards an algebro–geometric interpretation of the Neumann problem. – *Tohoku Math. J., 1984, v.36, p.407–426.*

[11] M.Adams, J.Harnad, and E.Previato. Isospectral Hamiltonian flows in finite and infinite dimensions. – *Commun. Math. Phys., 1988, v.117, p. 451–500.*

[12] J.Avan, M.Talon. Integrability and Poisson brackets for the Neumann–Moser–Uhlenbeck model. – *Int. J. Mod. Phys. A, 1990, v.5, p.4477–4485;*
Alternative Lax structures for the classical and quantum Neumann model. – *Phys. Lett. B, 1991, v.268, p.209–216.*

[13] A.G.Reyman, M.A.Semenov-Tian-Shansky. Group–theoretical method in the theory of finite–dimensional integrable systems. – *In: Encyclopaedia of Math.Sciences, V.16, Dynamical systems VII. Springer, 1993.*

[14] Yu.B.Suris. On the bi–Hamiltonian structure of Toda and relativistic Toda lattices. – *Physics Letters A, 1993, v.180, p.419–429.*

[15] A discrete Neumann system. – *Physics Lett. A, 1992, v.167, p.165–171;*
Dynamical r–matrices for integrable maps. – *Physics Lett. A, 1995, v.198, p.295–305;*
O.Ragnisco, S.Rauch–Wojciechowski. Integrable maps for the Garnier and for the Neumann systems. – *Preprint Linköping Univ., 1994;*
O.Ragnisco, C.Cao, Y.Wu. On the relation of the stationary Toda equation and the symplectic maps. – *J. Phys. A: Math. and Gen., 1995, v.28, p.573–588.*

[16] A.P.Veselov. Integrable systems with discrete time and difference operators. – *Funct. Anal. Appl., 1988, v.22, p.1–13;*
J.Moser, A.P.Veselov. Discrete versions of some classical integrable systems and factorization of matrix polynomials. – *Commun. Math. Phys., 1991, v.139, p.217–243;*
P.Deift, L.-Ch. Li, C.Tomei. Loop groups, discrete versions of some classical integrable systems, and rank 2 extensions. – *Mem. Amer. Math. Soc., 1992, No 479.*

[17] Yu.B.Suris. A discrete–time Garnier system. – *Physics Letters A, 1994, v.189, p.281–289;*
A family of integrable standard–like maps related to symmetric spaces. – *Physics Letters A, 1994, v.192, p.9–16;*
Discrete–time analogs of some nonlinear oscillators in an inverse-square potential. – *J. Phys. A: Math. and Gen., 1994, v.27, p.8161–8169.*

[18] E.K.Sklyanin. Separation of variables in the classical integrable $SL(3)$ magnetic chain. – *Commun. Math. Phys., 1992, v.142, p.123–132.*

Orlando Ragnisco   Yuri. B. Suris
Dipartimento di Fisica   Centre for Complex Systems
Universita di Roma TRE   and Visualizaton
00146 Roma, Italy   University of Bremen
ragnisco@roma1.infn.it   28334 Bremen, Germany
                                           suris@mathematik.uni-bremen.de

Received December, 1995

Multiscale Expansions, Symmetries and the Nonlinear Schrödinger Hierarchy

Paolo Maria Santini

Abstract

We study the propagation of quasi-monocromatic, nondissipative and weakly nonlinear waves, modelled by partial differential equations in $1 + 1$ dimensions, using a multitime expansion. In the case of pure radiation, we show that the asymptotic character of this expansion is guaranted by requiring that the modulation of the leading amplitude of the waves satisfy the nonlinear Schrödinger hierarchy of evolution equations with respect to the slow space-time variables characteristic of the problem. The theory of the symmetries of integrable systems plays a crucial role in the derivation of this result.

1. Introduction

In this paper we review results concerning the propagation of quasi-monocromatic, nondissipative and weakly nonlinear waves, modelled by partial differential equations in $1+1$ dimensions. Most of the presentation is based on the results presented in the work [1], co-authored by Degasperis and Manakov.

It is well-known that some integrable nonlinear PDE's are **universal**, i.e. they can be derived through asymptotically exact techniques of physical significance, in the limit of weak nonlinearity, from large classes of PDE's [2]–[10]. For instance, consider the following general nonlinear PDE in $1+1$ (one spatial and one temporal) dimensions:

$$L(\partial_t, \partial_x)u = N(u), \qquad u = u(x,t) \in \mathcal{R}, \qquad (1)$$

where L is a linear dispersive differential operator with constant coefficients and $N(u)$, the nonlinear part of the equation, is an entire function of u and of its x-derivatives. The linearized equation: $Lu = 0$ admits the monocromatic wave solution:

$$u = Ae^{i\theta} + c.c., \qquad \theta = kx - \omega(k)t, \qquad k \in \mathcal{R} \qquad (2)$$

for an arbitrary complex constant amplitude A, where the frequency $\omega = \omega(k)$ is given in terms of the wave number k through the dispersion relation

$$L(-i\omega, ik) = 0. \tag{3}$$

If one is interested in small solutions: $u = O(\epsilon)$, $\epsilon \ll 1$ of (1), one can show that the effect of the weak nonlinearity on the approximate solution (2) occurs at large space-time scales:

$$\xi = \epsilon x, \qquad t_n = \epsilon^n t, \quad n \in \mathcal{N} \tag{4}$$

and determines a redistribution of the energy to the higher harmonics $e^{i\alpha\theta}$, $\alpha \neq \pm 1$, and a dependence of the coefficients of such harmonics on the slow space and time variables (4) (the modulation of the wave packet). This leads to the following ansatz:

$$u = \sum_{\alpha=-\infty}^{\infty} u^{(\alpha)} e^{i\alpha\theta}, \qquad (u \in \mathcal{R} \Rightarrow u^{(-\alpha)} = u^{(\alpha)^*}) \tag{5}$$

where the amplitude $u^{(\alpha)}$ of each harmonics is expanded in powers of ϵ:

$$u^{(\alpha)} = \sum_{n=n_\alpha}^{\infty} \epsilon^n u_n^{(\alpha)}, \tag{6}$$

with $n_\alpha = |\alpha|$, $\alpha \neq 0$ and $n_0 = 1$, and the coefficients $u_n^{(\alpha)}$ depend only on the slow variables (4). We have therefore the following "multiscale expansion" of the solution of equation (1):

$$\begin{aligned}
u(x,t) &= \sum_{\alpha=-\infty}^{\infty} \sum_{n=n_\alpha}^{\infty} \epsilon^n e^{i\alpha\theta} u_n^{(\alpha)}(\xi, t_1, t_2, t_3, ..) \\
&= \sum_{n=1}^{\infty} \sum_{\alpha=-n}^{n} \epsilon^n e^{i\alpha\theta} u_n^{(\alpha)}(\xi, t_1, t_2, t_3, ..).
\end{aligned} \tag{7}$$

It is well-known [4]–[9] that, if the following "no-resonance condition"

$$L(-i\alpha\omega, i\alpha k) \neq 0, \quad |\alpha| \neq 1, \tag{8}$$

is satisfied, i.e., if no higher harmonics, different from the leading one, solves the linearized equation: $Lu = 0$, then the dependence of the leading amplitude $\psi = u_1^{(1)}$ of $e^{i\theta}$ on the first two stretched times t_1, t_2 is given by

$$u(x,t) = \epsilon\psi(\xi - \omega_1 t_1, t_2) e^{i\theta} + c.c. + O(\epsilon^2), \tag{9}$$

$$\psi_{t_2} = i\omega_2(\psi_{\xi\xi} - 2c|\psi|^2\psi) =: K_2(\psi), \tag{10}$$

where

$$\omega_n = \frac{1}{n!}\frac{d^n\omega(k)}{dk^n}, \qquad n \in \mathcal{N}. \tag{11}$$

Namely: **the leading amplitude ψ translates w.r.t. t_1 with the group velocity and evolves in t_2 according to the celebrated cubic nonlinear Schrödinger (NLS) equation,** which therefore describes the modulation of the amplitude of a monocromatic wave, due to a weak nonlinearity, at the time scale $t = O(\epsilon^{-2})$. The constant coefficient c (in general complex) contains some information on the original equation (1).

In a similar way one could show that:

i) perturbing equation (1) around a finite number M of monocromatic waves

$$u = \sum_{j=0}^{M} \epsilon\psi_j e^{i(k_j x - \omega(k_j)t)} + c.c. + O(\epsilon^2), \tag{12}$$

one obtains, generically, the system of the "nonresonant M-wave interaction" (M-WI)

$$\psi_{j_{t_1}} + \omega'(k_j)\psi_{j_{\xi}} = i\psi_j \sum_{l=1}^{M} \alpha_{jl}|\psi_l|^2, \qquad j = 1,..,n, \tag{13a}$$

and, if $M = 3$ and the following resonance condition

$$\sum_{j=0}^{3} k_j = 0, \qquad \sum_{j=1}^{3} \omega(k_j) = 0 \tag{13b}$$

is satisfied, then one obtains, generically, the system of the "resonant 3-wave interaction" (3-WI) [6],[7], [10] equations:

$$\psi_{j_{t_1}} + \omega'(k_j)\psi_{j_{\xi}} = \sum_{k,l=1}^{n} \gamma_{jkl}\psi_k^*\psi_l^*, \qquad k \neq l \neq j, \qquad j = 1,..,n. \tag{13c}$$

Therefore equations (13a), (13c) describe the nonlinear interaction of a finite number of dispersive waves.

ii) Given a hyperbolic system with weak nonlinearity and dispersion, their effect on the evolution of the Riemann invariants along the characteristics of the hyperbolic system is generically described, on a suitably large space-time scale, by the Korteweg–de Vries (KdV) equation [3],[5]-[7]:

$$u_t = u_{xxx} + uu_x. \tag{14}$$

Such a striking universality makes the NLS, M-WI and KdV equations **ubiquous** and **applicable** in several areas of the Natural Sciences, from

Fluid Dynamics to Plasma Physics, from Nonlinear Optics to Biology, etc....
[5], [6], [7].

This universality is also responsable for the very special and beautiful mathematical properties possessed by these systems since, as pointed out in [9], it is enough that in the large class of PDE's which reduce to the above universal equations there is an integrable one, to garanty that the universal model be **integrable**. Indeed, the main ingredients of integrability, like the existence of an associated Lax pair, the existence of ∞-many commuting symmetries and of ∞-many constants of the motion in involution, are preserved by the above reductive - perturbative procedure. It is not surprising, therefore, that the universal equation (10) is integrable if $c \in \mathcal{R}$ [11].

Concentrating our attention on the NLS equation (10) (but all the considerations we make here extend naturally to any other universal model obtained through a multiple scale expansion, like the KdV and the n-WI equations), the following natural questions arise at this point.

1) What is the evolution of the envelope ψ at larger time scales $t = O(\epsilon^{-k})$, $k > 2$, i.e. with respect to the time variables t_k, $k > 2$?

2) Are the equations describing the evolution of ψ w.r.t. the higher times as universal and integrable as the NLS equation?

To answer these two questions we are obliged to investigate, at each order in ϵ, the asymptotic character of the series generated by the ansatz (7).

Before starting this investigation, which is the goal of the following sections, we give immediately a qualitative answer to these questions, showing how the higher symmetries of the NLS equation (10) must play an important role in this theory.

If the asymptotic expansion can be carried over to higher orders, defining higher dynamics of the envelope ψ:

$$\psi_{t_j} = K_j(\psi), \quad j \in \mathcal{N}, \tag{15}$$

the independence between the different times $t_1, t_2, ...$, which is the essence of the multiscale method, implies necessarily the commutation of the corresponding flows:

$$\psi_{t_j t_m} = \psi_{t_m t_j} \quad \Leftrightarrow \quad K'_j(\psi) \cdot K_m(\psi) = K'_m(\psi) \cdot K_j(\psi), \quad j, m \in \mathcal{N}, \tag{16}$$

where $K'(\psi) \cdot f$ is the so-called "Frechet derivative of K with respect to ψ in the direction f":

$$K'(\psi) \cdot f := \partial_\epsilon K(\psi + \epsilon f)|_{\epsilon=0}. \tag{17}$$

Therefore the higher flows $K_j(\psi)$, $j \in \mathcal{N}$, must be (commuting) symmetries of the NLS equation (10), i.e. (commuting) solutions σ of the linearized (about a solution ψ) NLS equation:

$$[\partial_{t_2} - K'_2(\psi)\cdot]\sigma = 0, \tag{18}$$

where

$$K_2'(\psi) \cdot \sigma = \partial_\epsilon K_2(\psi + \epsilon\sigma)|_{\epsilon=0} = iw_2[\sigma_{\xi\xi} - 2c(\psi^2\bar{\sigma} + 2|\psi|^2\sigma)]. \quad (19)$$

On the other hand, the existence of (higher order) symmetries of a given PDE is a very exceptional fact, and it is typical of integrable PDE's only [12], [13]. It is well-known, for instance, that the NLS equation (10), for $c \in \mathcal{R}$, is the second member (corresponding to $n = 2$ and $d_2 = w_2$) of the following hierarchy of integrable commuting flows:

$$\psi_{t_n} = d_n\sigma_n(\psi), \quad n \in \mathcal{N}, \quad (20a)$$

$$\sigma_n(\psi) := \Phi^n\sigma_0, \quad \sigma_0 := -i\psi, \quad (20b)$$

where

$$\Phi f := i[f_\xi - 4c\psi\partial_\xi^{-1}Re(\bar{\psi}f)] \quad (20c)$$

is the so-called "recursion operator" [14]–[17]. The first few commuting symmetries of equation (10), for $c \in \mathcal{R}$, read:

$$\sigma_0 = -i\psi, \quad \sigma_1 = \psi_\xi, \quad \sigma_2 = i(\psi_{\xi\xi} - 2c|\psi|^2\psi), \quad \sigma_3 = -(\psi_{\xi\xi\xi} - 6c|\psi|^2\psi_\xi),$$

$$\sigma_4 = -i[\psi_{\xi\xi\xi\xi} - 2c(3\psi_\xi^2\bar{\psi} + 4|\psi|^2\psi_{\xi\xi} + 2\psi|\psi_\xi|^2 + \psi^2\bar{\psi}_{\xi\xi}) + 6c^2\psi|\psi|^4]. \quad (21)$$

We conclude that:

i) if equation (10) is integrable (i.e., if $c \in \mathcal{R}$) and if the asymptotic expansion can be carried over to higher orders, then the corresponding dynamics (15) **must** be described by elements of the NLS hierarchy (20). Furthermore, a dimensionality argument implies that: $K_n(\psi) = d_n\sigma_n(\psi)$, for some real constants d_n.

ii) If equation (10) is not integrable, then higher commuting flows do not exist and we expect either that the asymptotic expansion break or that the higher dynamics be essentially described by trivial symmetries of equation (10).

These qualitative considerations are made rigorous in the following section, in which we will show that, for localized purely radiative solutions of the NLS equation (10), a necessary and sufficient condition to keep the expansion (7) asymptotic is that the dependence of the leading amplitude ψ on the slow times t_n, $n \in \mathcal{N}$ is described by the NLS hierarchy (20), with $d_n = (-)^n w_n$.

Therefore the theory of the symmetries of integrable systems in 1+1 dimensions, whose development owes much to Irene Dorfman [16], plays a crucial role in the derivation of the above results. This paper is dedicated to Irene Dorfman.

We end this introductory section remarking that the first attempts to compute the higher terms of the perturbation expansion at all orders have been presented in [3],[4]. The main difference with respect to our approach is the artificial limitation to the dependence on t_1 and t_2 only, and the restriction of the analysis to the pure soliton solution. For an approach similar to ours, but applied to the weakly dispersive limit (which gives rise to the KdV hierarchy) and restricted to the one soliton solution only, the interested reader is referred to [18].

2. Multiscale Expansion and the NLS Hierarchy

In this section we apply the multiscale method to a scalar PDE of the type (1), concentrating our attention on the following two classes of equations:

$$L_+u = [\partial_t + i\omega(-i\partial_x)]u = N(u), \tag{22a}$$

$$L_+L_-u = [\partial_t^2 - \omega^2(-i\partial_x)]u = N(u), \tag{22b}$$

where

$$L_\pm := \partial_t \pm i\omega(-i\partial_x) \tag{22c}$$

and $\omega(\cdot)$ is an odd function, having always in mind the following two concrete examples:

$$A: \qquad u_t - u_{xxx} = (au^3 + bu^5)_x, \tag{23a}$$

$$B: \qquad u_{tt} - u_{xx} + \mu^2 u = au^3 + bu^5, \tag{23b}$$

where a, b, μ are arbitrary real constants. Equation (23a) is a modified KdV equation with a quintic nonintegrable correction; equation (23b) is a nonlinear wave equation.

Inserting the asymptotic ansatz (7) into equations (22), one obtains a double family of equations for the coefficients of the n-th power of ϵ and of the α-th harmonics. For instance, for equations (23a) and (23b) we have, respectively:

$$\sum_{m=1}^{n} L_{+(n-m)}^{(\alpha)} u_m^{(\alpha)} = H_n^{(\alpha)}, \tag{24a}$$

$$H_n^{(\alpha)} = \sum_{\alpha_j=-\infty}^{\infty} \sum_{n_j=1}^{\infty} \{a\Delta(\alpha_1 + \alpha_2 + \alpha_3 - \alpha)$$

$$[i\alpha k\Delta(n_1 + n_2 + n_3 - n)u_{n_1}^{(\alpha_1)}u_{n_2}^{(\alpha_2)}u_{n_3}^{(\alpha_3)}$$

$$+\Delta(n_1 + n_2 + n_3 + 1 - n)(u_{n_1}^{(\alpha_1)}u_{n_2}^{(\alpha_2)}u_{n_3}^{(\alpha_3)})_\xi]$$

$$+b\Delta(\alpha_1 + \alpha_2 + \alpha_3 + \alpha_4 + \alpha_5 - \alpha)$$

$$\cdot[i\alpha k\Delta(n_1 + n_2 + n_3 + n_4 + n_5 - n)u_{n_1}^{(\alpha_1)}u_{n_2}^{(\alpha_2)}u_{n_3}^{(\alpha_3)}u_{n_4}^{(\alpha_4)}u_{n_5}^{(\alpha_5)}$$

$$+\Delta(n_1 + n_2 + n_3 + n_4 + n_5 + 1 - n)(u_{n_1}^{(\alpha_1)}u_{n_2}^{(\alpha_2)}u_{n_3}^{(\alpha_3)}u_{n_4}^{(\alpha_4)}u_{n_5}^{(\alpha_5)})_\xi]\}, \quad (24b)$$

and

$$\sum_{m=1}^{n} L_{+(n-m)}^{(\alpha)} \sum_{m'=1}^{m} L_{-(m-m')}^{(\alpha)} u_{m'}^{(\alpha)} = H_n^{(\alpha)}, \quad (25a)$$

$$\begin{aligned}
H_n^{(\alpha)} &= \sum_{\alpha_j=-\infty}^{\infty} \sum_{n_j=1}^{\infty} \{a\Delta(\alpha_1 + \alpha_2 + \alpha_3 - \alpha) \\
&\quad \cdot \Delta(n_1 + n_2 + n_3 - n)u_{n_1}^{(\alpha_1)}u_{n_2}^{(\alpha_2)}u_{n_3}^{(\alpha_3)} \\
&\quad + b\Delta(\alpha_1 + \alpha_2 + \alpha_3 + \alpha_4 + \alpha_5 - \alpha) \\
&\quad \Delta(n_1 + n_2 + n_3 + n_4 + n_5 - n) \\
&\quad \cdot u_{n_1}^{(\alpha_1)}u_{n_2}^{(\alpha_2)}u_{n_3}^{(\alpha_3)}u_{n_4}^{(\alpha_4)}u_{n_5}^{(\alpha_5)}\},
\end{aligned} \quad (25b)$$

where

$$L_{\pm\ n}^{(\alpha)} := \partial_{t_n} \mp (-i)^{n+1}\frac{\omega^{(n)}(\alpha k)}{n!}\partial_\xi^n \quad (26)$$

and $\Delta(n) = 0$, if $n \neq 0$, $\Delta(0) = 1$.

The no-resonance condition (8) implies that equations (24), (25) for $\alpha \neq \pm 1$ are algebraic and can be solved explicitly to yield the higher harmonics coefficients in terms of the amplitude $\psi = u_1^{(1)}$ and its higher corrections $u_n^{(1)}$, $n > 1$:

$$u_n^{(\alpha)} = u_n^{(\alpha)}(\psi, u_2^{(1)}, u_3^{(1)}, ..), \quad |\alpha| \neq 1. \quad (27)$$

In the following we shall concentrate on the equations for the coefficients $u_n^{(1)}$, $n \in \mathcal{N}$ of the first (and leading) harmonics $e^{i\theta}$. When, in these equations, there appear the coefficients of higher harmonics, we will use the corresponding equations (26) to express them in terms of the $u^{(1)}$'s. At $O(\epsilon)$ we obtain: $L(-i\omega, ik)\psi = 0$, $\psi := u_1^{(1)}$, which gives the dispersion relation (3). For our two examples:

$$A: \qquad \omega = k^3, \quad (28a)$$

$$B: \qquad \omega = (k^2 + 1)^{\frac{1}{2}}. \quad (28b)$$

At $O(\epsilon^2)$ we obtain (using also (3)):

$$L_{+\ 1}^{(1)}\psi = 0 \qquad \Rightarrow \qquad \psi = \psi(\xi - \omega_1 t_1, t_2, t_3, ..). \quad (29)$$

At $O(\epsilon^3)$:

$$L_{+\ 1}^{(1)}u_2^{(1)} = -[\psi_{t_2} - K_2(\psi)], \quad (30)$$

where $K_2(\psi)$ is defined in (10); for our two examples:

$$A: \qquad c = -\frac{a}{2}, \quad (31a)$$

$$B: \qquad c = -\frac{3a}{4\omega\omega_2} = -\frac{3a}{2}(1 + \frac{k^2}{\mu^2}). \qquad (31)$$

Since, from (29), the RHS of equation (30) is in the null space of $L_1^{(1)}$, the solution $u_2^{(1)}$ of equation (30) would blow linearly in the variable $\xi + \omega_1 t_1$, unless the RHS of equation (30) is zero. Therefore, to avoid this secularity, the t_2 - dependence of ψ must be such that the RHS of (30) is zero, i.e. ψ **must** evolves w.r.t. t_2 according to the NLS equation (10). Consequently

$$L_{+\ 1}^{(1)} u_2^{(1)} = 0 \quad \Rightarrow \quad u_2^{(1)} = u_2^{(1)}(\xi - \omega_1 t_1, \ t_2, t_3, ..). \qquad (32)$$

At the next orders in ϵ, the above secularity mechanism fixes recursively the t_1 dependence of all the coefficients $u_n^{(\alpha)}$ of the expansion in the same form:

$$u_n^{(\alpha)} = u_n^{(\alpha)}(\xi - \omega_1 t_1, t_2, t_3, ..), \qquad (33)$$

i.e.: **the whole solution translates with w.r.t. t_1 with the group velocity $\omega'(k)$.**

One is left with the following equations for the corrections $u_n^{(1)}$, $n \geq 2$ of the NLS field $\psi = u_1^{(1)}$:

$$O(\epsilon^{n+2}): \qquad [\partial_{t_2} - K_2'(\psi)]u_n^{(1)} = F_n(\psi, u_2^{(1)}, .., u_{n-1}^{(1)}), \qquad n \geq 2, \ (34)$$

where the forcing F_n depends on the fields ψ, $u_2^{(1)}, .., u_{n-1}^{(1)}$ determined in the previous iterations. Equation (34) has to be viewed as a **forced linear evolution equation for the field $u_n^{(1)}$**. We remark that its homogeneous version is equation (18) defining the NLS symmetries.

It is convenient to extract from the forcing F_n its linear part, which exhibits the following universal character:

$$F_n = -L_{+\ n+1}^{(1)}\psi - \sum_{k=2}^{n-1} L_{+\ n+2-k}^{(1)} u_k^{(1)} + G_n(\psi, u_2^{(1)}, .., u_{n-1}^{(1)}). \qquad (35)$$

Also the nonlinear part G_n of the forcing has a universal character, consisting of the sum of typical blocks that can be guessed a priori through a dimensionality argument; only the coefficients of such blocks depend on the original equation (22). For instance:

$$A: \quad G_2 = 3a(|\psi|^2\psi)_\xi, \qquad (36a)$$

$$B: \quad G_2 = -\frac{3a\omega_1}{2\omega^2}(|\psi|^2\psi)_\xi. \qquad (36b)$$

The general philosophy remains the same: at each order in ϵ we have to guarantee the asymptotic character of the multiscale expansion (7); therefore:

i) we have to identify in each forcing F_n the (potential) secular terms;

ii) we have to fix the (unknown at $O(\epsilon^{n+2})$) dependence of ψ on t_{n+1} in order to eliminate, if possible, such a secular terms.

One can prove by iteration the following result [1].

Let $\psi(\xi, t_2)$ be a solution of the NLS equation (10) for a rapidly decaying initial condition corresponding to pure radiation (the no-soliton case). Then:

i) the secular terms of the forcing F_n are the linear ones;

ii) such secular terms are eliminated imposing that the t_n-dependence of the leading amplitude ψ be described by the NLS hierarchy of evolution equations:

$$\psi_{t_n} = (-)^n \omega_n \sigma_n(\psi) =: (-i)^{n+1} \omega_n \frac{\partial^n \psi}{\partial \xi^n} + V_n(\psi), \quad n \in \mathcal{N}, \quad (37)$$

where the symmetries $\sigma_n(\psi)$ are defined in (20). Consequently, equations (34),(35) take the following secularity free form:

$$[\partial_{t_2} - K_2'(\psi)]u_n^{(1)} = \tilde{F}_n(\psi, u_2^{(1)}, .., u_{n-1}^{(1)}), \quad (38a)$$

$$\tilde{F}_n := -V_n(\psi) - \sum_{k=2}^{n-1} L^{(1)}_{+\ n+2-k} u_k^{(1)} + G_n(\psi, u_2^{(1)}, .., u_{n-1}^{(1)}), \quad n \geq 2,$$
$$(38b)$$

Referring to [1] for the details of the proof by iteration, here we sketch only the first step of this iteration.

First, we remind that an arbitrary localized initial condition $\psi(\xi, t_2 = 0)$ of the NLS equation (10) decomposes asymptotically into a finite number N of solitons plus radiation [11]. In the focusing case ($c < 0$), solitons dominate on the radiation and are generically well separated asymptotically [5]:

$$\psi \sim \psi_{sol}^{(j)}(\xi, t_2) := \frac{2p_j}{\sqrt{|c|}} \frac{e^{i\varphi_j(\xi, t_2)}}{\cosh z_j(\xi, t_2)}, \quad p_j > 0, \ z_j = O(1), \ t_2 1, \ j = 1, .., N,$$
$$(39a)$$

$$\varphi_j(\xi, t_2) := -2s_j \xi + 4\omega_2 t_2(p_j{}^2 - s_j{}^2) + \alpha_j, \quad z_j(\xi, t_2) := 2p_j(\xi + 4s_j \omega_2 t_2) + \beta_j.$$
$$(39b)$$

In the absence of solitons (like in the defocusing ($c > 0$) case), only radiation is present asymptotically:

$$\psi(\xi, t_2) = \psi_{rad}(\xi, t_2)(1 + O(\frac{\ln t_2}{t_2})), \quad t_2 1, \quad (40a)$$

$$\psi_{rad}(\xi, t_2) := \sqrt{\frac{\rho(y)}{\omega_2 t_2}} e^{i[\frac{y^2}{4\omega_2} t_2 - 2c\rho(y)\ln(\omega_2 t_2) + \delta(y)]}, \qquad y = \frac{\xi}{t_2} = O(1),$$

(40b)

where the real slowly varying functions $\rho(y)$, $\delta(y)$ are expressed in terms of (the spectral data of) the initial condition $\psi(\xi, t_2 = 0)$ [19],[6].

With this in mind, we start investigating the first forcing F_2 (see (35),(36)), which appears at $O(\epsilon^4)$.

It was observed in [4] that, in the pure soliton case, only the linear term of F_2:

$$\psi_{sol\xi\xi\xi}^{(j)} = [8is_j(s_j^2 - 3p_j^2) + 8p_j(3s_j^2 - p_j^2)\tanh z_j$$
$$+ 48p_j^2(is_j + p_j\tanh z_j)\cosh^{-2}z_j]\psi_{sol}^{(j)}$$

(41)

contains the two solutions:

$$\Sigma_1 = \frac{ie^{i\varphi_j}}{\cosh z_j}, \qquad \Sigma_2 = \frac{e^{i\varphi_j}\tanh z_j}{\cosh z_j}$$

(42)

of the linearized equation (18). Therefore $\omega_3\psi_{sol\xi\xi\xi}^{(j)}$ is the only secular term, since it forces the solution $u_2^{(1)}$ of equation (34) to contain contributions of the type: $z_j\Sigma_1$ and $z_j\Sigma_2$, which would make the terms $\epsilon u_2^{(1)}$ and ψ comparable for $t_2 = O(\epsilon^{-1})$, destroying the asymptotic character of the expansion (7).

The multitime formalism gives an effective way to deal with such a secularity, using the fact that the t_3-dependence of ψ, unknown at this stage, can be chosen to eliminate this secular terms. This implies the following evolution equations for the the spectral parameters of the soliton:

$$p_{j t_3} = s_{j t_3} = 0, \quad \alpha_{j t_3} = 8\omega_3 s_j(s_j^2 - 3p_j^2), \quad \beta_{j t_3} = 8\omega_3 p_j(p_j^2 - 3s_j^2),$$

(43)

which corresponds to the t_3- dependence described by the complex modified KdV (cmKdV) equation:

$$\psi_{t_3} = \omega_3[\psi_{\xi\xi\xi} - 6c|\psi|^2\psi_\xi].$$

(44)

Even in the absence of solitons one is lead to the t_3-dependence described by (44). Indeed, from (40), it follows that only the linear term of F_2:

$$\omega_3\psi_{\xi\xi\xi} = \omega_3(\frac{iy}{2})^3\psi_{rad}(\xi, t_2)(1 + O(\frac{\ln t_2}{t_2})), \quad y = \frac{\xi}{t_2} = O(1), \quad t_2 \gg 1 \quad (45)$$

contains $O(1/\sqrt{t_2})$ terms which are secular, since they would force the solution $u_2^{(1)}$ of (34) to blow like $O(\sqrt{t_2})$, destroying the asymptotic character of the expansion (7). To eliminate such a secular term, the t_3-dependence of $\rho(y)$ and $\delta(y)$ **must** be as follows:

$$\rho_{t_3} = 0, \qquad \delta_{t_3} = -(\frac{iy}{2})^3 \omega_3, \tag{46}$$

corresponding to equation (44).

Replacing (44) into the forcing F_2, one obtains the secularity free forcing:

$$\tilde{F}_2 = 6c\omega_3 |\psi|^2 \psi_\xi + G_2(\psi). \tag{47}$$

For our two examples:

$$A: \quad \tilde{F}_2 = -6c\psi(|\psi|^2)_\xi, \tag{48a}$$

$$B: \quad \tilde{F}_2 = \frac{3ak}{2\omega^3}\psi(\psi^*\psi_\xi - \psi_\xi^*\psi). \tag{48b}$$

For the pure radiation case, this procedure can be iterated successfully to all orders in ϵ [1]. For the pure one soliton case, one can follow the analysis in [4], implemented by the formalism of [1], which includes all the slow times (4). The interesting problem of investigating the above perturbative series in the generic case (solitons plus radiation) is still open. It is plausible that, if equation (1) is not integrable, the perturbative expansion will loose its asymptotic character at some order in ϵ; up to that order, the dependence on the higher times: t_n, $n \in \mathcal{N}$ of the soliton parameters $p_j, s_j, \alpha_j, \beta_j$ and of the radiation slowly varying functions ρ, δ is the one described by the NLS hierarchy (37).

This theory, although developed in [1] for scalar PDE's of the type (22), has a general validity; in particular:

1) it has been successfully applied to the following two physical situations [20],[21]:

   i) the one-dimensional propagation of surface waves in an ideal nonrotational fluid;

   ii) the one-dimensional propagation of ion-acoustic waves in a Plasma.

2) it can be extended to the case partial differential equations in 2+1 dimensions [20].

3. Concluding Remarks

As we have already pointed out, the above results can be obtained starting from large classes of nonlinear PDE's of the type (22). Differences between different equations (22) give rise only to different coefficients of the secularity free terms of the forced equations (34), and do not seem to alter the asymptotic character of the multiscale expansion (7) in the pure radiation case. In order to appreciate such differences and, in particular, in order to distinguish between integrable and nonintegrable equations (22), it is necessary to implement properly the method. With this goal, we have recently discussed a method in which the above multiscale expansion is used to search for "approximate symmetries" of the original equation [22].

References

[1] A. Degasperis, S.V. Manakov and P.M. Santini, Multiple-scale perturbation beyond the nonlinear Schroedinger equation, preprint Dipartimento di Fisica, Università "La Sapienza", Roma (1995).

[2] T. Taniuti, *Suppl. Proc. Th. Phys.* **55** (1974), 1.

[3] Y. Kodama and T. Taniuti, *J. Phys. Soc. Japan* **45** (1978), 298.

[4] Y. Kodama, *J. Phys. Soc. Japan* **45** (1978), 311.

[5] S. P. Novikov, S.V. Manakov, L. P. Pitaeskij and V. E. Zakharov, *Theory of Solitons, the Inverse Scattering Method*, Consultant Burea, New York (1984).

[6] M. J. Ablowitz and H. Segur, *Solitons and the Inverse Scattering Transform*, Siam, Philadelphia (1981).

[7] A.C. Newell, *Solitons in Mathematics and Physics*, **45** SIAM, Philadelphia, PA (1985).

[8] V. E. Zakharov and E. A. Kuznetsov, *Physica* **18** D (1986), 455.

[9] F. Calogero and W. Eckhaus, *Inverse Problems* **3** (1987), 229.

[10] F. Calogero, *J. Math. Phys.* **30** (1989), 28.

[11] V. Zakharov and A. Shabat, *Sov. Phys. JETP* **34** (1972), 62.

[12] A. S. Fokas, *Stud. Appl. Math.* **77** (1987), 253.

[13] A.V. Mikhailov, A.B. Shabat and V.V. Sokolov, in *What is Integrability?*, edited by V.E. Zakharov Springer, Berlin (1990).

[14] M.J. Ablowitz, D. Kaup, A.C. Newell and H. Segur, *Stud. Appl. Math.* **53** (1974), 249.

[15] F. Magri, *J. Math. Phys.* **19** (1978), 1156; also in: *Nonlinear Evolution Equations and Dynamical Systems*, edited by M. Boiti, F. Pempinelli and G. Soliani, *Lectures Notes in Physics* **120** Springer, N.Y. (1980), 233.

[16] I. Gel'fand and I. Dorfman, *Funct. Anal Appl.* **13** (1979), 13; **14** (1980), 71.

[17] A.S. Fokas and B. Fuchssteiner, *Lett. Nuovo Cimento* **28** (1980), 299; Physica **4D** (1981), 47.

[18] R.A. Kraenkel, M.A. Manna and J.G. Pereira, *J. Math. Phys.* **36** (1995), 307.

[19] V.E. Zakharov, S.V. Manakov, *Sov. Phys. JETP* **44** (1976), 106.

[20] V. Catoni, Sviluppi a molti tempi in Fisica e la gerarchia dell'equazione di Schroedinger non lineare, Tesi di Laurea, Dipartimento di Fisica, Università La Sapienza, Roma, (1995).

[21] V. Catoni and P.M. Santini: "Multiscale expansions in Fluid Dynamics and Plasma Physics and the nonlinear Schrödinger hierarchy;" Preprint in preparation.

[22] A. Degasperis and P.M. Santini, Multiscale expansions of nonlinear partial differential equations and approximate symmetries, preprint in preparation.

Dipartimento di Fisica
Università di Catania
Corso Italia 57
95129 Catania, Italy

I. N. F. N., Sezione di Roma
Piazz.le A.Moro 2
00185 Roma, Italy
email: santini@catania.infn.it
          santini@roma1.infn.it

Revised February, 1996

On a Laplace Sequence of Nonlinear Integrable Ernst-Type Equations

W.K. Schief and C. Rogers

Abstract

Invariance under Laplace-Darboux-type transformations is established for the 2+1-dimensional Loewner-Konopelchenko-Rogers integrable system. This is exploited to derive a chain of novel, integrable Ernst-type equations which contain an arbitrary parameter.

1. Introduction

Laplace-Darboux-type transformations owe their origin to early work by Laplace [1] on form-invariance and reduction of linear hyperbolic equations. Later, Darboux [2] demonstrated that such transformations have interesting applications in the study of conjugate nets. In that context, they are applied iteratively to construct suites of surfaces on which the parametric curves constitute conjugate nets. Accounts of these geometric applications are to be found in the treatises of Eisenhart [3] and Lane [4].

Remarkably, it turns out that classical Laplace-Darboux transformations and their natural extensions have important applications in modern Soliton Theory. In particular, the concept of periodic fixed points of Laplace sequences leads to the celebrated integrable Toda lattice system [5]. Indeed, the two-dimensional version of the Toda lattice model was derived in the last century by Darboux.

In a recent development, Konopelchenko [6] has constructed generalised matrix Laplace transformations and derived thereby interesting invariance properties of a matrix-valued generalisation of the Boiti-Leon-Pempinelli (BLP) equation [7]. The latter represents a weak 2+1-dimensional integrable extension of the classical sinh-Gordon equation. It may be contrasted with the strong 2+1-dimensional extension of the sinh-Gordon equation embedded in the Loewner-Konopelchenko-Rogers (LKR) integrable system wherein the two spatial variables occur on an equal footing [8]–[10].

Here, matrix Laplace transformations are presented for the LKR system. In particular, application is made to a 2+1-dimensional integrable

Ernst-type equation recently introduced by Schief [10]. In a 2+0 dimensional reduction, a novel sequence of integrable Ernst-type equations involving an arbitrary parameter is constructed. An important subclass may be iteratively reduced to the classical Ernst equation by a sequence of Laplace transformations.

2. Matrix Laplace Transformations and the LKR System

It is readily established that the linear matrix equation

$$L\phi = \phi_{xt} + A\phi_x + B\phi_t + C\phi = 0 \tag{1}$$

where A, B and C are matrix-valued functions of the independent variables x and t is form-invariant under the Laplace-Darboux-type transformation

$$
\begin{aligned}
\phi_+ &= \mathcal{L}_+\phi = (\partial_t + A)\phi \\
A_+ &= HAH^{-1} - H_tH^{-1}, \quad B_+ = B \\
C_+ &= B_t - H + (HAH^{-1} - H_tH^{-1})B,
\end{aligned}
\tag{2}
$$

where the Laplace 'invariant'[1] H is given by

$$H = (\partial_x + B)(\partial_t + A) - L = A_x + BA - C. \tag{3}$$

Thus, the presence of non-zero H represents an impediment to the factorization of L and it is natural to seek iterative sequences of Laplace-Darboux transformations which result in vanishing H.

The investigation of form-invariance under Laplace-Darboux transformations is now extended to 2+1-dimensional LKR systems of the type [8, 9]

$$
\begin{aligned}
\phi_y &= S\phi_x \\
\phi_{xt} &= V\phi_x + W\phi \\
\phi_{yt} &= V\phi_y + SW\phi
\end{aligned}
\tag{4}
$$

with compatibility the integrable nonlinear system

$$
\begin{aligned}
S_t &= [V, S] \\
V_y - V_xS + [W, S] &= 0 \\
W_y - (SW)_x &= 0.
\end{aligned}
\tag{5}
$$

In the linear representation (4), the matrix ϕ now depends on x, t and the additional variable y.

---

[1]It is noted that H transforms as $H \rightarrow gHg^{-1}$ under the gauge transformation $\phi \rightarrow g\phi$

The identification $(A, B, C) = (-V, 0, -W)$ shows that equation $(4)_2$ of the above LKR representation is form-invariant under $\mathcal{L}_+$ where

$$\phi_+ = \mathcal{L}_+ \phi = \phi_t - V\phi. \tag{6}$$

Now, in general, (3) implies that

$$H\phi = (\partial_x + B)\phi_+ \tag{7}$$

while, in the case of the above LKR system, this yields

$$\phi = H^{-1}\phi_{+x} = W_+^{-1}\phi_{+x}. \tag{8}$$

The latter suggests the introduction of the Laplace-Darboux transformation $\mathcal{L}_-$ such that

$$\phi_- = \mathcal{L}_-\phi = W^{-1}\phi_x \tag{9}$$

whence,

$$
\begin{aligned}
(\mathcal{L}_- \circ \mathcal{L}_+)\phi = \mathcal{L}_-\phi_+ &= W_+^{-1}\phi_{+x} \\
&= W_+^{-1}\partial_x(\phi_t - V\phi) \\
&= W_+^{-1}(\phi_{xt} - V_x\phi - V\phi_x) \\
&= W_+^{-1}(W - V_x)\phi = W_+^{-1}W_+\phi = \phi
\end{aligned}
\tag{10}
$$

so that

$$\mathcal{L}_- \circ \mathcal{L}_+ = \text{id}. \tag{11}$$

In what follows, it proves convenient to focus attention on $\mathcal{L}_-$ which delivers

$$
\begin{aligned}
V_- &= W^{-1}VW - W^{-1}W_t \\
W_- &= W + (W^{-1}VW - W^{-1}W_t)_x.
\end{aligned}
\tag{12}
$$

Since the first two equations of the LKR representation (4) imply the third, modulo the conditions (5), it remains to establish that $(4)_1$ is form-invariant under $\mathcal{L}_-$. In fact, direct substitution readily shows that

$$\phi_{-y} = S_-\phi_{-x} \tag{13}$$

where

$$S_- = W^{-1}SW. \tag{14}$$

Accordingly, it is seen that the LKR system (4) is form-invariant under the Laplace-Darboux transformation (9) with $(S, V, W) \rightarrow (S_-, V_-, W_-)$.

It is observed that (9) may be brought into the form

$$\phi_- = \Phi_x^{-1}\phi_x \tag{15}$$

where Φ is the potential associated with $(5)_3$ and, accordingly, itself satisfies $(4)_1$. Now, bearing in mind that the latter is nothing but the 'scattering problem' for the multi-component modified Kadomtsev-Petviashvili (mKP) hierarchy, it is readily deduced that (15) is a composite gauge and Darboux transformation generated by the 'eigenfunction' Φ [11, 12].

In the next section, the form-invariance of the LKR representation (4) under the Laplace-Darboux transformation (9) is exploited in connection with Ernst-type equations which lie within the LKR integrable system.

3. Application to Integrable Ernst-Type Equations

In [10], a 2+1-dimensional Ernst-type equation was derived via the LKR system (4), namely

$$\left[\partial_t - \frac{\mathcal{E}_t}{\Re(\mathcal{E})} + i\Re(\rho) \right] \left[\mathcal{E}_{xx} + \mathcal{E}_{yy} - \frac{\mathcal{E}_x^2 + \mathcal{E}_y^2}{\Re(\mathcal{E})} \right]$$

$$+ \mathcal{E}_x \left[\frac{\mathcal{E}_t \bar{\mathcal{E}}_x - \bar{\mathcal{E}}_t \mathcal{E}_x}{2\Re(\mathcal{E})^2} + i\rho_x \right] + \mathcal{E}_y \left[\frac{\mathcal{E}_t \bar{\mathcal{E}}_y - \bar{\mathcal{E}}_t \mathcal{E}_y}{2\Re(\mathcal{E})^2} + i\rho_y \right] = 0$$

$$(16)$$

$$i(\rho_{xx} + \rho_{yy}) + \left[\frac{\mathcal{E}_t \bar{\mathcal{E}}_x - \bar{\mathcal{E}}_t \mathcal{E}_x}{2\Re(\mathcal{E})^2} \right]_x + \left[\frac{\mathcal{E}_t \bar{\mathcal{E}}_y - \bar{\mathcal{E}}_t \mathcal{E}_y}{2\Re(\mathcal{E})^2} \right]_y + \left[\frac{\mathcal{E}_x \bar{\mathcal{E}}_x + \mathcal{E}_y \bar{\mathcal{E}}_y}{2\Re(\mathcal{E})^2} \right]_t = 0.$$

This corresponds to a 2×2 reduction of the system (5) with the additional consistent constraints

$$S^2 = -\mathbf{1}, \quad \text{Tr}\, S = 0 \tag{17}$$

wherein the matrices S, V and W have been parametrized according to

$$S = \frac{1}{\mathcal{E} + \bar{\mathcal{E}}} \begin{pmatrix} i(\mathcal{E} - \bar{\mathcal{E}}) & -2\mathcal{E}\bar{\mathcal{E}} \\ 2 & -i(\mathcal{E} - \bar{\mathcal{E}}) \end{pmatrix}$$

$$V = \tfrac{1}{2}(\varrho - S_t)S + \tfrac{1}{2}\nu \tag{18}$$

$$W = \tfrac{1}{4}[(\varrho - S_t)S + \sigma]_x - \tfrac{1}{4}S[(\varrho - S_t)S + \sigma]_y.$$

Here, the quantities ϱ, σ and ν are defined via the decomposition

$$\rho = \varrho + i\sigma \tag{19}$$

and the relations

$$\nu_x = \varrho_y - i\frac{\mathcal{E}_t \bar{\mathcal{E}}_y - \bar{\mathcal{E}}_t \mathcal{E}_y}{2\Re(\mathcal{E})^2}, \quad \nu_y = -\varrho_x + i\frac{\mathcal{E}_t \bar{\mathcal{E}}_x - \bar{\mathcal{E}}_t \mathcal{E}_x}{2\Re(\mathcal{E})^2}. \tag{20}$$

It is noted that the integrability condition for the system (20) is satisfied modulo the 2+1-dimensional Ernst-type equation (16). The latter was shown in [10] to admit Darboux-Levi-type transformations and to possess integrable reductions akin to the 2+1-dimensional sine-Gordon system/equation given in [8, 9], a Darboux system descriptive of conjugate coordinate systems, as well as the well-known self-induced transparency (SIT) equations.

In the t-independent case, (16) adopts the form

$$\mathcal{E}_{xx} + \mathcal{E}_{yy} + \frac{\rho_x \mathcal{E}_x + \rho_y \mathcal{E}_y}{\Re(\rho)} = \frac{\mathcal{E}_x^2 + \mathcal{E}_y^2}{\Re(\mathcal{E})}, \qquad \rho_{xx} + \rho_{yy} = 0. \qquad (21)$$

This represents a new integrable extension of the well-known Ernst equation of General Relativity [13, 14]. The latter is retrieved in the case $\Im(\rho) = 0$.

Here, we focus on the action of the Laplace transformation $\mathcal{L}_-$ on the generalized Ernst equation (21). In this case, S, V and W transform according to

$$S_- = W^{-1}SW, \quad V_- = W^{-1}VW, \quad W_- = W + (W^{-1}VW)_x \qquad (22)$$

and (20) reduce to the Cauchy-Riemann equations

$$\nu_x = \varrho_y, \quad \nu_y = -\varrho_x. \qquad (23)$$

By virtue of the parametrization (18), it is seen that the traces of $(22)_2$ and of the latter multiplied by S_- read

$$\nu_- = \nu, \quad \varrho_- = \varrho \qquad (24)$$

respectively, whereas those of $(22)_3$ yield

$$\sigma_- = \sigma + 2\nu + \text{const.} \qquad (25)$$

Hence, we conclude, without loss of generality, that

$$\rho_- = \rho + 2i\nu \qquad (26)$$

which implies that the condition $\Im(\rho) = 0$ associated with the Ernst equation is not preserved under $\mathcal{L}_-$.

In view of (26), it is natural to specialize the imaginary part of ρ to be proportional to ν. Thus, we set

$$\rho = \varrho - in\nu, \qquad (27)$$

where n is an arbitrary constant, which takes the Ernst-type equation (21) to

$$\mathcal{E}_{z\bar{z}} + \frac{1}{2}(1-n)\frac{\varrho_{\bar{z}}}{\varrho}\mathcal{E}_z + \frac{1}{2}(1+n)\frac{\varrho_z}{\varrho}\mathcal{E}_{\bar{z}} = \frac{\mathcal{E}_z \mathcal{E}_{\bar{z}}}{\Re(\mathcal{E})}, \qquad \varrho_{z\bar{z}} = 0, \qquad (28)$$

where the complex variable $z = x + iy$ has been introduced. In this case, evaluation of the transformation formula $(22)_1$ for S in terms of $\mathcal{E}$ yields the Laplace transform

$$\mathcal{E}_- = \frac{(1-n)\mathcal{E}\Re(\mathcal{E})\varrho_{\bar{z}} + \varrho\bar{\mathcal{E}}\mathcal{E}_{\bar{z}}}{(1-n)\Re(\mathcal{E})\varrho_{\bar{z}} - \varrho\mathcal{E}_{\bar{z}}}$$

$$n_- = n - 2, \quad \varrho_- = \varrho.$$

(29)

The inverse of this transformation associated with $\mathcal{L}_+$ is readily obtained via the interchange $(\bar{z}, n) \leftrightarrow (z, -n)$, i.e.

$$\mathcal{E}_+ = \frac{(1+n)\mathcal{E}\Re(\mathcal{E})\varrho_z + \varrho\bar{\mathcal{E}}\mathcal{E}_z}{(1+n)\Re(\mathcal{E})\varrho_z - \varrho\mathcal{E}_z}$$

$$n_+ = n + 2, \quad \varrho_+ = \varrho.$$

(30)

In particular, it is important to note that the Ernst-type equation (28) may be reduced to the Ernst equation $(n = 0)$ via a sequence of Laplace transformations $\mathcal{L}_+$ iff n is an even integer.

We conclude with two remarks. Firstly, transformations which change parameters in integrable ordinary differential equations of Painlevé-type are well-established [15]. These are typically inherited from Bäcklund transformations for 1+1-dimensional integrable equations [16].

Secondly, it is readily seen that the linearized version of the Ernst-type equation (28), viz

$$\hat{\mathcal{E}}_{z\bar{z}} + \frac{1}{2}(1-n)\frac{\varrho_{\bar{z}}}{\varrho}\hat{\mathcal{E}}_z + \frac{1}{2}(1+n)\frac{\varrho_z}{\varrho}\hat{\mathcal{E}}_{\bar{z}} = 0, \quad \varrho_{z\bar{z}} = 0,$$

(31)

admits the classical Laplace-Darboux transformation

$$\hat{\mathcal{E}}_+ = [(1+n) + 2\frac{\varrho}{\varrho_z}\partial_z]\hat{\mathcal{E}}, \quad n_+ = n + 2, \quad \varrho_+ = \varrho.$$

(32)

In fact, in the limit $\mathcal{E} = \hat{\mathcal{E}} + c$, $c \to \infty$, where c is a real constant, the Ernst-type equation (28) linearizes to (31) while the Laplace-Darboux transformation (30) reduces to the classical one (32) up to an irrelevant constant factor.

Acknowledgement. The authors wish to express their gratitude to Professor Jack Hale and Professor Shui-Nee Chow for facilitating their visit to the Center for Dynamical Systems and Nonlinear Studies, Georgia Institute of Technology where this work was carried out. The support of the Australian Research Council is also gratefully acknowledged.

References

[1] P.S. Laplace, Recherches sur le Calcul Intégral aux Différences Partielles, *Mémoires de l'Academie Royale des Sciences de Paris* (1773).

[2] G. Darboux, *Leçons sur la Théorie Générale des Surfaces*, Gauthier-Villars, Paris (1887).

[3] L.P. Eisenhart, *Transformation of Surfaces*, Chelsea, New York (1962).

[4] E.P. Lane, *Projective Differential Geometry of Curves and Surfaces*, The University of Chicago Press, Chicago (1932).

[5] M. Toda, *J. Phys. Soc. Japan* **22** (1967), 431.

[6] B.G. Konopelchenko, *Phys. Lett. A* **156** (1991), 221.

[7] M. Boiti, J.J.P. Leon and F. Pempinelli, *Inverse Problems* **3** (1987), 25.

[8] B.G. Konopelchenko and C. Rogers, *Phys. Lett. A* **158** (1991), 391.

[9] B.G. Konopelchenko and C. Rogers, *J. Math. Phys.* **34** (1993), 214.

[10] W.K. Schief, *Proc. R. Soc. London A* **446** (1994), 381.

[11] W. Oevel and C. Rogers, *Rev. Math. Phys.* **5** (1993), 299.

[12] W. Oevel and W. Schief, *Rev. Math. Phys.* **6** (1994), 1301.

[13] F. Ernst, *Phys. Rev.* **167** (1968), 1175.

[14] F. Ernst, *Phys. Rev.* **168** (1968),1415.

[15] A.S. Fokas and M.J. Ablowitz, *J. Math. Phys.* **23** (1982), 2033.

[16] W.K. Schief, *J. Phys. A: Math. Gen.* **27** (1994), 547.

School of Mathematics
The University of New South Wales Sydney 2052, Australia
email: CRogers@unsw.edu.au

Received November, 1995

Classical and Quantum Nonultralocal Systems on the Lattice

Michael Semenov-Tian-Shansky and Alexey Sevostyanov

Abstract

We classify nonultralocal Poisson brackets for 1-dimensional lattice systems and describe the corresponding regularizations of the Poisson bracket relations for the monodromy matrix. A nonultralocal quantum algebras on the lattices for these systems are constructed. For some class of such algebras an ultralocalization procedure is proposed. The technique of the modified Bethe–Anzatz for these algebras is developed and is applied to the nonlinear sigma model problem.

Introduction

This article is devoted to an old problem, which arose in the beginning of the development of the Classical Inverse Scattering Method (CISM) [1]. An important point of CISM is the calculation of the Poisson brackets relations for the monodromy matrix of an auxiliary linear problem. This calculation is usually performed under the technical assumption of 'ultralocality' of the Poisson brackets for local variables (this condition means simply that the Poisson operator defining the bracket is a multiplication operator and does not contain any derivations). In many interesting models this condition is violated, and in this case getting consistent Poisson brackets relations for the monodromy becomes nontrivial. Technically, the trouble is that the Frechet derivative of the monodromy has a discontinuity, and so one has to extend a differential operator to functions with a jump. It is easy to observe that Poisson operators are nonultralocal precisely for the models with non-skew-symmetric r-matrices. A naive calculation of the Poisson brackets for the monodromy in this case gives

$$\{M_1, M_2\} = aM_1M_2 - M_1M_2a,$$
$$a = \tfrac{1}{2}\left(r - r^*\right). \tag{1}$$

This bracket does not satisfy the Jacobi identity, since the skew part of r usually does not satisfy the Yang–Baxter identity (in fact, the bracket (1) is inconsistent even if it does). A natural way to regularize the monodromy brackets in this case has been proposed in [19]. This method allows to regularize some (though not all) of the Poisson brackets of the type (1). The idea is that to extend the Poisson operator to functions with a jump one has to add to it a boundary form sensitive to the jump, which is well in the spirit of the operator extensions theory.

In this article we classify all regularized r-matrices and all regularizations of this kind using the Belavin–Drinfeld classification theorem for the modified Yang–Baxter equation [5]. Unfortunately, our classification is given in an implicit form because the Belavin–Drinfeld classification theorem describes solutions of the modified Yang–Baxter equation only up to automorphisms of the corresponding affine Lie algebra. This fact doesn't enable to write all regularizations in the explicit form. But we give a natural way to find all regularizations. We define the corresponding quantum algebras by means of the Faddeev-Reshetikhin-Takhtajan approach [10]. The same class of Poisson structures and of the corresponding quantum algebras has been recently studied in a slightly different way by J.M. Maillet and L. Freidel [22] and by S. Parmentier [21]; to describe them we use the unified approach based on the notion of the twisted double (cf.[18], [20])

The second goal of the present work is to construct quantum nonultralocal systems on the lattice, which possess infinite series of conservation laws and to calculate the spectrum of the corresponding commuting operators. For this calculation we develop a generalization of the Bethe–Ansatz construction.

About the contents of this paper:

In Section 1 we review the construction of Poisson algebras on the lattice arising in the study of Lax equations on the lattice with non-ultralocal Poisson brackets.

In Section 2 we remind the main construction of [19]. We reformulate the Belavin–Drinfeld classification theorem [5] in terms of the affine root systems. This reformulation is convenient for our purposes. We reduce the classification of regularizations to the search of some class of solutions of the Yang–Baxter equation on the square of a finite-dimensional Lie algebra.

In Section 3 we discuss the main examples of regularizations and the corresponding nonultralocal algebras and investigate their algebraic properties. In particular, we determine their centers; under some additional conditions it is possible to find a new system of generators of these algebras which already satisfy *local* commutation relations This ultralocalization procedure has been discussed earlier in [13]. We present new examples of ultralocalization; the new system of generators is related to the original one by an appropriate quantum lattice gauge transformation. At the end of Section 3 we describe a generalization of the algebraic Bethe Ansatz for nonultralocal algebras.

In Section 4 we apply the technique developed in the previous sections to the nonlinear sigma model problem. It is well known that integrable models usually admit several different Poisson structures; the simplest one for the nonlinear sigma model is associated with its standard Lagrangian formulation. We were unable to find a regularization of this Poisson structure; however, the general scheme introduced in section 2 may be applied to another, and a fairly natural Poisson structure which we introduce in this section for a nonlinear sigma model with values in an arbitrary Riemannian symmetric space. We explicitly describe the corresponding quantum lattice systems. For the n-field (i.e., the sigma model with values in the unit sphere S^2) we get a representation of the local quantum lattice Lax operator via the canonical Weyl pairs. It turns out that the n-field with this Poisson structure is gauge equivalent to the lattice Sine–Gordon model.

In the conclusion we discuss some open problems.

1. General Construction of Lattice Algebras

It is natural to assume that the phase space of a mechanical system associated with a 1-dimensional lattice $\Gamma = \mathbf{Z}/N\mathbf{Z}$ is the direct product $\mathcal{M}^N$ of "1-particle spaces." In applications to integrable systems these "elementary" phase spaces are parametrized by Lax matrices and hence are modeled on submanifolds of an appropriate Lie group (usually, a loop group associated with a finite-dimensional semisimple Lie group). In simple cases the Poisson structure on $\mathcal{M}^N$ is the product structure. (The corresponding Poisson bracket is called *ultralocal*.) The auxiliary linear problem associated with Lax equations on the lattice is

$$\psi_{n+1} = L^n \psi_n. \tag{2}$$

The associated monodromy map is the product map

$$M : G^N \to G : (L^1, ..., L^N) \mapsto \prod_{n=1}^{N} L^n. \tag{3}$$

It is natural to demand that M be a Poisson map. In the ultralocal case this condition means that G should be a Poisson Lie group. It is interesting (and also important for applications) to study the most general Poisson structures on G^N which are compatible with this property of the monodromy. The corresponding Poisson algebras are referred to as *lattice algebras*. First examples of nonultralocal lattice algebras appeared in [18]; further examples and a classification (for finite dimensional semisimple Lie algebras) appeared in [22], [4], [21]. In this section we briefly recall the construction of lattice algebras using the approach proposed in [18], [19].

Fix an affine Lie algebra $\mathfrak{g}$ with a normalized invariant bilinear form $\langle \cdot, \cdot \rangle$. It is well known that $\mathfrak{g}$ admits the structure of a quasitriangular Lie bialgebra (the corresponding classical r-matrices are listed in [5]). Put $\mathfrak{d} = \mathfrak{g} \oplus \mathfrak{g}$. We define the bilinear invariant form on the square of $\mathfrak{g}$ in the following way:

$$\langle\langle (X_1, Y_1), (X_2, Y_2) \rangle\rangle = \langle X_1, X_2 \rangle - \langle Y_1, Y_2 \rangle, \tag{4}$$

so that the diagonal subalgebra is isotropic. As a Lie algebra, $\mathfrak{d}$ is isomorphic to the double of $\mathfrak{g}$. (This isomorphism does not depend on a particular choice of the r-matrix.) Hence $\mathfrak{d}$ carries a natural r-matrix, the r-matrix of the double; for our present goals, however, we shall need *arbitrary* classical r-matrices on $\mathfrak{d}$ which define on it the structure of a quasitriangular Lie bialgebra. In other words, we are interested in r-matrices which are skew with respect to(4) and satisfy the modified classical Yang–Baxter equation on $\mathfrak{g} \oplus \mathfrak{g}$.

Let $\mathcal{R}$ be such a solution; it may be written in the block form

$$\mathcal{R} = \begin{pmatrix} A & B \\ B^* & D \end{pmatrix}, \quad A^* = -A, \; D^* = -D. \tag{5}$$

For $\varphi \in Fun(D), D = G \times G$ let $\mathrm{D}\varphi, \mathrm{D}'\varphi \in (\mathfrak{g} \oplus \mathfrak{g})^*$ be the left and right derivatives of φ:

$$\begin{aligned} \langle\langle \mathrm{D}\varphi(g), X \rangle\rangle &= \tfrac{d}{dt}\big|_{t=0}\, \varphi\left(e^{tX} g\right), \\ \langle\langle \mathrm{D}'\varphi(g), X \rangle\rangle &= \tfrac{d}{dt}\big|_{t=0}\, \varphi\left(g e^{tX}\right), \\ g \in D, X &\in \mathfrak{g} \oplus \mathfrak{g}. \end{aligned} \tag{6}$$

It is well known that for any two solutions $\mathcal{R}_1, \mathcal{R}_2$ of the modified classical Yang–Baxter equation the bracket

$$\{\varphi, \psi\}_{\mathcal{R}_1, \mathcal{R}_2} = \langle\langle \mathcal{R}_1 \mathrm{D}\varphi, \mathrm{D}\psi \rangle\rangle + \langle\langle \mathcal{R}_2 \mathrm{D}'\varphi, \mathrm{D}'\psi \rangle\rangle \tag{7}$$

satisfies the Jacobi identity. Let us take, in particular, $\mathcal{R}_1 = \mathcal{R}, \mathcal{R}_2 = \pm\mathcal{R}$; we get the following important brackets

$$\{\varphi, \psi\}_{D_\pm} = \langle\langle \mathcal{R}\mathrm{D}\varphi, \mathrm{D}\psi \rangle\rangle \pm \langle\langle \mathcal{R}\mathrm{D}'\varphi, \mathrm{D}'\psi \rangle\rangle. \tag{8}$$

We denote by $D_\pm$ the group D with the bracket $\{\cdot, \cdot\}_{D_\pm}$. The bracket $\{,\}_{D_-}$ equips D with the structure of a Poisson-Lie group, while the "+" sign corresponds to an almost nondegenerate Poisson structure on D_+. (It is symplectic on an open cell in D containing the unit element, see [23] for the description of the symplectic leaves of D_+.)

The multiplication map $D \times D \to D$ defines a Poisson group action $D_- \times D_+ \to D_+$; its restriction to the diagonal subgroup $G \subset D$ is *admissible* [18], and hence it is possible to perform Poisson reduction over the action of G. The quotient space is canonically identified with G itself; in fact, it is clear that the map $\pi : D \to G : (g_1, g_2) \mapsto g_1 g_2^{-1}$ is constant on the right coset classes of G.

To calculate the explicit form of the quotient Poisson structure on G choose $\varphi \in Fun\,(G)$ and put $\hat{\varphi} = \pi^*\varphi$; let $\nabla\varphi, \nabla'\varphi$ be the left and right derivatives of φ.

$$\begin{aligned}
\langle \nabla\varphi\,(g)\,, X \rangle &= \tfrac{d}{dt}\,|_{t=0}\,\varphi\left(e^{tX}g\right), \\
\langle \nabla'\varphi\,(g)\,, X \rangle &= \tfrac{d}{dt}\,|_{t=0}\,\varphi\left(ge^{tX}\right), \\
g &\in G,\ X \in \mathfrak{g}.
\end{aligned} \tag{9}$$

Then

$$\mathrm{D}\hat{\varphi}\,(g_1, g_2) = \left(\nabla\varphi\left(g_1 g_2^{-1}\right), \nabla'\varphi\left(g_1 g_2^{-1}\right)\right). \tag{10}$$

After a short computation this yields

$$\{\varphi, \psi\}_G = \langle A\nabla\varphi, \nabla\psi\rangle - \langle D\nabla'\varphi, \nabla'\psi\rangle + \langle B\nabla'\varphi, \nabla\psi\rangle - \langle B^*\nabla\varphi, \nabla'\psi\rangle. \tag{11}$$

In general, this Poisson structure is degenerate.

Suppose that τ is an automorphism of $\mathfrak{g}$; then $\tau \oplus \tau$ is an automorphism of $\mathfrak{g} \oplus \mathfrak{g}$. Let us assume that $\tau \oplus \tau$ commutes with $\mathcal{R}$. To twist the r-matrix on $\mathfrak{d}$ we shall use another extension of τ to $\mathfrak{d}$; namely, we put $\hat{\tau} = \tau \oplus \tau^{-1}$. Put

$$\mathcal{R}^\tau = \hat{\tau}\mathcal{R}\hat{\tau}^{-1} = \begin{pmatrix} A & B\tau^{-1} \\ \tau B^* & D \end{pmatrix}. \tag{12}$$

$\mathcal{R}^\tau$ satisfies the Yang–Baxter equation. Put $\mathcal{R}_1 = \mathcal{R}^\tau, \mathcal{R}_2 = \mathcal{R}$ in (7) and denote by D_τ the group D with the corresponding Poisson structure:

$$\{\varphi, \psi\}_{D_\tau} = \langle\langle \mathcal{R}^\tau \mathrm{D}\varphi, \mathrm{D}\psi\rangle\rangle + \langle\langle \mathcal{R}\mathrm{D}'\varphi, \mathrm{D}'\psi\rangle\rangle. \tag{13}$$

(If $\mathcal{R}$ is the r-matrix of the double of $\mathfrak{g}$, the group D_τ is usually referred to as the *twisted double*.) This Poisson structure on D_τ also admits reduction with respect to the action of the diagonal subgroup; the quotient structure on $\mathrm{G}= \pi(D_\tau)$ is given by

$$\{\varphi, \psi\}_\tau = \langle A\nabla\varphi, \nabla\psi\rangle - \langle D\nabla'\varphi, \nabla'\psi\rangle + \langle B\tau^{-1}\nabla'\varphi, \nabla\psi\rangle - \langle \tau B^*\nabla\varphi, \nabla'\psi\rangle. \tag{14}$$

In particular, let us apply this construction to the group $\mathrm{G} = G^N$; in this case $D = (G \times G)^N$, and τ is the cyclic shift in the direct sum $\bigoplus_1^N \mathfrak{g}$. Let

$$r = \begin{pmatrix} A & B \\ B^* & D \end{pmatrix} \tag{15}$$

be a solution of the modified classical Yang–Baxter equation on $\mathfrak{g} \oplus \mathfrak{g}$; put $\mathcal{R} = \bigoplus_{1}^{N} r$. Evidently, $\mathcal{R}$ commutes with $\tau \oplus \tau$. To describe the resulting lattice Poisson algebra it is convenient to introduce tensor notations. Fix an exact matrix representation ρ_V of G and denote

$$L^n = \rho_V(g_n), L_1^n = L^n \otimes I, L_2^n = I \otimes L^n,$$
$$g = (g_1,...,g_n) \in G^N. \tag{16}$$

The reduced Poisson brackets of the matrix coefficients of L^n have the form

$$\{L_1^n, L_2^n\} = -A L_1^n L_2^n + L_1^n L_2^n D,$$
$$\{L_1^n, L_2^{n+1}\} = L_1^n B^* L_2^{n+1}, \tag{17}$$
$$\{L_1^n, L_2^m\} = 0, |n - m| \geq 2.$$

For shortness, we keep writing A, B,etc instead of $(\rho_V \otimes \rho_V) A$,etc. The main property of the Poisson bracket (17) is given by the following assertion:

Theorem . [20] *Equip* $\mathsf{G} = G^N$ *with the Poisson structure (11); then the monodromy map*

$$M : \mathsf{G}_\tau \to G, \quad M(g_1, \ldots, g_N) = g_1 \cdot \ldots \cdot g_N$$

is a Poisson mapping if and only if the r-matrix (15) satisfies the additional constraint

$$A + B = B^* + D. \tag{18}$$

In that case the Poisson structure in the target space of the monodromy map is given by (11).

In tensor notations we have the following brackets for M :

$$\{M_1, M_2\} = -A M_1 M_2 + M_1 M_2 D + M_1 B^* M_2 - M_2 B M_1 \tag{19}$$

Later we shall describe symplectic leaves of the bracket (17) in the main examples. In the particular case when $B = 0, A = D$ the bracket is ultralocal. The reader may keep in mind this possibility as a degenerate case.

The remainder of this section is devoted to the quantization of the Poisson brackets (17), (19). It may be easily performed on the lines of [10] provided that we know the quantum R-matrix which corresponds to the chosen classical r-matrix on $\mathfrak{d}$. More precisely, let $U_q(\mathfrak{d}; \mathcal{R})$ be the quantized universal enveloping algebra of $\mathfrak{d}$ which corresponds to $\mathcal{R}$ [6]; note that its description is not quite obvious since in the existing literature only the standard algebras $U_q(\mathfrak{d}; \mathcal{R})$ which correspond to simplest solutions of the classical Yang–Baxter equation are usually considered. It is widely believed that all solutions from the Belavin–Drinfeld list [5] give rise to quasitriangular Hopf algebras. (Examples in Section 3 below give evidence to support this belief.) Assuming that the algebra $U_q(\mathfrak{d};\mathcal{R})$ exists, let

$$\mathcal{R}_q = \begin{pmatrix} A_q & B_q \\ B_q^* & D_q \end{pmatrix} \in U_q(\mathfrak{d};\mathcal{R}) \otimes U_q(\mathfrak{d};\mathcal{R}) \tag{20}$$

be its universal quantum R-matrix. We omit the explicit form of the relations in the algebra $U_q(\mathfrak{d};R)$. Let ρ_V be some representation of the algebra $U_q(\mathfrak{d};\mathcal{R})$ in the space V and let

$$\mathcal{R}_q^{VV} = (\rho_V \otimes \rho_V)\mathcal{R}_q = \begin{pmatrix} A_q & B_q \\ B_q^* & D_q \end{pmatrix}. \tag{21}$$

The following theorem is parallel to the description of the twisted quantum double and of the lattice current algebra [20], [4].

Theorem . *The free algebra $Fun_q^{\mathcal{R}}(G_\tau)$ generated by the matrix elements of the matrices $L^n \in Fun_q^{\mathcal{R}}(G_\tau) \otimes End(V)$, satisfying the following relations:*

$$\begin{aligned} A_q L_1^n L_2^n &= L_2^n L_1^n D_q \\ L_1^n B_q^{*^{-1}} L_2^{n+1} &= L_2^{n+1} L_1^n \end{aligned} \tag{22}$$

is the quantization of the Poisson algebra (19).

Theorem . *The free algebra $Fun_q^{\mathcal{R}}(G)$ generated by the matrix elements of the matrix $M \in Fun_q^{\mathcal{R}}(G) \otimes End(V)$, satisfying the relations:*

$$A_q M_1 B_q^{*^{-1}} M_2 = M_2 B_q^{-1} M_1 D_q \tag{23}$$

is the quantization of the Poisson algebra (19).

Finally, we formulate the quantum version of Theorem 1.1

Theorem . *The map*

$$\begin{aligned} M : Fun_q^{\mathcal{R}}(G_\tau) &\to Fun_q^{\mathcal{R}}(G), \\ \left(L^1, \ldots, L^N\right) &\mapsto L^1 \cdot \ldots \cdot L^N \end{aligned}$$

is an homomorphism of algebras.

The algebras (22), (23) are the principal objects of our investigation.

2. Regularization of Nonultralocal Poisson Brackets

The goal of this section is to link the construction of lattice algebras with Hamiltonian systems on coadjoint orbits of current algebras. This approach is outlined in [19] where a regularization procedure for the Poisson brackets of the monodromy matrix is proposed which matches naturally with lattice

Poisson algebras described in Section 1. This will also allow us to construct consistent lattice approximations of nonultralocal systems on the circle.

We recall some details of the construction of dynamical systems on coadjoint orbits [17], [1].

Let $\mathfrak{G} = C^\infty\left(S^1, \mathfrak{g}\right)$ be a current algebra with the values in some affine Lie algebra $\mathfrak{g}$. Let us define the invariant scalar product on $\mathfrak{G}$:

$$(X, Y) = \int\limits_0^{2\pi} \langle X, Y \rangle \, dz, \qquad (24)$$

where $X, Y \in \mathfrak{G}, \langle \cdot, \cdot \rangle$ is a invariant bilinear form on $\mathfrak{g}$. Let $\widehat{\mathfrak{G}}$ be the central extension of the algebra $\mathfrak{G}$ which corresponds to the 2-cocycle

$$\omega\left(X, Y\right) = \left(X, \partial_z Y\right). \qquad (25)$$

By definition, $\widehat{\mathfrak{G}}$ is the set of pairs (X, a), $X \in \mathfrak{G}, a \in \mathbf{C}$ with the commutator

$$[(X.a), (Y, b)] = ([X, Y], \omega\left(X, Y\right)). \qquad (26)$$

If r is a solution of the modified classical Yang–Baxter equation on $\mathfrak{g}$, we put as usual

$$[X, Y]_r = [rX, Y] + [X, rY].$$

Let $\mathfrak{g}_r$ be the algebra $\mathfrak{g}$ equipped with this bracket. Put $\mathfrak{G}_r = C^\infty\left(S^1, \mathfrak{g}_r\right)$; it is easy to see that

$$\omega_r\left(X, Y\right) = \omega\left(rX, Y\right) + \omega\left(X, rY\right)$$

is a 2-cocycle on $\mathfrak{G}_r$; thus we may define the second structure of a Lie algebra on $\widehat{\mathfrak{G}}$

$$[(X, a), (Y, b)]_r = ([X, Y]_r, \omega_r\left(X, Y\right)). \qquad (27)$$

In this formula it is not assumed that r is skew-symmetric with respect to the scalar product $\langle \cdot, \cdot \rangle$. (In fact, if it is, the cocycle ω_r vanishes identically.)

Let $\widehat{\mathfrak{G}}^*$ be the dual space of $\widehat{\mathfrak{G}}$; using the inner product (24) we may identify it with $\mathfrak{G} \oplus \mathbf{C}$. The Poisson bracket used in the CISM is the Lie-Poisson bracket which corresponds to the commutator (27). The variable $e \in \mathbf{C}$ is central with respect to this bracket. If $X_\varphi \in \widehat{\mathfrak{G}}$ is a derivative of a function $\varphi \in Fun\left(\widehat{\mathfrak{G}}^*\right)$:

$$((X_\varphi, X)) = \tfrac{d}{dt}\big|_{t=0}\, \varphi\left(L + tX\right),$$
$$X, L \in \widehat{\mathfrak{G}}^*, \qquad (28)$$

here $((\cdot, \cdot))$ is a natural pairing between $\widehat{\mathfrak{G}}$ and $\widehat{\mathfrak{G}}^*$. Then

$$\{\varphi, \psi\} (L, e) = \left(((L, e), [X_\varphi, X_\psi]_r) \right). \tag{29}$$

Without loss of generality we may assume that $e = 1$ and suppress it in the notations. The bracket (29) may be represented as the bilinear form of the Poisson operator:

$$\mathcal{H} = adL \circ r + r^* \circ adL - (r + r^*) \partial_z, \tag{30}$$

$$\{\varphi, \psi\} (L) = (\mathcal{H} X_\varphi, X_\psi). \tag{31}$$

The operator $\mathcal{H}$ is unbounded, so the formula (31) requires some caution when the gradients are not smooth on the circle. This is precisely the case for the Poisson brackets of the monodromy matrix. Let ψ be the fundamental solution of the equation:

$$\partial_z \psi = L\psi \tag{32}$$

normalized by $\psi(0) = I$; then the monodromy matrix is equal to

$$M = \psi(2\pi) \in G. \tag{33}$$

Fix $\Phi \in Fun(G)$. According to [1], the Frechet derivative of the functional $L \mapsto \Phi(M[L])$ is given by

$$X_\Phi(z) = \psi(z) \nabla' \Phi(M) \psi(z)^{-1} \tag{34}$$

and in general is discontinuous at $z = 0$:

$$\begin{aligned} X_\Phi(0) &= \nabla' \Phi(M), \\ X_\Phi(2\pi) &= \nabla \Phi(M). \end{aligned} \tag{35}$$

To regularize Poisson brackets of the monodromy, we shall use an idea borrowed from the theory of self-adjoint extensions [19]. Let $\triangle :$ $C^\infty([0, 2\pi]; \mathfrak{g}) \to \mathfrak{g} \oplus \mathfrak{g}$ be the map which associates to a function on $[0, 2\pi]$ its boundary values,

$$\triangle X_\varphi = \begin{pmatrix} X_\varphi(0) \\ X_\varphi(2\pi) \end{pmatrix}. \tag{36}$$

Choose $B \in End \left(\overset{\circ}{\mathfrak{g}} \oplus \overset{\circ}{\mathfrak{g}} \right)$, here $\overset{\circ}{\mathfrak{g}} \subset \mathfrak{g}$ is the spreading finite-dimensional Lie algebra which corresponds to an affine Lie algebra $\mathfrak{g}$; we extend operator B to the space $\mathfrak{g} \oplus \mathfrak{g}$ as a zero operator outside $\overset{\circ}{\mathfrak{g}} \oplus \overset{\circ}{\mathfrak{g}}$ and define the regularized Poisson bracket in the following way:

$$\{\varphi, \psi\} (L, 1) = \frac{1}{2} ((\mathcal{H} X_\varphi, X_\psi) - (\mathcal{H} X_\psi, X_\varphi)) + \langle\langle B \triangle X_\varphi, \triangle X_\psi \rangle\rangle. \tag{37}$$

The bracket (37) must coincide with the bracket (31) on smooth functions, hence B must satisfy the condition

$$\left\langle\!\!\left\langle B\left(\begin{array}{c} X \\ X \end{array}\right),\left(\begin{array}{c} Y \\ Y \end{array}\right)\right\rangle\!\!\right\rangle = 0. \tag{38}$$

The additional restriction on B imposed in [19] follows from the study of the linearized bracket for the monodromy (37) for $M \to 1$; it is natural to demand that this linearized bracket should coincide with the one defined by r. This gives, after a short computation:

$$\{\Phi, \Psi\}(M) = \left\langle\!\!\left\langle \mathcal{R}\left(\begin{array}{c} \nabla\Phi \\ \nabla'\Phi \end{array}\right),\left(\begin{array}{c} \nabla\Psi \\ \nabla'\Psi \end{array}\right)\right\rangle\!\!\right\rangle, \tag{39}$$

where

$$\mathcal{R} = \left(\begin{array}{cc} -a+\alpha & -\alpha-s \\ \alpha-s & -a-\alpha \end{array}\right), \alpha^* = -\alpha, a = \tfrac{1}{2}(r-r^*), s = \tfrac{1}{2}(r+r^*),$$
$$\text{where } \alpha \in \overset{\circ}{\mathfrak{g}} \wedge \overset{\circ}{\mathfrak{g}} \text{ because } B \in End\left(\overset{\circ}{\mathfrak{g}} \oplus \overset{\circ}{\mathfrak{g}}\right),$$
$$\tag{40}$$

and our choice of B supposes that $s \in \overset{\circ}{\mathfrak{g}} \otimes \overset{\circ}{\mathfrak{g}}$. The Jacobi identity for this bracket will be valid if $\mathcal{R}$ satisfies the modified Yang–Baxter equation. In tensor notations a Poisson brackets of monodromy matrix have the form

$$\begin{aligned} \{M_1, M_2\} &= (a-\alpha)\, M_1 M_2 \text{-} M_1 M_2\,(a+\alpha)+ \\ &+ M_1\,(\alpha-s)\, M_2 + M_2\,(\alpha+s)\, M_1. \end{aligned} \tag{41}$$

The corresponding lattice Poisson algebra for which the monodromy matrix has the brackets (41) is

$$\begin{aligned} \{L_1^n, L_2^n\} &= (a-\alpha)\, L_1^n L_2^n - L_1^n L_2^n\,(a+\alpha), \\ \{L_1^n, L_2^{n+1}\} &= L_1^n\,(\alpha-s)\, L_2^{n+1}, \\ \{L_1^n, L_2^m\} &= 0, |n-m| \geq 2. \end{aligned} \tag{42}$$

Our next step is the classification of the Poisson brackets of type (31) for which the Poisson brackets of the monodromy matrix may be regularized. It is difficult to classify all non-skew solutions of the Yang–Baxter equation for which there exists an $\alpha \in \overset{\circ}{\mathfrak{g}} \wedge \overset{\circ}{\mathfrak{g}}$ such that a matrix $\mathcal{R}$ in (40) is a solution of the Yang–Baxter equation. But we can easily construct all solutions of the Yang–Baxter equation for $\mathfrak{g} \oplus \mathfrak{g}$ according to the Belavin–Drinfeld classification theorem [5], and then choose solutions of the form (40). To realize this program we start with an easy theorem:

Theorem . *If $\mathcal{R}$ is a solution of the modified Yang–Baxter equation for $\mathfrak{g} \oplus \mathfrak{g}$ of the type (40), then $a+s$ is a solution of the modified Yang–Baxter equation for $\mathfrak{g}$.*

Proof. Let $r_1 = -a + \alpha, r_2 = -\alpha - s$, then $r = -(r_1 + r_2)$. From the Yang–Baxter equation for $\mathcal{R}$ it follows:

$$
\begin{aligned}
&[r_1 X, r_2 Y] - r_1 \left([r_1 X, Y] + [X, r_1 Y]\right) = -[X, Y], \\
&[r_2 X, r_2 Y] - r_2 \left([(r_1 - 2\alpha) X, Y] + [X, (r_1 - 2\alpha) Y]\right) = 0, \\
&[r_1 X, r_2 Y] - r_1 [X, r_2 Y] - r_2 \left[(r_2 + 2\alpha) X, Y\right] = 0, \\
&\qquad\qquad X, Y \in \mathfrak{g}.
\end{aligned}
\tag{43}
$$

It is easy to check that the (43) implies the Yang–Baxter equation for r:

$$
[rX, rY] - r \left([rX, Y] + [X, rY]\right) = -[X, Y].
\tag{44}
$$

Let us turn now to the detailed study of the case of affine Lie algebras. We use the terminology and notations of the book [2].

Let $\mathfrak{g}$ be an affine Lie algebra, Δ_+ the set of its positive roots, $\overset{\circ}{\mathfrak{g}}$ the corresponding spreading simple Lie algebra, $\overset{\circ}{\Delta}_+$ the set of its positive roots. Let $\Delta_{++} = \Delta_+ \backslash \overset{\circ}{\Delta}_+$. Using this notations we formulate some version of the Belavin–Drinfeld classification theorem.

Theorem . [5] *Up to an automorphism any solution of the modified classical Yang–Baxter equation for an affine Lie algebra $\mathfrak{g}$ has the form:*

$$
R = \sum_{\alpha \in \Delta_{++}} e_\alpha \wedge e_{-\alpha} + r,
$$

where r is a solution of the modified Yang–Baxter equation for $\overset{\circ}{\mathfrak{g}}$.

For an explicit form of such solutions see [5]. In [5] such solutions are called trigonometric. Thus from theorem 2.2 we have the following ansatz for a:

$$
a = \sum_{\alpha \in \Delta_{++}} e_\alpha \wedge e_{-\alpha} + a_0, \quad a_0 \in \overset{\circ}{\mathfrak{g}} \wedge \overset{\circ}{\mathfrak{g}}.
\tag{45}
$$

and we reduce our problem to the Yang–Baxter equation on the square of a finite dimensional Lie algebra $\overset{\circ}{\mathfrak{g}}$. Namely, $\mathcal{R}$ is a solution of the Yang–Baxter equation iff

$$
\begin{pmatrix}
-a_0 + \alpha & -\alpha - s \\
\alpha - s & -a_0 - \alpha
\end{pmatrix}
\tag{46}
$$

is a solution of the Yang–Baxter equation for $\overset{\circ}{\mathfrak{g}} \oplus \overset{\circ}{\mathfrak{g}}$.

Theorem . *Let*

$$
\begin{pmatrix}
A & B \\
B^* & D
\end{pmatrix}, A^* = -A, D^* = -D
$$

be a solution of the Yang–Baxter equation for $\overset{\circ}{\mathfrak{g}} \oplus \overset{\circ}{\mathfrak{g}}$. *It has the form* (46) *iff*

$$A + B = B^* + D. \tag{47}$$

Under this condition

$$\alpha = \frac{B^* - B}{2}, \quad a_0 = -\frac{A + D}{2}, \quad s = -\frac{B + B^*}{2}. \tag{48}$$

Notice that we again come to the condition (18).

Remark. It may be showed that it is not necessary to impose the condition $B \in End\left(\overset{\circ}{\mathfrak{g}} \oplus \overset{\circ}{\mathfrak{g}}\right)$ a priori. Actually, this condition follows from the generalization of Theorem 2.2 for a direct sum of two copies of an affine Lie algebra because $\alpha \in \overset{\circ}{\mathfrak{g}} \wedge \overset{\circ}{\mathfrak{g}}$, $s \in \overset{\circ}{\mathfrak{g}} \otimes \overset{\circ}{\mathfrak{g}}$ for every solution of the modified Yang–Baxter equation for $\mathfrak{g} \oplus \mathfrak{g}$ of the form (40).

Unfortunately, condition (47) is not stable under the automorphisms of $\overset{\circ}{\mathfrak{g}}$. So we cannot use the Belavin–Drinfeld theorem to classify all regularizations. But this theorem gives a possibility to construct sufficiently general examples of such regularizations. These examples will be presented in the next section. In the $\widehat{sl(2)}$ case we shall able to classify all regularizations.

3. Main Examples of Regularizations

Now we are ready to discuss examples of regularizations using the results of the previous sections. We shall consider affine Lie algebras of type $X_N^{(1)}$ and $X_N^{(2)}$ in the loop realization. We shall describe the corresponding lattice quantum algebras and their Casimir elements. In the $\widehat{sl(2)}$ case we shall explain the Algebraic Bethe Ansatz construction for such algebras.

Example 1. Nontwisted loop algebras. The first example is connected with the r-matrix of the double of a finite dimensional Lie algebra $\overset{\circ}{\mathfrak{g}}$ equipped with the structure of a quasitriangular Lie bialgebra. Let $\mathfrak{g}$ be an affine Lie algebra of type $X_N^{(1)}$, $\overset{\circ}{\mathfrak{g}}$ the corresponding finite-dimensional Lie algebra. To apply Theorem 2.3 consider the r-matrix of its double $\overset{\circ}{\mathfrak{d}} = \overset{\circ}{\mathfrak{g}} \oplus \overset{\circ}{\mathfrak{g}}$; we have

$$r = \begin{pmatrix} \alpha & -2\alpha_+ \\ 2\alpha_- & -\alpha \end{pmatrix}, \tag{49}$$

where α is some solution of the modified Yang–Baxter equation for $\overset{\circ}{\mathfrak{g}}$ and $\alpha_\pm = \frac{1}{2}(\alpha \pm I)$. ($I$ is the identity operator in $\overset{\circ}{\mathfrak{g}}$; its kernel is the Casimir element t.) According to Theorem 2.3, in this case one gets:

$$s = I, a_0 = 0. \tag{50}$$

In this case $r = a + I$ is the rational r-matrix for $\mathfrak{g}$. We choose for $\mathfrak{g}$ the non-twisted loop realization [2]. We remind that in this realization $\mathfrak{g} = \overset{\circ}{\mathfrak{g}} \otimes \mathbf{C} \left[\lambda, \lambda^{-1} \right]$, and the invariant bilinear form is given by

$$\langle X(\lambda), Y(\lambda) \rangle = Res\, tr\, (X(\lambda) Y(\lambda)) \frac{d\lambda}{\lambda}, \tag{51}$$

where tr is an invariant bilinear form on $\overset{\circ}{\mathfrak{g}}$. The kernel of a in this realization is

$$a(\lambda, \mu) = -t \frac{\lambda + \mu}{\lambda - \mu}, \tag{52}$$

where we identify $\mathfrak{g} \otimes \mathfrak{g}$ with $\overset{\circ}{\mathfrak{g}} \otimes \overset{\circ}{\mathfrak{g}} \otimes \mathbf{C} \left[\lambda, \lambda^{-1} \right] \otimes \mathbf{C} \left[\mu, \mu^{-1} \right]$. Thus we have the following formulas for $-a \pm \alpha$:

$$\begin{aligned}
-a + \alpha &= \frac{\lambda}{\lambda - \mu} 2\alpha_+ - \frac{\mu}{\lambda - \mu} 2\alpha_-, \\
-a - \alpha &= \frac{\mu}{\lambda - \mu} 2\alpha_+ - \frac{\lambda}{\lambda - \mu} 2\alpha_-.
\end{aligned} \tag{53}$$

Let $\mathcal{R}_\pm$ be the finite-dimensional quantum R-matrix in the fundamental representation, which corresponds to $2\alpha_\pm$ after quantization. In the classical limit

$$\mathcal{R}_\pm = I + 2\alpha_\pm h + o(h), \tag{54}$$

where h is the deformation parameter.
We have:

$$\mathcal{R}_- = P\left(\mathcal{R}_+^{-1}\right), \tag{55}$$

where P is the permutation operator in the tensor square. Using these data we may construct the quantum r-matrices corresponding to $-a \pm \alpha$. If we denote the quantum r-matrix corresponding to $-a - \alpha$ by $\mathcal{R}(\lambda, \mu)$, then

$$\mathcal{R}(\lambda, \mu) = \frac{\lambda}{\lambda - \mu} \mathcal{R}_-^{-1} - \frac{\mu}{\lambda - \mu} \mathcal{R}_+^{-1}, \tag{56}$$

and the quantum r-matrix $\mathcal{R}(\lambda, \mu)^T$ corresponds to $-a + \alpha$:

$$\mathcal{R}(\lambda, \mu)^T = \frac{\lambda}{\lambda - \mu} \mathcal{R}_+ - \frac{\mu}{\lambda - \mu} \mathcal{R}_-, \tag{57}$$

here T is the conjugation with respect to the scalar product tr.
Finally, we have the quantum R-matrix on the square of $\mathfrak{g}$:

$$\mathcal{R}_q = \begin{pmatrix} \mathcal{R}(\lambda, \mu)^T & \mathcal{R}_+^{-1} \\ \mathcal{R}_- & \mathcal{R}(\lambda, \mu) \end{pmatrix}. \tag{58}$$

According to Theorem 1.2 one can get the relations in the quantum lattice algebra $Fun_q^{\mathcal{R}}(\mathbf{G}_r)$, which gives a lattice approximation of the continuous system:

$$
\begin{aligned}
\mathcal{R}(\lambda,\mu)^T L_1^n(\lambda) L_2^n(\mu) &= L_2^n(\mu) L_1^n(\lambda) \mathcal{R}(\lambda,\mu), \\
L_1^n(\lambda) \mathcal{R}_-^{-1} L_2^{n+1}(\mu) &= L_2^{n+1}(\mu) L_1^n(\lambda), \\
L^n(\lambda) &\in Fun_q^{\mathcal{R}}(\mathbf{G}_r) \otimes End(V).
\end{aligned}
\tag{59}
$$

Here V is the fundamental representation space.

From Theorem 1.3 we get the following commutation relations for the monodromy matrix:

$$
\mathcal{R}(\lambda,\mu)^T M_1(\lambda) \mathcal{R}_-^{-1} M_2(\mu) = M_2(\mu) \mathcal{R}_+ M_1(\lambda) \mathcal{R}(\lambda,\mu).
\tag{60}
$$

The algebra (59) is connected with the Lattice Kac–Moody algebra $\mathcal{A}_{LC}$ [4]. Namely, the algebra (59) admits a family of representation for which there exists the limit

$$
L^n(\lambda) 0 \xrightarrow[\lambda \to 0]{} \lambda^{-k_n} \overset{\circ}{L}{}^n + o\left(\lambda^{-k_n}\right), k_n \in \mathbf{Z}.
\tag{61}
$$

From (56), (57) we have the asymptotic conditions for $\mathcal{R}(\lambda,\mu)$, $\mathcal{R}(\lambda,\mu)^T$:

$$
\begin{aligned}
\mathcal{R}(\lambda,\mu) 0 &\xrightarrow[\mu \to 0]{} \mathcal{R}_-^{-1}, \\
\mathcal{R}(\lambda,\mu)^T &\xrightarrow[\mu \to 0]{} \mathcal{R}_+.
\end{aligned}
\tag{62}
$$

Using (59),(62), one can get the relations for $\overset{\circ}{L}{}^n$:

$$
\begin{aligned}
\overset{\circ}{L}{}_2^n \overset{\circ}{L}{}_1^n &= \mathcal{R}_+ \overset{\circ}{L}{}_1^n \overset{\circ}{L}{}_2^n \mathcal{R}_- \\
\overset{\circ}{L}{}_1^n \mathcal{R}_-^{-1} \overset{\circ}{L}{}_2^{n+1} &= \overset{\circ}{L}{}_2^{n+1} \overset{\circ}{L}{}_1^n
\end{aligned}
\tag{63}
$$

These are the relations in the Lattice Kac–Moody algebra $\mathcal{A}_{LC}$. If $\alpha = \sum\limits_{\alpha \in \overset{\circ}{\Delta}_+} e_\alpha \wedge e_{-\alpha}$, then the monodromy matrix for this algebra

$$
\overset{\circ}{M} = \overset{\circ}{L}{}^1 \cdot \ldots \cdot \overset{\circ}{L}{}^N
\tag{64}
$$

satisfies the commutation relations for the quantum group $U_q\left(\overset{\circ}{\mathfrak{g}}\right)$:

$$
\mathcal{R}_+ \overset{\circ}{M}_1 \mathcal{R}_-^{-1} \overset{\circ}{M}_2 = \overset{\circ}{M}_2 \mathcal{R}_+ \overset{\circ}{M}_1 \mathcal{R}_-^{-1}.
\tag{65}
$$

This construction is useful for the computation of the center of the algebra (59).

Theorem . *For generic q and N odd the center of the algebra* (59) *contains the elements*

$$C_k = tr_q \left(\overset{\circ}{M} \right)^k = tr\ q^{2\rho} \left(\overset{\circ}{M} \right)^k, k = 1, \ldots, rk\ \overset{\circ}{\mathfrak{g}} . \qquad (66)$$

(Here ρ is half the sum of positive roots in $U_q \left(\overset{\circ}{\mathfrak{g}} \right)$.)

It is natural to expect that in an appropriate topology these elements generate the center of the algebra (59). The proof of this theorem is complete similar to the proof of such theorem for the algebra $\mathcal{A}_{LC}$ [4].

From this theorem one can deduce that the center of the algebra (59) coincides with the center of $\mathcal{A}_{LC}$ [4] and with the center of $U_q \left(\overset{\circ}{\mathfrak{g}} \right)$. Thus we have

Corollary 1. *The extensions*

$$centU_q \left(\overset{\circ}{\mathfrak{g}} \right) \subset U_q \left(\overset{\circ}{\mathfrak{g}} \right) \subset \mathcal{A}_{LC} \subset Fun_q^R (G_r)$$

are central.

Example 2. Twisted loop algebras. We are leaving for a short time our first example to consider the second one. Let us consider an affine Lie algebra $\mathfrak{g}$ of a type $X_N^{(r)}, r = 1, 2, 3$. Let

$$a_0 = \sum_{\alpha \in \overset{\circ}{\Delta}_+} e_\alpha \wedge e_{-\alpha}, \alpha = 0, s \in End \left(\overset{\circ}{\mathfrak{h}} \right) . \qquad (67)$$

It is not difficult to verify that the corresponding r-matrix of the type (46) satisfies the Yang–Baxter equation for $\overset{\circ}{\mathfrak{g}} \oplus \overset{\circ}{\mathfrak{g}}$. In this case $r = P_+ - P_- + s$, where $P_\pm$ are projection operators onto the opposite Borel subalgebras of $\mathfrak{g}$. We shall use a twisted loop realization for $\mathfrak{g}$ with invariant bilinear form (51) [2], [5]. In this realization $\mathfrak{g}$ is the set of stable points of an automorphism of an affine Lie algebra $X_{N'}^{(1)}$ in a non-twisted loop realization. Here N' does not in general coincide with N. The automorphism $\widehat{C}$ of $X_{N'}^{(1)}$ is given by

$$\widehat{C} X (\lambda) = C X \left(\lambda e^{-\frac{2\pi i}{r \cdot h}} \right), \qquad (68)$$

where h is the Coxeter number of the algebra $X_N^{(r)}$, C is the Coxeter automorphism of $\overset{\circ'}{\mathfrak{g}}$ which corresponds to the affine Lie algebra $X_{N'}^{(1)}$.

If $\mathfrak{g} = X_N^{(1)}$, N' coincides with N and $\overset{\circ'}{\mathfrak{g}}$ coincides with $\overset{\circ}{\mathfrak{g}}$. For simplicity we restrict ourselves to this case. The automorphism C has degree h and the algebra $\overset{\circ}{\mathfrak{g}}$ is decomposed into the direct sum $\overset{\circ}{\mathfrak{g}} = \overset{h-1}{\underset{j=0}{\bigoplus}} \overset{\circ}{\mathfrak{g}}_j$, $\overset{\circ}{\mathfrak{g}}_0 = \mathfrak{h}$, where $\overset{\circ}{\mathfrak{g}}_j$ is the eigenspace of C corresponding to the eigenvalue $e^{\frac{2\pi i}{h}j}$. The Casimir element of $\overset{\circ}{\mathfrak{g}}$ is decomposed into the sum $t = \overset{h-1}{\underset{j=0}{\sum}} t_j$, $t_j \in \overset{\circ}{\mathfrak{g}}_j \otimes \overset{\circ}{\mathfrak{g}}_{-j}$, $t_0 \in \mathfrak{h} \otimes \overset{\circ}{\mathfrak{h}}$. The kernel of the operator a in this realization is

$$a(\lambda,\mu) = -\frac{\overset{h-1}{\underset{j=0}{\sum}} t_j \left(\lambda^{h-j}\mu^j + \lambda^j \mu^{h-j}\right)}{\lambda^h - \mu^h}. \tag{69}$$

We suppose that $s = t_0$. Now we are ready to describe the quantization. Let $\mathcal{R}(\lambda,\mu)$ be the quantum r-matrix in the fundamental representation V which corresponds to $a(\lambda,\mu)$ after quantization. Evidently, such quantum r-matrix exists because the classical r-matrix a coincides with the classical r-matrix $a + \alpha$ from example 1 if $\alpha = \underset{\alpha \in \overset{\circ}{\Delta}_+}{\sum} e_\alpha \wedge e_{-\alpha}$. For the latter classical r-matrix the corresponding quantum r-matrix exists (see example 1). Let $\mathcal{R}_0 = e^{hs}$. We have the following quantum r-matrix on the square:

$$\mathcal{R}_q = \begin{pmatrix} \mathcal{R}(\lambda,\mu)^{-1} & \mathcal{R}_0^{-1} \\ \mathcal{R}_0^{-1} & \mathcal{R}(\lambda,\mu)^{-1} \end{pmatrix}. \tag{70}$$

In the standard way we built the quantum lattice algebra $Fun_q^{\mathcal{R}}(G_\tau)$ which gives a lattice counterpart for our continuous system. According to Theorem 1.2, the commutation relations for it have the form

$$\begin{aligned} \mathcal{R}(\lambda,\mu) L_1^n(\lambda) L_2^n(\mu) &= L_2^n(\mu) L_1^n(\lambda) \mathcal{R}(\lambda,\mu), \\ L_1^n(\lambda) \mathcal{R}_0 L_2^{n+1}(\mu) &= L_2^{n+1}(\mu) L_1^n(\lambda), \\ L^n(\lambda) &\in Fun_q^{\mathcal{R}}(G_\tau) \otimes End(V). \end{aligned} \tag{71}$$

For the monodromy matrix we have the relation:

$$\mathcal{R}(\lambda,\mu) M_1(\lambda) \mathcal{R}_0 M_2(\mu) = M_2(\mu) \mathcal{R}_0 M_1(\lambda) \mathcal{R}(\lambda,\mu). \tag{72}$$

Similarly to Theorem 3.1, there exists a description of the center of the algebra (71).

Theorem . *Let*

$$M0 \underset{\lambda \to 0}{\longrightarrow} \lambda^{-k} \overset{\circ}{M} + o\left(\lambda^{-k}\right). \tag{73}$$

For a generic q and N odd the center of the algebra (71) contains the elements:

$$C_k = tr\left(\overset{\circ}{M}\right)^k, k = 1, \ldots, rk \overset{\circ}{\mathfrak{g}}. \tag{74}$$

As in Theorem 3.1, it is natural to expect that in an appropriate topology these elements generate the center of the algebra (71).

The algebra (71), as opposed to the algebra (59), possesses a remarkable property; namely, it admits an ultralocalization. This property is important for the construction of representations of lattice algebras. The exact statement is given by the following theorem.

Theorem . *Let ρ be the fundamental representation of the group $\overset{\circ}{H}$ corresponding to the algebra $\overset{\circ}{\mathfrak{h}}$, and V the space of its fundamental representation used for the definition of the algebra (71). Then ρ acts naturally in V. Let A be the free algebra generated by the matrix coefficients of the matrices $\widehat{L}^n \in A \otimes EndV, G^n \in A \otimes \rho$, satisfying the relations:*

$$\mathcal{R}\left(\lambda,\mu\right)\widehat{L}_1^n\left(\lambda\right)\widehat{L}_2^n\left(\mu\right) = \widehat{L}_2^n\left(\mu\right)\widehat{L}_1^n\left(\lambda\right)\mathcal{R}\left(\lambda,\mu\right),$$
$$G_1^n\widehat{L}_2^{n+1}\left(\mu\right) = \mathcal{R}_0^{\frac{1}{2}}\widehat{L}_2^{n+1}\left(\mu\right)G_1^n, \tag{75}$$
$$G_1^n\widehat{L}_2^n\left(\mu\right) = \widehat{L}_2^n\left(\mu\right)G_1^n\mathcal{R}_0^{\frac{1}{2}}.$$

Then there exist the homomorphism of the algebras:

$$Fun_q^{\mathcal{R}}\left(\mathsf{G}_\tau\right) \to A,$$
$$L^n \mapsto G^{n-1}\widehat{L}^nG^{n-1},$$
$$M \mapsto G^N\widehat{M}G^{N-1}; \tag{76}$$
$$here\ \widehat{M} = \widehat{L}^1 \cdot \ldots \cdot \widehat{L}^N.$$

Remark. This homomorphism has the form of a lattice gauge transformation: the generators L^n of the nonultralocal algebra $Fun_q^{\mathcal{R}}\left(\mathsf{G}_\tau\right)$ are connected by a lattice gauge transformation with the generators $\widehat{L}^n$ satisfying ultralocal relations. Therefore this homomorphism is called ultralocalization. The monodromy matrices M and $\widehat{M}$ are conjugate and hence nonultralocality is equivalent to including some quasiperiodic boundary conditions. This is natural in the spirit of the operator extension theory used in the previous section. In particular examples it is easier to construct representations of the algebra A than those of the algebra (71). It is the main motivation for the definition of the algebra A.

The proof of this theorem consists in a straightforward check of the relations (71) for the generators $G^{n-1}\widehat{L}^nG^{n-1}$. A similar theorem on ultralocalization exists for the algebra $\mathcal{A}_{LC}$ [13].

Now we are ready to describe all regularized r-matrices and their regularizations for the algebra $A_1^{(1)}$. According to the Belavin–Drinfeld theorem for finite-dimensional semisimple Lie algebras [5], we may construct all solutions of the modified Yang–Baxter equation on the square of corresponding finite-dimensional Lie algebra $\overset{\circ}{\mathfrak{g}} = sl\left(2\right)$. Modulo some freedom in the choice of the Cartan components of R-matrix on the skew-diagonal, there exist only two solutions of the Yang–Baxter equation on $sl\,(2) \oplus sl\,(2)$ considered in examples 1 and 2 in this section. Moreover, the r-matrix of the double (49) in example 1 is connected with the unique structure of a bialgebra on $sl\,(2)$ for which

$$\alpha = \sum_{\alpha \in \overset{\circ}{\Delta}_+} e_\alpha \wedge e_{-\alpha}. \tag{77}$$

For the algebra $A_1^{(1)}$ the Coxeter number h equals 2, and the Coxeter automorphism has the form:

$$CX = DXD^{-1},$$
$$D = \begin{pmatrix} 1 & 0 \\ 0 & -1 \end{pmatrix}. \tag{78}$$

We summarize all results for the algebra $A_1^{(1)}$ in the following theorem. We formulate at once the quantum version of all formulas. The reader may reproduce the corresponding classical formulas by a quasiclassical limit.

Theorem . *For the algebra $A_1^{(1)}$ there exists only two nonultralocal Lie-Poisson brackets of the type (31) which may be regularized in the way (37). Namely, let $b_\pm$ be the opposite Borel subalgebras in $A_1^{(1)}$ and $\overset{\circ}{b}_\pm$ the opposite Borel subalgebras in $sl\,(2)$. Then the regularized r-matrices and the corresponding lattices algebras are:*

1. $r = P_{b_+\backslash b_+} - P_{b_-\backslash b_-} + I$ *is the regularized r-matrix, here* $P_{b_+\backslash b_+}, P_{b_-\backslash b_-}$ *are projectors onto the corresponding subspaces, I is the identity operator in $sl\,(2)$. It is precisely the rational r-matrix. If we use the non-twisted current realization $A_1^{(1)} = sl\,(2) \otimes \mathbf{C}\left[\lambda, \lambda^{-1}\right]$ and the scalar product (51), its kernel is given by*

$$r\,(\lambda, \mu) = -\frac{\mu t}{\lambda - \mu}, \tag{79}$$

where t is the Casimir element for $sl\,(2)$.
   The relations in the corresponding lattice algebra are given by formulas (59), where for the fundamental representation $A_1^{(1)}$ in $\mathbf{C}^2\left[\lambda, \lambda^{-1}\right]$:

$$\mathcal{R}_+ = q^{-\frac{1}{2}} \begin{pmatrix} q & 0 & 0 & 0 \\ 0 & 1 & q - q^{-1} & 0 \\ 0 & 0 & 1 & 0 \\ 0 & 0 & 0 & q \end{pmatrix}, \tag{80}$$
$$\mathcal{R}_- = P\left(\mathcal{R}_+^{-1}\right),$$

$$R(\lambda, \mu) = \begin{pmatrix} \frac{\lambda q - \mu q^{-1}}{\lambda - \mu} & 0 & 0 & 0 \\ 0 & 1 & \frac{\mu}{\lambda - \mu}(q - q^{-1}) & 0 \\ 0 & \frac{\lambda}{\lambda - \mu}(q - q^{-1}) & 1 & 0 \\ 0 & 0 & 0 & \frac{\lambda q - \mu q^{-1}}{\lambda - \mu} \end{pmatrix}.$$

$$(81)$$

2. $r = P_{b_+} - P_{b_-} + \xi P_{\overset{\circ}{h}}$ is the regularized r-matrix, $P_{\overset{\circ}{h}}$ is the projection operator onto the Cartan subalgebra of $sl(2)$. For simplicity we consider only the case $\xi = 1$. Fix the twisted current realization

$$A_1^{(1)} = \left\{ X(\lambda) \in sl(2) \otimes \mathbf{C}\left[\lambda, \lambda^{-1}\right] : X(-\lambda) = D X(\lambda) D^{-1} \right\} \qquad (82)$$

and the scalar product (51). Then the kernel of r is given by

$$r(\lambda, \mu) = -\frac{\mu^2 t_0}{\lambda^2 - \mu^2} - \frac{\lambda \mu t_1}{\lambda^2 - \mu^2}, \qquad (83)$$

where t_0 is the $\overset{\circ}{\mathfrak{h}}$ – component of the $sl(2)$ Casimir element and $t_1 \in \overset{\circ}{\mathfrak{g}}_1$ $\otimes \overset{\circ}{\mathfrak{g}}_{-1}, \overset{\circ}{\mathfrak{g}}_1 \in \{X \in sl(2), D X D^{-1} = -X\}$. The relations in the corresponding lattice algebra are given by the formulas (71), where for the fundamental representation $A_1^{(1)}$ in $\mathbf{C}^2\left[\lambda, \lambda^{-1}\right]$:

$$R_0 = q^{-\frac{1}{2}} \begin{pmatrix} q & 0 & 0 & 0 \\ 0 & 1 & 0 & 0 \\ 0 & 0 & 1 & 0 \\ 0 & 0 & 0 & q \end{pmatrix}, \qquad (84)$$

$$R(\lambda, \mu) = \begin{pmatrix} \frac{\lambda^2 q - \mu^2 q^{-1}}{\lambda^2 - \mu^2} & 0 & 0 & 0 \\ 0 & 1 & \frac{2\lambda\mu}{\lambda^2 - \mu^2}(q - q^{-1}) & 0 \\ 0 & \frac{2\lambda\mu}{\lambda^2 - \mu^2}(q - q^{-1}) & 1 & 0 \\ 0 & 0 & 0 & \frac{\lambda^2 q - \mu^2 q^{-1}}{\lambda^2 - \mu^2} \end{pmatrix}.$$

$$(85)$$

Remark. The quantum R-matrices (81) and (85) are in fact the ordinary trigonometric quantum r-matrices in different realizations. To see this just notice that the classical r-matrices $a + \alpha$ from the example 1 (see (50),(45),(77)) and a from the example 2 (see (45), (67) coincide.

We conclude this section with the Bethe–Ansatz construction for algebras with relations (59), (71) in the $A_1^{(1)}$ case.

We start with example 1. The "generating function" producing an infinite series of quantum commuting integrals of motion [8], [9] is

$$
\begin{aligned}
tr_q M(\lambda) &= qA(\lambda) + q^{-1}D(\lambda), \\
\text{where } M(\lambda) &= \begin{pmatrix} A(\lambda) & B(\lambda) \\ C(\lambda) & D(\lambda) \end{pmatrix} \\
\text{and } [tr_q M(\lambda), tr_q M(\mu)] &= 0.
\end{aligned}
\tag{86}
$$

We suppose that there exists a reference state Ω:

$$
A(\lambda)\Omega = a(\lambda)\Omega, D(\lambda)\Omega = d(\lambda)\Omega, C(\lambda)\Omega = 0, B(\lambda)\Omega \neq 0.
\tag{87}
$$

We shall try to seek a representation of the algebra (60) in which $tr_q M(\lambda)$ is diagonal. According to the standard Bethe Ansatz technique we try to find the eigenvectors of $tr_q M(\lambda)$ in the form:

$$
\psi(\lambda_1, \ldots, \lambda_n) = B(\lambda_1) \ldots B(\lambda_n)\Omega.
\tag{88}
$$

The relations between $A(\lambda), B(\lambda), C(\lambda), D(\lambda)$ which are essential to calculate the spectrum of $tr_q M(\lambda)$ are:

$$
\begin{aligned}
[B(\lambda), B(\mu)] &= 0, \\
A(\lambda)B(\mu) &= \tfrac{\mu - \lambda q^{-2}}{\mu - \lambda} B(\mu)A(\lambda) - \tfrac{\lambda}{\mu - \lambda}\left(1 - q^{-2}\right)B(\lambda)A(\mu) - \\
&\quad - \left(1 - q^{-2}\right)B(\lambda)D(\mu), \\
D(\lambda)B(\mu) &= \tfrac{\mu}{\mu - \lambda}\left(q^2 - 1\right)B(\lambda)D(\mu) + \tfrac{\mu - \lambda q^2}{\mu - \lambda}B(\mu)D(\lambda).
\end{aligned}
\tag{89}
$$

The vector $\psi(\lambda_1, \ldots, \lambda_n)$ will be an eigenvector of $tr_q M(\lambda)$ with the eigenvalue

$$
qa(\lambda)\prod_{i=1}^{n}\frac{\lambda q^{-2} - \lambda_i}{\lambda - \lambda_i} + q^{-1}d(\lambda)\prod_{i=1}^{n}\frac{\lambda q^2 - \lambda_i}{\lambda - \lambda_i}
\tag{90}
$$

iff λ_i satisfy the equations [8],[9]:

$$
\frac{d(\lambda_j)}{a(\lambda_j)} = \prod_{i \neq j}\frac{\lambda_j q^{-2} - \lambda_i}{\lambda_j q^2 - \lambda_i}.
\tag{91}
$$

Similarly, in example 2, the element

$$
\begin{aligned}
tr M(\lambda) &= A(\lambda) + D(\lambda), \\
\text{where } M(\lambda) &= \begin{pmatrix} A(\lambda) & B(\lambda) \\ C(\lambda) & D(\lambda) \end{pmatrix}
\end{aligned}
\tag{92}
$$

produces an infinite series of commuting conservation laws. Under the assumption (87) we look for the eigenvectors of $tr M(\lambda)$ which have the form (88). The essential commutation relations for this procedure are (72):

$$
\begin{aligned}
[B(\lambda), B(\mu)] &= 0, \\
A(\lambda)B(\mu) &= \tfrac{\mu^2 - \lambda^2 q^{-2}}{\mu^2 - \lambda^2} B(\mu)A(\lambda) - \tfrac{2\lambda\mu}{\mu^2 - \lambda^2}\left(1 - q^{-2}\right)B(\lambda)A(\mu), \\
D(\lambda)B(\mu) &= \tfrac{2\lambda\mu}{\mu^2 - \lambda^2}\left(q^2 - 1\right)B(\lambda)D(\mu) + \tfrac{\mu^2 - \lambda^2 q^2}{\mu^2 - \lambda^2}B(\mu)D(\lambda).
\end{aligned}
\tag{93}
$$

The vector $\psi(\lambda_1, \ldots, \lambda_n)$ is an eigenvector of $tr M(\lambda)$ with the eigenvalue

$$a(\lambda) \prod_{i=1}^{n} \frac{\lambda^2 q^{-2} - \lambda_i^2}{\lambda^2 - \lambda_i^2} + d(\lambda) \prod_{i=1}^{n} \frac{\lambda^2 q^2 - \lambda_i^2}{\lambda^2 - \lambda_i^2} \tag{94}$$

iff λ_i satisfy the equations:

$$q^2 \frac{d(\lambda_j)}{a(\lambda_j)} = \prod_{i \neq j} \frac{\lambda_j^2 q^{-2} - \lambda_i^2}{\lambda_j^2 q^2 - \lambda_i^2}. \tag{95}$$

We omit the standard Bethe Ansatz calculations leading to these formulas (90), (91), (94), (95) [16]. It is easy to verify that the modification of the commutation relations (89), (93) (as compared to the standard XXZ relations) only weakly influences the Bethe Ansatz construction, so that the principal idea [16] may be applied as before.

4. Application to the Nonlinear Sigma Model

We recall some facts about chiral fields with values in Riemannian symmetric spaces. Let $(\mathfrak{k}, \mathfrak{p})$ be a Riemannian symmetric pair for a semisimple finite-dimensional Lie algebra $\overset{\circ}{\mathfrak{g}}$ [15]. It means that $\overset{\circ}{\mathfrak{g}} = \mathfrak{k} \dotplus \mathfrak{p}$ as a linear space, and

$$[\mathfrak{k}, \mathfrak{k}] \subset \mathfrak{k}, [\mathfrak{k}, \mathfrak{p}] \subset \mathfrak{p}, [\mathfrak{p}, \mathfrak{p}] \subset \mathfrak{k}. \tag{96}$$

so that $\mathfrak{k}$ is a subalgebra in $\overset{\circ}{\mathfrak{g}}$. We suppose that the decomposition $\overset{\circ}{\mathfrak{g}} = \mathfrak{k} \dotplus \mathfrak{p}$ is orthogonal with respect to the standard scalar product on $\overset{\circ}{\mathfrak{g}}$, so that the projectors $P_{\mathfrak{p}}$ and $P_{\mathfrak{k}}$ onto the subspaces $\mathfrak{p}$ and $\mathfrak{k}$ are orthogonal. Let $\overset{\sim}{\overset{\circ}{\mathfrak{g}}} = C^\infty \left(S^1, \overset{\circ}{\mathfrak{g}} \right)$ and let us introduce the left currents $l_x, l_t \in \overset{\sim}{\overset{\circ}{\mathfrak{g}}}$,

$$l_x = g^{-1} \partial_x g, l_t = g^{-1} \partial_t g, g \in C^\infty \left(S^1, \overset{\circ}{G} \right) = \overset{\sim}{\overset{\circ}{G}}, \tag{97}$$

where $\overset{\circ}{G}$ is a Lie group corresponding to $\overset{\circ}{\mathfrak{g}}$. Let A_μ, B_μ be the two projections of the current:

$$A_\mu = P_{\mathfrak{k}} l_\mu, B_\mu = P_{\mathfrak{p}} l_\mu. \tag{98}$$

In these notations the action of the nonlinear sigma model has the form [14]:

$$S(g) = \tfrac{1}{2} \int tr \left(B_x B_x - B_t B_t \right) dx dt,$$
$$S(g) \in Fun \left(\overset{\sim}{\overset{\circ}{G}} \right). \tag{99}$$

The action functional is unchanged under the right gauge action of the group $\tilde{K} = C^{\infty}\left(S^1, K\right)$:

$$g \mapsto gk, k \in \tilde{K}, g \in \overset{\circ}{\tilde{G}}, \tag{100}$$

so that this is a well defined function on the symmetric space $\overset{\circ}{\tilde{G}}/\,\tilde{K}$. The equations of motion which follow from the action (99) are:

$$\begin{aligned} \partial_t B_t + [A_t, B_t] &= \partial_x B_x + [A_x, B_x], \\ \partial_x A_t - \partial_t A_x + [A_x, A_t] + [B_x, B_t] &= 0, \\ \partial_x B_t - \partial_t B_x + [A_x, B_t] - [A_t, B_x] &= 0. \end{aligned} \tag{101}$$

The two last equations are zero curvature conditions which serve to restore the group variable $g \in \overset{\circ}{\tilde{G}}/\,\tilde{K}$ from A_μ, B_μ. For this model there exists a Lax pair:

$$\begin{aligned} L &= -\left(A_x + \tfrac{\lambda}{2}\left(B_x + B_t\right) + \tfrac{1}{2\lambda}\left(B_x - B_t\right)\right), \\ T &= -\left(A_t + \tfrac{\lambda}{2}\left(B_x + B_t\right) - \tfrac{1}{2\lambda}\left(B_x - B_t\right)\right). \end{aligned} \tag{102}$$

The equations of motion (101) are expressed as the zero curvature condition:

$$[\partial_x - L, \partial_t - T] = 0. \tag{103}$$

Now we define a natural Poisson structure connected with the Lax pair (102). Let $\mathfrak{g} = \overset{\circ}{\mathfrak{g}} \otimes \mathbf{C}\left[\lambda, \lambda^{-1}\right]$ be the current Lie algebra with the scalar product (51). Let $\mathfrak{g}^\sigma$ be the twisted current Lie algebra which corresponds to the Cartan automorphism σ associated with $(\mathfrak{k}, \mathfrak{p})$:

$$\sigma\mid_{\mathfrak{k}} = id, \sigma\mid_{\mathfrak{p}} = -id. \tag{104}$$

Thus

$$\mathfrak{g}^\sigma = \left\{X\left(\lambda\right) \in \mathfrak{g} : \sigma X\left(-\lambda\right) = X\left(\lambda\right)\right\}. \tag{105}$$

For our model we may choose the standard r-matrix Lie-Poisson structure bracket (31) using the r-matrix

$$r = P_+ - P_-; \tag{106}$$

where P_- is the projection operator onto the negative part of a Laurent series and P_+ is the complementary projection operator. The projection operator P_- has the kernel:

$$P_-\left(\lambda, \mu\right) = t_A \frac{\mu^2}{\lambda^2 - \mu^2} + t_B \frac{\lambda\mu}{\lambda^2 - \mu^2}, \tag{107}$$

where $t_A = (P_\mathfrak{k} \otimes P_\mathfrak{k})\, t, t_B = (P_\mathfrak{p} \otimes P_\mathfrak{p})\, t$ are the $\mathfrak{k}-$ and $\mathfrak{p}-$ components of the Casimir element, respectively. The r-matrix (106) has a symmetric part with the kernel:

$$s(\lambda, \mu) = t_A. \tag{108}$$

Example 3. The principal chiral field. Let $\overset{\circ}{\mathfrak{g}} = \mathfrak{a} \oplus \mathfrak{a}$ be the direct sum of two copies of a simple Lie algebra $\mathfrak{a}$. Let $\mathfrak{k}$ be the diagonal subalgebra in $\overset{\circ}{\mathfrak{g}}$ and $\mathfrak{p}$ be the anti-diagonal subspace:

$$
\begin{aligned}
\mathfrak{k} &= \{(x, x)\,, x \in \mathfrak{a}\}\,, \\
\mathfrak{p} &= \{(x, -x)\,, x \in \mathfrak{a}\}\,, \\
\sigma(x, y) &= (y, x)\,.
\end{aligned}
\tag{109}
$$

This is a Riemannian symmetric pair. The symmetric space for the corresponding current group is $\tilde{A} \times \tilde{A}/\tilde{A}$, where $\tilde{A}$ is the current group of the Lie group corresponding to the Lie algebra $\mathfrak{a}$. In this case $\tilde{A} \simeq \tilde{P}$ and acts on $\tilde{A} \times \tilde{A}$ according to (100):

$$
\begin{aligned}
\tilde{A} \times \left(\tilde{A} \times \tilde{A} \right) &\to \tilde{A} \times \tilde{A}, \\
g \circ (g_1, g_2) &\mapsto (g_1 g, g_2 g)
\end{aligned}
$$

So in this case the action (99) is a well-defined function on $\tilde{A}$. If we define the projection

$$
\begin{aligned}
\pi &: \tilde{A} \times \tilde{A} \to \tilde{A}, \\
\pi &: (g_1, g_2) \mapsto g_1 g_2^{-1} = g
\end{aligned}
\tag{110}
$$

onto the quotient space $\tilde{A} \simeq \tilde{A} \times \tilde{A}/\tilde{A}$ and define the currents

$$l_\mu^k = g_k^{-1} \partial_\mu g_k, k = 1, 2, l_\mu = g^{-1} \partial_\mu g, \tag{111}$$

then the variables A_μ and B_μ have the form:

$$
\begin{aligned}
A_\mu &= \left(\tfrac{1}{2} \left(l_\mu^1 + l_\mu^2 \right), \tfrac{1}{2} \left(l_\mu^1 + l_\mu^2 \right) \right), \\
B_\mu &= \left(\tfrac{1}{2} \left(l_\mu^1 - l_\mu^2 \right), -\tfrac{1}{2} \left(l_\mu^1 - l_\mu^2 \right) \right),
\end{aligned}
\tag{112}
$$

and the action (99) takes the form:

$$S(g) = \int (l_x l_x - l_t l_t)\, dx dt. \tag{113}$$

This is the action of the principal chiral field model on $\tilde{A}$, so that we deal with a realization of this model on the symmetric space $\tilde{A} \times \tilde{A}/\tilde{A}$. The Lax operator (102) is a direct sum of the two ones:

$$L(\lambda) = \left(L^1(\lambda), L^1(-\lambda) \right), \tag{114}$$

and the r-matrix (106) is a direct sum of a two copies of the rational r-matrix:

$$r = \begin{pmatrix} P_+ - P_- & 0 \\ 0 & P_+ - P_- \end{pmatrix}. \tag{115}$$

Let us return to the situation of example 3.1. Our model is a direct sum of two copies of this example. The quantum lattice algebra is a direct sum of two copies of the algebra (59), where L^n is a quantum version of a lattice approximation of the Lax operator $L^1(\lambda)$. If

$$M(\lambda) = \left(M^1(\lambda), M^1(-\lambda) \right) \tag{116}$$

is the quantum monodromy matrix for our model, then $M^1(\lambda)$ satisfies the relation (60). If $\mathfrak{a} = sl(2)$ then the generating function for the integrals of motion is:

$$tr_q M^1(\lambda) + tr_q M^1(-\lambda), \tag{117}$$

and we may apply the technique of the modified Bethe Ansatz developed in Section 3 to calculate the spectrum.

Example 4. The n-field. Let $\overset{\circ}{\mathfrak{g}} = su(2)$ and let $\mathfrak{k} = \overset{\circ}{\mathfrak{h}}$ be the Cartan subalgebra of $su(2)$, and $\mathfrak{p}$ the complementary subspace. This is a Riemannian symmetric pair and the corresponding automorphism σ is given by (cf. 78):

$$\sigma X = DXD^{-1}, X \in su(2), D = \begin{pmatrix} 1 & 0 \\ 0 & -1 \end{pmatrix}. \tag{118}$$

Let

$$A_x = \begin{pmatrix} ia & 0 \\ 0 & -ia \end{pmatrix}, B_x = \begin{pmatrix} 0 & A_1 + A_{-1} \\ -(\overline{A}_1 + \overline{A}_{-1}) & 0 \end{pmatrix},$$
$$B_t = \begin{pmatrix} 0 & A_1 - A_{-1} \\ -(\overline{A}_1 - \overline{A}_{-1}) & 0 \end{pmatrix} \tag{119}$$

then in this case the Lax operator (102) has the form:

$$L = -\begin{pmatrix} ia & \lambda A_1 + \frac{1}{\lambda} A_{-1} \\ -\lambda \overline{A}_1 - \frac{1}{\lambda} \overline{A}_{-1} & -ia \end{pmatrix}. \tag{120}$$

Since the automorphism σ coincides with the Coxeter automorphism (78) for $sl(2)$, the twisted affine Lie algebra $\mathfrak{g}^\sigma$ (105) coincides with the compact real form of the twisted realization of the algebra $A_1^{(1)}$. To define a Poisson structure we may use the r-matrix (106) which in this case is precisely the r-matrix from example 3.1 So that we are returning to the situation investigated in this example. The lattice quantum algebra is given by Theorem 3.4, part 2, where L^n is a quantum version of a lattice approximation of the Lax operator (120). According to Theorem 3.3 for this algebra there exists an ultralocalization. Let

$$\widehat{L}^n(\lambda) = -\begin{pmatrix} A_0^n & \lambda A_1^n + \frac{1}{\lambda} A_{-1}^n \\ -\lambda A_1^{n^*} - \frac{1}{\lambda} A_{-1}^{n^*} & A_0^{n^*} \end{pmatrix},$$

$$G^n = \begin{pmatrix} a_n & 0 \\ 0 & a_n^* \end{pmatrix},$$

(121)

where we use the notations of theorem 3.3 and $*$ is the anti-involution in this algebra (it is supposed that $| q |= 1$). We shall give an explicit realization of this algebra via the canonical Weyl pairs

$$U_i^n V_i^n = q V_i^n U_i^n, i = 1, 2; n = 1, \dots, N,$$

$$U_i^{n^*} = U_i^{n^{-1}}, V_i^{n^*} = V_i^{n^{-1}}.$$

(122)

We omit the complicated calculations and give only the result:

$$A_0^n = \left(1 - q^{-1} V_1^{n^{-2}}\right) U_1^n U_2^n,$$

$$A_1^n = V_1^n U_2^n, A_{-1}^n = V_1^{n^{-1}} U_2^n,$$

$$a_n = V_1^{n^{-\frac{1}{2}}} V_2^{n^{\frac{1}{4}}} V_2^{n+1^{-\frac{1}{4}}}.$$

(123)

The Lax operator $\widehat{L}^n(\lambda)$ coincides up to an unessential factor with the one of the Lattice Sine–Gordon model [9]:

$$\widehat{L}^n = L_{LSG}^n \otimes U_2^n.$$

(124)

Since U_2^n is the shift operator in the standard Weyl representation of the algebra (122), we may choose the reference state in the form:

$$\Omega = \Omega_{LSG} \otimes 1,$$

$$1 = \underbrace{1 \otimes \dots \otimes 1}_{N},$$

(125)

where Ω_{LSG} is the reference state of the Lattice Sine–Gordon model. U_2^n acts trivially on 1, so that the spectrum of our model coincides with the one for the Lattice Sine–Gordon model. Here we are using the gauge equivalence of these two models and the connection between generating functions for the integrals of motion:

$$tr M = q^{\frac{1}{2}} tr \widehat{M},$$

$$M = L^1 \cdot \dots \cdot L^N,$$

$$\widehat{M} = \widehat{L}^1 \cdot \dots \cdot \widehat{L}^N.$$

(126)

5. Conclusion

In this section we formulate some problems related to the subject of this paper. The first problem is connected with a generalization of our construction of nonultralocal algebras to an arbitrary graph. It is not difficult to construct a nonultralocal algebra on a graph using a solution of the Yang–Baxter equation on the square of some Lie algebra and the 'polyuble construction' [11]. In the finite-dimensional case these algebras were introduced in [11] in connection with the quantization of the moduli space. The lattice algebra on a graph similar to one considered in example 1 of this paper was investigated in [3].

A more interesting problem is the regularization of the Poisson bracket relations for the monodromy   for a more wide class of scalar products on current algebras, e.g. when

$$\langle X(\lambda), Y(\lambda) \rangle = Restr\left(X(\lambda) Y(\lambda)\right) \phi(\lambda) d\lambda,$$

where $\phi(\lambda)$ is some rational function. The Belavin–Drinfeld classification theorem does not allow to define regularization in the way considered in Section 2, because there exist no nontrivial solutions of the Classical Yang–Baxter equation on $\mathfrak{g} \oplus \mathfrak{g}$ of the type (40) which are skew-symmetric with respect to this scalar product . One of the possible ways to tackle with this situation is to use Quasi-Hopf algebras [7]. This requires the study of r-matrices which do not satisfy the Yang–Baxter equation. However, the physical motivation to introduce such algebras is not clear.

Another problem is the construction of representations of nonultralocal algebras. These representations are important for the definition of reference states Ω and of the Bethe Ansatz construction. A natural way to solve this problem is the ultralocalization procedure. As we have seen, ultralocalization does not exist for all lattice algebras. It is interesting to describe all representations of nonultralocal algebras and to choose from them those which arise from an ultralocalization procedure. For finite-dimensional algebras this question is connected with the anomaly problem in the quantum field theory [4].

The last problem is to construct so called fundamental Lax operator for nonultralocal algebras [9]. The first step to its solution was made in [12].

The authors would like to thank L.D. Faddeev, F.A. Smirnov, and A. Yu. Alekseev for numerous helpful discussions.

References

[1] L.D. Faddeev, L.A. Takhtajan (1987), *Hamiltonian Methods in the Soliton Theory*, Berlin: Springer.

[2] V.G. Kac (1983), *Infinite dimensional Lie algebras*, Boston, Basel, Stuttgart: Birkhauser.

[3] A. Alekseev (1994), Integrability of the Chern–Simons theory in Hamiltonian, *Algebra and Analysis* **6** (2), 53–66.

[4] A. Alekseev, L. Faddeev, M. Semenov-Tian-Shansky (1992), Hidden quantum groups Inside Kac–Moody algebra, *Comm. Math. Phys.* **149**, 335–345.

[5] A.A. Belavin, V.G. Drinfeld (1984), Triangle equation and simple Lie algebras, *Sov. Sci. Rev. Sect. C* **4**, 93–170.

[6] V.G. Drinfeld (1986), *Quantum Groups, Proc. Internat. Congress of Mathematicians (Berkeley, 1986)*, **1**, Amer. Math. Soc. (1987), 798–820.

[7] V.G. Drinfeld (1990), Quasi-Hopf algebras, *Leningrad Math. J.* **1**, 1419–1457.

[8] L.D.Faddeev (1984), *Integrable models in (1+1) dimensional quantum field theory /Recent advances in field theory and statistical mechanics (Lectures in Les Houches, 1982)*, Elsevier Science Publishers, pp. 563-608.

[9] L.D. Faddeev (1993), *The Bethe–Ansatz*, SFB 288, preprint no. 70, Berlin.

[10] L.D. Faddeev, N.Yu. Reshetikhin, L.A. Takhtajan (1990), Quantization of Lie groups and Lie algebras, *Leningrad Math. J.* **1**, 193–226.

[11] V.V. Fock, A.A. Rosly (1993), Flat connections and polyubles, *Theor. Math. Phys.* **95** (2), 228–238.

[12] L.D. Faddeev, A. Yu. Volkov (1993), Abelian current algebra and the Virasoro algebra on the lattice, preprint HU-TFT-93-29.

[13] L.D. Faddeev, A. Yu. Volkov (1994), private communications.

[14] L.D. Faddeev, M.A. Semenov-Tian-Shansky (1977), On the theory of nonlinear chiral fields, *Vestnik Leningr. Universiteta* **13** (3).

[15] S. Helgason (1978), *Differential Geometry, Lie Groups, and Symmetric Spaces*, New-York: Academic Press.

[16] E.K. Sklyanin, L.A. Takhtajan, L.D. Faddeev (1979), Quantum inverse problem method 1, *Theor. Math. Phys.* **40** (2), 194–220.

[17] M.A. Semenov-Tian-Shansky (1983), What is the classical r-matrix, *Funct. Anal. i ego Pril.* **17** (4), 17–33.

[18] M.A. Semenov-Tian-Shansky (1985), *Publ. RIMS, Kyoto University* **21** (6), 1237–1260.

[19] M.A. Semenov-Tian-Shansky (1992), Monodromy map and classical r-matrices, *Zapiski Nauchn. Semin. LOMI* **200**, 156–166.

[20] M.A. Semenov-Tian-Shansky (1992), Poisson-Lie groups, quantum duality principle and the twisted quantum double, *Theor. Math. Phys.* **93** (2), 302–329.

[21] S. Parmentier (1991), Twisted affine Poisson structures, decompositions of Lie algebras, and the classical Yang–Baxter equation, M. Planck Institute preprint, no. 82, Bonn.

[22] L. Freidel, J.M. Maillet (1991), Quadratic algebras and integrable systems, *Phys. Lett. B* **262** (2, 3), 278.

[23] A. Yu. Alakseev, A.Z. Malkin (1994), Symplectic structures associated to Lie-Poisson groups, *Comm. Math. Phys.* **162** (1), 147.

Michael Semenov-Tian-Shansky
Université de Bourgogne
Physique Mathématique
B.P. 138 21004 Dijon CEDEX France
email: semenov@u-bourgogne.fr
and Steklov Mathematical Institute
St. Petersburg, Russia

Alexey Sevostyanov
Steklov Mathematical Institute
St. Petersburg, Russia

Received March, 1996.